Building the World

Building the World

An Encyclopedia of the Great Engineering Projects in History

Volume 2

Frank P. Davidson and
Kathleen Lusk Brooke

Greenwood Press
Westport, Connecticut • London

Library of Congress Cataloging-in-Publication Data

Building the world : an encyclopedia of the great engineering projects in history / by
Frank P. Davidson and Kathleen Lusk Brooke.

      p.   cm.

  Includes bibliographical references and index.

  ISBN 0–313–33354–8 (set: alk. paper)—0–313–33373–4 (vol 1: alk. paper)—0–313–33374–2
(vol 2: alk. paper)

  1. Industrial engineering. I. Title: Encyclopedia of the great engineering projects in
history. II. Davidson, Frank Paul, 1918– III. Lusk Brooke, Kathleen.

  T56.I44 1997

  620—dc22   2005037902

British Library Cataloguing in Publication Data is available.

Copyright © 2006 by Frank P. Davidson and Kathleen Lusk Brooke

Library of Congress Catalog Card Number: 2005037902
ISBN: 0–313–33354–8 (set)
          0–313–33373–4 (vol 1)
          0–313–33374–2 (vol 2)

First published in 2006

Greenwood Press, 88 Post Road West, Westport, CT 06881
An imprint of Greenwood Publishing Group, Inc.
www.greenwood.com

Printed in the United States of America

∞™

The paper used in this book complies with the
Permanent Paper Standard issued by the National
Information Standards Organization (Z39.48–1984).

10 9 8 7 6 5 4 3 2 1

CONTENTS

Volume 1

Contents

Volume 2

Contents

Map:

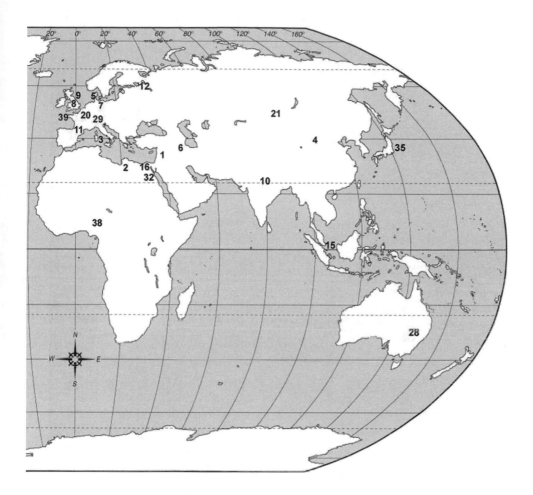

25
The Tennessee Valley Authority

United States

DID YOU KNOW . . . ?

➤ The Tennessee River basin covers 41,000 square miles (106,200 square kilometers).

➤ The Tennessee Valley Authority is the largest public power company in the United States, with 30,000 megawatts of generating capacity.

➤ It marks one of the first uses of a logo for a public project.

➤ It created a new professional community, the town of Norris, a showcase of consumer possibilities for electricity.

➤ Its low rate/high use policy accelerated the growth of the electric industry.

In 1941, the year of Pearl Harbor, when the United States entered World War II, the Tennessee Valley Authority (TVA) was the largest producer of electricity in the United States. Electricity was fundamental to war production, and President Franklin Roosevelt's long preparation for a federal role in producing electrical power was a great strategic decision.

The TVA used a logo. Its theme is "Power to the People," and the graphic shows a raised fist gripping a thunderbolt. The tag line is "Electricity for All."

HISTORY

The TVA was a pioneering attempt at managing the resource-development needs of an entire region of the United States, with the focus on a major river controlled through a unified, integrated network of large and small dams,

Aerial view of Wilson Dam and power house, Muscle Shoals, Alabama, 1926.
Courtesy of the Library of Congress.

hydropower plants, and flood-control measures. While such a wide-ranging, socially conscious intervention in regional affairs might seem a typical example of federal policy amid the hard times of the Great Depression, the TVA's origins can be traced back years earlier, to a World War I–era scheme to take advantage of rapids near Muscle Shoals, Alabama, to generate hydropower and manufacture nitrates used in explosives and fertilizer.

Following the war, the completion of the Wilson Dam and two nitrates plants provoked furious debate over whether the projects should be run by private industry. Senator George W. Norris, a Republican from Nebraska, who would later be remembered as "the father of the TVA," emerged as the main champion of public control. Upon assuming office in 1933, President Roosevelt pushed for public management and an expanded package of Tennessee Valley projects. He asked Congress to establish "a corporation clothed with the power of government but possessed of the flexibility and initiative of a private enterprise" (Roosevelt Library Archives, April 10, 1933). As a result, the Tennessee Valley Authority was established by a 1933 act of Congress (reproduced here). Subsequent court challenges of the TVA's constitutionality by competing private companies failed.

A public corporation run by a board of three presidential appointees, the TVA was assigned a broad array of functions relating to the Tennessee River basin, an area covering 41,000 square miles (106,200 square kilometers) and extending over much of the state of Tennessee and parts of Alabama, Georgia,

Kentucky, Mississippi, North Carolina, and Virginia. In addition to providing flood control and improved navigation on the river, it promoted agricultural and industrial development of the Tennessee Valley, long an economically underachieving area hurt by devastating floods, deforestation, substantial land erosion, and soil depletion. The TVA was mandated to produce inexpensive electricity on a massive scale, and supply power to rural areas previously unserved. Today the TVA's activities are funded primarily by the sale of electric power and by federal appropriations.

Over the years, the TVA grew by acquiring private land and facilities through condemnation as well as through construction of its own facilities. By 1944 it managed nine dams on the Tennessee River, plus many other dams on the river's tributaries; it also provided power for the federal atomic-energy program at Oak Ridge, Tennessee. Besides producing fertilizer and electricity, the authority boosted navigation on the river by building locks, deepening channels, and supporting port-expansion projects; river-trade tonnage surged to 161 million ton-miles in 1942, up from 32 million ton-miles in 1933. Recreational opportunities burgeoned on the so-called Great Lakes of the South created by the TVA's dams.

CULTURAL CONTEXT

Both coasts of the United States were well developed by the 1930s, but the center of the country was left behind: poverty was rife in less populated regions and living conditions were difficult. Upton Sinclair wrote popular novels describing the poverty of the Midwest. Born in 1878, Sinclair grew up in an extremely poor family to become a well-known socialist novelist. An activist who once ran for public office, he swayed many more with his pen. Before his death in 1968, Sinclair had published more than 90 books.

As part of the New Deal, President Roosevelt was determined to use the power of the federal government to relieve unemployment and the economic crises that created poverty. Millions of people were displaced from their homes by both economic conditions and drought. Sinclair's novels described the fate of those who had to leave their ranches and farms because of the Depression. Millions of displaced farmers hoped to find financial salvation in California. Roosevelt's philosophy differed markedly from that of Herbert Hoover and the Republican Party, who wanted to treat the Depression as a series of bad moments that would eventually dissipate. Roosevelt correctly understood that he needed to reverse the downward economic spiral.

Among the first programs Roosevelt initiated was the Civilian Conservation Corps (CCC), which offered jobs to unemployed young men to preserve forests, protect the soil, and fight natural hazards. Many "CCC boys" worked in Department of Agriculture programs for terracing land and fighting erosion. Vice President Henry A. Wallace was the advocate for these programs; he came from a family known for developing specialized agricultural crops. Drawing from his family lineage and extensive knowledge of agriculture, Wallace developed the idea of an "ever-normal granary." Part of this approach was an assurance that electric power would be cheap enough that poor families in rural areas could afford it. One of the TVA's missions supports this policy—that the federal government would develop power through hydroelectricity generated from the Tennessee River.

Hydroelectric technology was key, but harnessing that force required large-scale building and management skills to coordinate the work of many engineers and construction experts. Also needed were people skilled at managing large amounts of money and pacing the work according to available funding. Equally critical was learning to work with the large number of companies that participated in the project. Although large-scale management skills were developing in the culture, it would be many years before they were given a name: macroengineering.

Did America become a society driven by electricity because of the TVA and other related projects like the Bonneville Dam or the Hoover Dam? Peak/off-peak and low rate/high use policies may have had a direct impact on research to develop all things electric, from kitchen appliances to children's toys, computers, televisions, and home lighting. Can an entire culture be motivated by one policy, low rate/high use? It was a revolutionary idea at the time.

A Tennessee Valley Authority worker looking over books in a tool box, a part of the libraries brought to the men on projects in remote areas, circa 1935. Courtesy of the Library of Congress.

The electricity-centered culture of the United States has created industries, such as entertainment and home appliances, which have raised standards far beyond U.S. borders. For example, televisions in Bangkok, air conditioners in Singapore, rice cookers in Beijing, and kitchen appliances in Germany all run on electricity. During the Thai economic boom of 1986–94, the Electricity Generating Authority of Thailand (EGAT) grew more quickly and built more plants than any other sector in the economy. That country's need for electricity is enormous; for instance, there are thousands of elevators in tall buildings, one solution to exorbitant land prices, which increased 600 percent during Bangkok's economic boom. In cities and villages around the world, economic growth parallels electricity use. Of course, these changes cannot be attributed to the TVA, but it cannot be denied that the scale of the TVA at a time when electrical use was modest and the TVA's policy of low rate/high use fostered accelerated electrical research and development.

PLANNING

A three-member board planned and oversaw the operation of the TVA: Arthur Morgan, who believed the TVA's main role was to reduce poverty through national and community planning; Harcourt Morgan (no relation), who was an advocate for southern commercial farming and a firm nongovernment, non-public-works believer; and David Lilienthal, an advocate of publicly owned electric power. Arthur Morgan was the first president of the TVA, guiding a cooperative relationship between government and business. Many stories chronicle the uneasy relations that existed among the three members. In 1938, Harcourt Morgan and David Lilienthal joined forces and succeeded in ousting Arthur Morgan. That allowed Harcourt Morgan to lead the board for three years from 1938 to 1941, when Lilienthal became president. It was during Lilienthal's presidency that the TVA reached its pinnacle as the largest source of electricity in the United States.

When Lilienthal was later appointed chairman of the Atomic Energy Commission, the question arose as to who should invest in breeder reactors—the government or private industry. Lilienthal strongly favored government involvement. Another question arose: should atomic energy be secretive and handled by the military, or peaceful and provide economic value to many? Once again, the thrust of Roosevelt's New Deal was to open everything for civilian benefit—a policy that had its origins in the planning of the TVA. This policy, as it turned out, was also beneficial from a military point of view.

BUILDING

While one usually associates the TVA with electrical supply, the authorizing document presented here opens with the stated purposes of the project.

Mentioned first: "To improve the navigability and to provide for the flood control of the Tennessee River." Additional benefits are stated: reforestation and use of marginal lands along the river—both serving environmental purposes—followed by industrial and agricultural development. The final purpose is "To provide for national defense by the creation of a corporation for the operation of Government properties at and near Muscle Shoals in the State of Alabama."

Founded in 1923, Muscle Shoals is the birthplace of the Tennessee Valley Authority and the site of the Wilson Dam. The TVA's Environmental Research Center, the Public Power Institute, and the International Fertilizer Development Center, all nationally and internationally recognized institutions, are located in Muscle Shoals. The Shoals area is known for its music-recording industry, as well as for being the birthplace of Helen Keller and W. C. Handy, father of the blues.

After stating the purposes of the project, section 2 establishes a three-person board. Why three members rather than a single head? Because the TVA was launched as a corporation, and the desire was to maintain a small board. In addition, board members may not "be engaged in any other business" and may not "have financial interest in any public-utility corporation engaged in the business of distributing and selling power to the public, nor in any corporation engaged in the manufacture, selling, or distribution of fixed nitrogen or fertilizer." These noncompete clauses are important because the TVA was launched in competition with private power companies.

Section 4(i) authorizes the corporation to acquire the necessary real estate to construct dams and other buildings that would be needed. It asserts that landowners will be paid fair price, but if they refuse to sell, the government has the right "to condemn all property that it deems necessary." In the end, there was considerable dislocation (as there has been in China for the Three Gorges Dam). While some may be tempted to criticize the Chinese for dislocating thousands of people for the Three Gorges Dam project, in 1933 the TVA required the same actions. Similarly, the Quabbin Reservoir (built in the 1930s in Massachusetts) completely covers towns that were evacuated so the entire area could be flooded for a public water supply.

Section 5 of the act is a full discussion of fertilizer, and it is interesting and revealing in its implications for the future. Because nitrate results from hydroelectric activities, clause (d) authorizes that "in order to improve and cheapen the production of fertilizer," the corporation will sell fertilizer and build plants "for the fixation of atmospheric nitrogen or the cheapening of the production of fertilizer." This certainly adheres to one of the TVA goals: to improve agriculture.

Most surprising is section 5(g), which demands that the board maintain "in stand-by condition nitrate plant numbered 2 for the fixation of atmospheric nitrogen for the production of explosives in the event of war or for a national emergency"—fast-forward to the twenty-first century and its overarching concerns about terrorism, especially the use of fertilizer in conjunction with making explosives.

In 1933, the United States was aware of the military implications of harnessing the Tennessee River. Such awareness led to sections 5(j) and 5(k), which specify the obligations of the TVA to the secretary of war, the secretary of the navy, and the War Department. Section 5(m) stipulates that "no products of the Corporation shall be sold for use outside of the United States, its Territories and possessions except to the United States Government for the use of its Army or Navy, or to its allies in case of war."

It is not until section 10 that the production of electricity is discussed: the board is authorized to "sell the surplus power ... to States, counties, municipalities, corporations, partnerships, and individuals." True to its origin as a public-works project, the TVA is instructed to give "preference to States, counties, municipalities, and cooperative organizations of citizens or farmers not organized or doing business for profit, but primarily for the purpose of supplying electricity to its own citizens or members." Here, too, is an indication of the TVA's vision: "To promote and encourage the fullest possible use of electric light and power," especially on farms, and to "make better use of electric power for agricultural and domestic use, or for small or local industries."

Sections 13–15 deal with finances, apportioning revenues to Alabama and Tennessee from dams to be constructed there. There is also discussion of financing through 50-year bonds. In 1959, bonds were issued, and in the 1990s the TVA paid back $2.5 billion to the U.S. Treasury. In comparison, the Paris subway was built with 80-year bonds. If there is a precedent for bonds of such length, it might be possible to build projects of considerable scope.

There is a good summary in section 14 of the multitiered purposes of TVA: flood control, navigation, fertilizer, national defense, and development of power.

No discussion of the TVA and the implications for the future of the world of appliances and lighting is complete without mentioning the town of Norris. Senator George W. Norris of Nebraska is known as "the father of the TVA," and the Norris Dam is his namesake. In 1933, the TVA planned and built the town, designed to showcase the modern marvels of electricity. Set in the Tennessee hills 20 miles north of Knoxville, the immediate purpose of the town was to house workers building the Norris Dam four miles away on the Clinch River. Norris was more than a planned settlement to house workers; it was a window to the world of what life could be like with abundant electricity. Norris featured state-of-the-electrical-art homes and co-op businesses ranging from poultry farms to canneries and creameries. The well-lighted public school provided classes for both children and adults and specialized classes for farmers. Every aspect of Norris, right down to the gas station, had the TVA touch.

Unfortunately, African American workers who were employed at the TVA were not invited to live in Norris. Although arguing to the contrary, the fact was that in other rural areas, black and white people dwelled in the region.

Eventually the National Association for the Advancement of Colored People (NAACP) had to bring inquiries to the TVA regarding its policies toward African Americans, not only for housing but for jobs and training. After the Norris Dam was built, the character of the town itself changed, and the community became an enclave of professionals who worked at the TVA and later in Knoxville.

In 1948, the government sold the town of Norris to a private contractor, who repackaged everything into individual lots and sold them to families, often the residents themselves, who were currently living in the houses that once showed the world the marvels of what cheap electricity could do to improve living conditions.

Low rates/high use—the means of providing "electricity for all"—changed the face of America and the world. Showcased in the town of Norris, and highlighted in the policies implemented by Roosevelt, energy became the clay of invention, spawning a stream of innovations for industrial, military, and private use based on easy-to-use electricity found in so many applications and forms. Perhaps no other single project can be said to have affected daily life as greatly as the TVA.

Recreational structures at Big Ridge Lake, 1937. This recreation area was developed by the National Park Service and the CCC (Civilian Conservation Corps) in cooperation with the Tennessee Valley Authority. Courtesy of the Library of Congress.

As of the late 1990s, the TVA was the biggest public power company in the United States, with 30,000 megawatts of generating capacity that produces electricity used by eight million people in and beyond the Tennessee River basin. Coal-burning steam plants account for the majority of the TVA's installed capacity, but it also has 50 dams with an installed hydroelectric power capacity exceeding 6,000 megawatts. Other sources of power include nuclear plants, a pumped-storage plant, and (as of 2000) the sun and the wind.

The TVA is well documented by two photographers, Lewis Hine and Charles Krutch. TVA historian Patricia Bernard Ezzell has collected their dramatic photos showing workers high atop scaffolds of the Norris Dam as well as many other images. Hine also documented the construction of the Empire State Building, and his depictions of life in the CCC and other events of the times are legendary. Krutch was an amateur photographer who devoted himself to the TVA, capturing famous images of its transformation into a prosperous and technologically advanced leading region of the country from what was an economically disadvantaged region.

IMPORTANCE IN HISTORY

When the TVA was established, it ushered in discussions about the uses of power, starting with electrical and eventually moving to nuclear. The archives at the University of Arizona, organized by former senator Morris K. Udall, detail aspects of these issues that were considered by the Interior Committee.

The TVA had a large planning and research unit to explore various energy issues. One source of power for desert regions could be solar. The world has not yet made a decision as to whether solar-power satellites can be sufficiently supervised so as to pose no appreciable health or defense risks. However, this research unit supported an agricultural project that would bring deserts all over the world to bloom because it would be so inexpensive to desalinate water with electricity, suggesting one cent per kilowatt-hour as the cost of electricity. The TVA's hopes for desalination were too optimistic, but their discussions led to a range of literature on the possibilities of agro-industrial coastal complexes. Think of the effects of such projects on the Middle East if such complexes turned out to be buildable! But how long will the development of desalination technology take? Meanwhile, we may have neglected older and simpler technologies such as aqueducts and gravity-flow water—technologies at which the Romans excelled. Now engineers can give gravity a little push with electric pumps.

One could speculate about how an act like the one authorizing the TVA could be modified if solar-power satellite transmission was authorized. Could the TVA act given here be used as a basis for structuring future energy contracts and transmission agreements? With microwave transmission of electrical power, one could transmit it anywhere on earth. Could solar-power satellites supply the energy needed for building in space?

One outcome of the TVA's activity was the assembling of a core of specialists who evolved into a research center staffed with personnel who made themselves available as experts. When Oak Ridge was built, was it because the TVA had seeded the area with experts? When the U.S. Department of Defense established the RAND (Research and National Development) Corporation, it soon became a think tank that focused on many policy areas. When governments make a decision to start a project that requires the presence of professionals for extended periods of time, and make provisions for those individuals to bring their families, permanent relocation can be anticipated. Thus a new professional base is created and can be encouraged to launch related businesses through entrepreneurial and industrial tax incentives.

For future historians the importance of TVA may lie more in its demonstration of organizational technique than in its embodiment of large-scale technology. Once the political world begins to debate the implications of an ever-normal energy supply, politicians may decide to review the legislation that led to organizing the TVA and the interstate compact that spawned the Hoover Dam.

FOR FURTHER REFERENCE

Books and Articles

Callahan, North. *TVA: Bridge over Troubled Waters*. Cranbury, NJ: Cornwall Books, 1981.

Chandler, W. U. *The Myth of TVA: Conservation and Development in the Tennessee Valley, 1933–1983*. Boston: Ballinger Publishing Company, 1984.

Clark, Wilson. *Energy for Survival*. New York: Anchor Press, 1974.

Culver, J.C., and J. Hyde. *American Dreamer: A Life of Henry A. Wallace*. New York: Norton, 2000.

Ezzell, Patricia Bernard. *TVA Photography: Thirty Years of Life in the Tennessee Valley*. Jackson: University Press of Mississippi, 2003.

Hubbard, P. J. *Origins of the TVA: The Muscle Shoals Controversy, 1920–1932*. Nashville, TN: Vanderbilt University Press, 1961.

Moore, John R., ed. *The Economic Impact of TVA*. Knoxville: University of Tennessee Press, 1967.

Internet

For general information on TVA, see http://www.tva.gov.

Roosevelt, Franklin D. *Message to Congress suggesting the Tennessee Valley Authority*. From FDR Library homepage, http://www.fdrlibrary.marist.edu/odtvacon.html. This Web site contains the complete text of FDR's speech.

For a complete source on FDR, including extensive materials on TVA, see http://newdeal.feri.org/.

For more information about the town of Norris, see http://tva.gov/heritage/norris/.

On Morris Udall's 1970–84 archives of the TVA, with a complete listing of the files and notations from all reports and committee documents, see http://dizzy.library.arizona.edu/branches/spc/udall/udallfindingaid/.

On the triumvirate of the TVA, Morgan, Morgan, and Lilienthal, see http://newdeal.
feri.org/tva/tva05.htm.

For a collection of references from magazines, newspapers, and press releases on the
TVA, plus pictures and maps, see http://www.encyclopedia.com/html/section/
TennVA_Bibliography.asp. (*Note:* HighBeam Research, LLC. Copyright 2004.)

To see the TVA logo, a fist grabbing the thunderbolt with the slogan "Electricity for
All," see http://newdeal.feri.org/tva/index.htm.

For a library of the many cartoons depicting the TVA, see http://newdeal.feri.org/toons/
toon10.htm.

Music

Armstrong, Louis. *Louis Armstrong Plays W. C. Handy.* Sony 64925, 1997.

For the music of W. C. Handy and the W. C. Handy Festival, see http://www.
wchandyfest.com/.

Documents of Authorization

AN ACT

To improve the navigability and to provide for the flood control of the Tennessee River; to provide for reforestation and the proper use of marginal lands in
the Tennessee Valley; to provide for the agricultural and industrial development
of said valley; to provide for the national defense by the creation of a corporation for the operation of Government properties at and near Muscle Shoals in
the State of Alabama, and for other purposes.

*Be it enacted by the Senate and House of Representatives of the United States
of America in Congress assembled,* That for the purpose of maintaining and operating the properties now owned by the United States in the vicinity of Muscle
Shoals, Alabama, in the interest of the national defense and for agricultural and
industrial development, and to improve navigation in the Tennessee River and to
control the destructive flow waters in the Tennessee River and Mississippi River
Basins, there is hereby created a body corporate by the name of the "Tennessee
Valley Authority" (hereinafter referred to as the "Corporation"). The board of
directors first appointed shall be deemed the incorporators, and the incorporation shall be held to have been effected from the date of the first meeting of the
board. This Act may be cited as the "Tennessee Valley Authority Act of 1933".

SEC. 2. (a) The board of directors of the Corporation (hereinafter referred
to as the "board") shall be composed of three members, to be appointed by the
President, by and with the advice and consent of the Senate. In appointing the
members of the board, the President shall designate the chairman. All other officials, agents, and employees shall be designated and selected by the board.

(b) The terms of office of the members first taking office after the
approval of this Act shall expire as designated by the President at the time of
nomination, one at the end of the third year, one at the end of the sixth year,

and one at the end of the ninth year, after the date of approval of this Act. A successor to a member of the board shall be appointed in the same manner as the original members and shall have a term of office expiring nine years from the date of the expiration of the term for which his predecessor was appointed.

(c) Any member appointed to fill a vacancy in the board occurring prior to the expiration of the term for which his predecessor was appointed shall be appointed for the remainder of such term.

(d) Vacancies in the board so long as there shall be two members in office shall not impair the powers of the board to execute the functions of the Corporation, and two of the members in office shall constitute a quorum for the transaction of the business of the board.

(e) Each of the members of the board shall be a citizen of the United States, and shall receive a salary at the rate of $10,000 a year, to be paid by the Corporation as current expenses. Each member of the board, in addition to his salary, shall be permitted to occupy as his residence one of the dwelling houses owned by the Government in the vicinity of Muscle Shoals, Alabama, the same to be designated by the President of the United States. Members of the board shall be reimbursed by the Corporation for actual expenses (including traveling and subsistence expenses) incurred by them in the performance of the duties vested in the board by this Act. No member of said board shall, during his continuance in office, be engaged in any other business, but each member shall devote himself to the work of the Corporation.

(f) No director shall have financial interest in any public-utility corporation engaged in the business of distributing and selling power to the public nor in any corporation engaged in the manufacture, selling, or distribution of fixed nitrogen or fertilizer, or any ingredients thereof, nor shall any member have any interest in any business that may be adversely affected by the success of the Corporation as a producer of concentrated fertilizers or as a producer of electric power.

(g) The board shall direct the exercise of all the powers of the Corporation.

(h) All members of the board shall be persons who profess a belief in the feasibility and wisdom of this Act.

SEC. 3. The board shall without regard to the provisions of Civil Service laws applicable to officers and employees of the United States, appoint such managers, assistant managers, officers, employees, attorneys, and agents, as are necessary for the transaction of its business, fix their compensation, define their duties, require bonds of such of them as the board may designate, and provide a system or organization to fix responsibility and promote efficiency. Any appointee of the board may be removed in the discretion of the board. No regular officer or employee of the Corporation shall receive a salary in excess of that received by the members of the board.

All contracts to which the Corporation is a party and which require the employment of laborers and mechanics in the construction, alteration, maintenance, or repair of buildings, dams, locks, or other projects shall contain a provision that not less than the prevailing rate of wages for a work of similar nature prevailing in the vicinity shall be paid to such laborers or mechanics.

In the event any dispute arises as to what are the prevailing rates of wages, the question shall be referred to the Secretary of Labor for determination, and his decision shall be final. In the determination of such prevailing rate or rates, due regard shall be given to those rates which have been secured through collective agreement by representatives of employers and employees.

Where such work as is described in the two preceding paragraphs is done directly by the Corporation the prevailing rate of wages shall be paid in the same manner as though such work had been let by contract.

Insofar as applicable, the benefits of the Act entitled "An Act to provide compensation for employees of the United States suffering injuries while in the performance of their duties, and for other purposes," approved September 7, 1916, as amended, shall extend to persons given employment under the provisions of this Act.

SEC. 4. Except as otherwise specifically provided in this Act, the Corporation—

(a) Shall have succession in its corporate name.

(b) May sue and be sued in its corporate name.

(c) May adopt and use a corporate seal, which shall be judicially noticed.

(d) May make contracts, as herein authorized.

(e) May adopt, amend, and repeal bylaws.

(f) May purchase or lease and hold such real and personal property as it deems necessary or convenient in the transaction of its business, and may dispose of any such personal property held by it.

The board shall select a treasurer and as many assistant treasurers as it deems proper, which treasurer and assistant treasurers shall give such bonds for the safe-keeping of the securities and moneys of the said Corporation as the board may require: *Provided,* That any member of said board may be removed from office at any time by a concurrent resolution of the Senate and the House of Representatives.

(g) Shall have such powers as may be necessary or appropriate for the exercise of the powers herein specifically conferred upon the Corporation.

(h) Shall have power in the name of the United States of America to exercise the right of eminent domain, and in the purchase of any real estate or the acquisition of real estate by condemnation proceedings, the title to such real estate shall be taken in the name of the United States of America, and thereupon all such real estate shall be intrusted to the Corporation as the agent of the United States to accomplish the purposes of this Act.

(i) Shall have power to acquire real estate for the construction of dams, reservoirs, transmission lines, power houses, and other structures, and navigation projects at any point along the Tennessee River, or any of its tributaries, and in the event that the owner or owners of such property shall fail and refuse to sell to the Corporation at a price deemed fair and reasonable by the board, then the Corporation may proceed to exercise the right of eminent domain, and to condemn all property that it deems necessary for carrying out the purpose of this Act, and all such condemnation proceedings shall be had pursuant to the provisions and requirements hereinafter specified, with reference to any and all condemnation proceedings.

(j) Shall have power to construct dams, reservoirs, power houses, power structures, transmission lines, navigation projects, and incidental works in the Tennessee River and its tributaries, and to unite the various power installations into one or more systems by transmission lines.

SEC. 5. The board is hereby authorized—

(a) To contract with commercial producers for the production of such fertilizers or fertilizer materials as may be needed in the Government's program of development and introduction in excess of that produced by Government plants. Such contracts may provide either for outright purchase of materials by the board or only for the payment of carrying charges on special materials manufactured at the board's request for its program.

(b) To arrange with farmers and farm organizations for large-scale practical use of the new forms of fertilizers under conditions permitting an accurate measure to the economic return they produce.

(c) To cooperate with National, State, district, or county experimental stations or demonstration farms, for the use of new forms of fertilizer or fertilizer practices during the initial or experimental period of their introduction.

(d) The board in order to improve and cheapen the production of fertilizer is authorized to manufacture and sell fixed nitrogen, fertilizer, and fertilizer ingredients at Muscle Shoals by the employment of existing facilities, by modernizing existing plants, or by any other process or processes that in its judgment shall appear wise and profitable for the fixation of atmospheric nitrogen or the cheapening of the production of fertilizer.

(e) Under the authority of this Act the board may make donations or sales of the product of the plant or plants operated by it to be fairly and equitably distributed through the agency of county demonstration agents, agricultural colleges, or otherwise as the board may direct, for experimentation, education, and introduction of the use of such products in cooperation with practical farmers so as to obtain information as to the value, effect, and best methods of their use.

(f) The board is authorized to make alterations, modifications, or improvements in existing plants and facilities, and to construct new plants.

(g) In the event it is not used for the fixation of nitrogen for agricultural purposes or leased, then the board shall maintain in stand-by condition nitrate plant numbered 2, or its equivalent, for the fixation of atmospheric nitrogen, for the production of explosives in the event of war or a national emergency, until the Congress shall by joint resolution release the board from this obligation, and if any part thereof be used by the board for the manufacture of phosphoric acid or potash, the balance of nitrate plant numbered 2 shall be kept in stand-by condition.

(h) To establish, maintain, and operate laboratories and experimental plants, and to undertake experiments for the purpose of enabling the Corporation to furnish nitrogen products for military purposes, and nitrogen and other fertilizer products for agricultural purposes in the most economical manner and at the highest standard of efficiency.

(i) To request the assistance and advice of any officer, agent, or employee of any executive department or of any independent office of the United States, to enable the Corporation the better to carry out its powers successfully, and as far as practical shall utilize the services of such officers, agents, and employees, and the President shall, if in his opinion, the public interest, service, or economy so require, direct that such assistance, advice, and service be rendered to the Corporation, and any individual that may be by the President directed to render such assistance, advice, and service shall be thereafter subject to the orders, rules, and regulations of the board: *Provided* That any invention or discovery made by virtue of and incidental to such service by an employee of the Government of the United States serving under this section, or by any employee of the Corporation, together with any patents which may be granted thereon, shall be the sole and exclusive property of the Corporation, which is hereby authorized to grant such licenses thereunder as shall be authorized by the board: *Provided further,* That the board may pay to such inventor such sum from the income from sale of licenses as it may deem proper.

(j) Upon the requisition of the Secretary of War or the Secretary of the Navy to manufacture for and sell at cost to the United States explosives or their nitrogenous content.

(k) Upon the requisition of the Secretary of War the Corporation shall allot and deliver without charge to the War Department so much power as shall be necessary in the judgment of said Department for use in operation of all locks, lifts, or other facilities in aid of navigation.

(l) To produce, distribute, and sell electric power, as herein particularly specified.

(m) No products of the Corporation shall be sold for use outside of the United States, its Territories and possession except to the United States Government for the use of its Army and Navy, or to its allies in case of war.

(n) The President is authorized, within twelve months after the passage of this Act, to lease to any responsible farm organization or to any corporation organized by it nitrate plant numbered 2 and Waco Quarry, together with the railroad connecting said quarry with nitrate plant numbered 2, for a term not exceeding fifty years at a rental of not less than $1 per year, but such authority shall be subject to the express condition that the lessee shall use said property during the term of said lease exclusively for the manufacture of fertilizer and fertilizer ingredients to be used only in the manufacture of fertilizer by said lessee and sold for use as fertilizer. The said lessee shall covenant to keep said property in first-class condition, but the lessee shall be authorized to modernize said plant numbered 2 by the installation of such machinery as may be necessary, and is authorized to amortize the cost of said machinery and improvements over the term of said lease or any part thereof. Said lease shall also provide that the board shall sell to the lessee power as it charges all other customers for power in the same class and quantity. Said lease shall also provide that, if the said lessee does not desire to buy power of the publicly owned plant, it shall have the right to purchase its power for the operation of said plant of the Alabama Power Company or any other publicly or privately owned corporation engaged in the generation and sale of electric power, and in such case the lease shall provide further that the said lessee shall have a free right of way to build a transmission line over Government property to said plant paying the actual expenses and damages, if any, incurred by the Corporation on account of such line. Said lease shall also provide that the said lessee shall covenant that during the term of said lease the said lessee shall not enter into any illegal monopoly, combination, or trust with any privately owned corporation engaged in the manufacture, production, and sale of fertilizer with the object or effect of increasing the price of fertilizer to the farmer.

SEC. 6. In the appointment of officials and the selection of employees for said Corporation, and in the promotion of any such employees or officials, no political test or qualification shall be permitted or given consideration, but all such appointments and promotions shall be given and made on the basis of merit and efficiency. Any member of said board who is found by the President of the United States to be guilty of a violation of this section shall be removed from office by the President of the United States, and any appointee of said board who is found by the board to be guilty of a violation of this section shall be removed from office by said board.

SEC. 7. In order to enable the Corporation to exercise the powers and duties vested in it by this Act—

(a) The exclusive use, possession, and control of the United States nitrate plans numbered 1 and 2, including steam plants, located, respectively, at Sheffield, Alabama, and Muscle Shoals, Alabama, together with all real estate and buildings connected therewith, all tools and machinery, equipment, accessories, and materials belonging thereto, and all laboratories and plants used as auxiliaries thereto; the fixed-nitrogen research laboratory, the Waco limestone quarry, in Alabama, and Dam Numbered 2, located at Muscle Shoals, its power house, and all hydroelectric and operating appurtenances (except the locks), and all machinery, lands, and buildings in connection therewith, and all appurtenances thereof, and all other property to be acquired by the Corporation in its own name or in the name of the United States of America, are hereby instructed to the Corporation for the purposes of this Act.

(b) The President of the United States is authorized to provide for the transfer to the Corporation of the use, possession, and control of such other real or personal property of the United States as he may from time to time deem necessary and proper for the purposes of the Corporation is herein stated.

SEC. 8. (a) The Corporation shall maintain its principal office in the immediate vicinity of Muscle Shoals, Alabama. The Corporation shall be held to be an inhabitant and resident of the northern judicial district of Alabama within the meaning of the laws of the United States relating to the venue of civil suits.

(b) The Corporation shall at all times maintain complete and accurate books of accounts.

(c) Each member of the board, before entering upon the duties of his office, shall subscribe to an oath (or affirmation) to support the Constitution of the United States and to faithfully and impartially perform the duties imposed upon him by this Act.

SEC. 9. (a) The board shall file with the President and with the Congress, in December of each year, a financial statement and a complete report as to the business of the Corporation covering the preceding governmental fiscal year. This report shall include an itemized statement of the cost of power at each power station, the total number of employees and the names, salaries, and duties of those receiving compensation at the rate of more than $1,500 a year.

(b) The Comptroller General of the United States shall audit the transactions of the Corporation at such times as he shall determine, but not less frequently than once each governmental fiscal year, with personnel of his selection. In such connection he and his representative shall have free and open access to all papers, books, records, files, accounts, plants, warehouses, offices, and all other things, property and places belonging to or under the control of or used or employed by the Corporation, and shall be afforded full facilities for counting all cash and verifying transactions with and balances in depositaries. He shall make report of each such audit in quadruplicate, one copy for the President of the United States, one for the chairman of the board, one for public inspection at the principal office of the corporation, and the other to be retained by him for the uses of the Congress. The expenses for each such audit may be paid from moneys advanced therefor by the Corporation, or from any appropriation or appropriations for the General Accounting Office, and appropriations so used shall be reimbursed promptly by the Corporation as billed by the Comptroller General. All such audit expenses shall be charged to operating expenses of the Corporation. The Comptroller General shall make special report to the President of the United States and to the Congress of any transaction or condition found by him to be in conflict with the powers or duties intrusted to the Corporation by law.

SEC. 10. The board is hereby empowered and authorized to sell the surplus power not used in its operations, and for operation of locks and other works generated by it, to States, counties, municipalities, corporations, partnerships, or individuals, according to the policies hereinafter set forth; and to carry out said authority, the board is authorized to enter into contracts for such sale for a term not exceeding twenty years, and in the sale of such current by the board it shall give preference to States, counties, municipalities, and cooperative organizations of citizens or farmers, not organized or doing business for profit, but primarily for the purpose of supplying electricity to its own citizens or members: *Provided,* That all contracts made with private companies or individuals for the sale of power, which power is to be resold for a profit, shall contain a provision authorizing the board to cancel said contract upon five years' notice in writing, if the board needs said power to supply the demands of States, counties, or municipalities. In order to promote and encourage the fullest possible use of electric light and power on farms within reasonable distance of any of its transmission lines the board in its discretion shall have power to construct transmission lines to farms and small villages that are not otherwise supplied with electricity at reasonable rates, and to make such rules and regulations governing such sale and distribution of such electric power as in its judgment may be just and equitable: *Provided further,* That the board is hereby authorized and directed to make better use of electric power for agricultural and domestic use, or for small or local industries, and it may cooperate with State governments, or their subdivisions or agencies, with educational or research institutions, and with cooperatives or other organizations, in the application of electric power to the fuller and better balanced development of the resources of the region.

SEC. 11. It is hereby declared to be the policy of the Government so far as practical to distribute and sell the surplus power generated at Muscle Shoals equitably among the States, counties, and municipalities within transmission distance. This policy is further declared to be that the projects herein provided for shall be considered primarily as for the benefit of the people of the section as a whole and particularly the domestic and rural consumers to whom the power can economically be made available, and accordingly that sale to and use by industry shall be a secondary purpose, to be utilized principally to secure a sufficiently high load factor and revenue returns which will permit domestic and rural use at the lowest possible rates and in such manner as to encourage increased domestic and rural use of electricity. It is further hereby declared to be the policy of the Government to utilize the Muscle Shoals properties so far as may be necessary to improve, increase, and cheapen the production of fertilizer ingredients by carrying out the provisions of this Act.

SEC. 12. In order to place the board upon a fair basis for making such contracts and for receiving bids for the sale of such power, it is hereby expressly authorized, either from appropriations made by Congress or from funds secured from the sale of such power, or from funds secured by the sale of bonds hereafter provided for, to construct, lease, purchase, or authorize the construction of transmission lines within transmission distance from the place where generated, and to interconnect with other systems. The board is also authorized to lease to any person, persons, or corporation the use of any transmission line owned by the Government and operated by the board, but no such lease shall be made that in any way interferes with the use of such transmission line by the board: *Provided,* That if any State, county, or municipality, or other public or cooperative organization of citizens or farmers, not organized or doing business for profit, but primarily for the purpose of supplying electricity to its own citizens or members, or any two or more of such municipalities or organizations, shall construct or agree to construct and maintain a properly designed and built transmission line to the Government reservation upon which is located a Government generating plant, or to a main transmission line owned by the Government or leased by the board and under the control of the board, the board is hereby authorized and directed to contract with such State, county, municipality, or other organization, or two or more of them, for the sale of electricity for a term not exceeding thirty years; and in any such case the board shall give to such State, county, municipality, or other organization ample time to fully comply with any local law now in existence or hereafter enacted providing for the necessary legal authority for such State, county, municipality, or other organization to contract with the board for such power: *Provided further,* That all contracts entered into between the Corporation and any municipality or other political subdivision or cooperative organization shall provide that the electric power shall be sold and distributed to the ultimate consumer without discrimination as between consumers of the same class, and such contract shall be voidable at the election of the board if a discriminatory rate, rebate, or other special concession is made or given to any consumer or user by the municipality or other political subdivision or cooperative organization: *And provided further,* That as to any surplus power not so sold as above provided to States, counties, municipalities, or other said organizations, before the board shall sell the same to any person or corporation engaged in the distribution and resale of electricity for profit, it shall require said person or corporation to agree that any resale of such electric power by said person or corporation shall be made to the ultimate consumer of such electric power at prices that shall not exceed a schedule fixed by the board from time to time as reasonable, just, and fair; and in case of any such sale, if an amount is charged the ultimate consumer which is in excess of the price so deemed to be just, reasonable, and fair by the board, the contract for such sale between the board and such distributor of electricity shall be voidable at the election of the board: *And provided further,* That the board is hereby authorized to enter into contracts with other power systems for the mutual exchange of unused excess power upon suitable terms, for the conservation of store water, and as an emergency or break-down relief.

SEC. 13. Five per centum of the gross proceeds received by the board for the sale of power generated at Dam Numbered 2, or from any other hydropower plant hereafter constructed in the State of Alabama, shall be paid to the State of Alabama; and 5 per centum of the gross proceeds from the sale of power generated at Cove Creek Dam, hereinafter provided for, or any other dam located in the State of Tennessee, shall be paid to the State of Tennessee. Upon the completion of said Cove Creek Dam the board shall ascertain how much additional power is thereby generated at Dam Numbered 2 and at any other dam hereafter constructed by the Government of the United States on the Tennessee River, in the State of Alabama, or in the State of Tennessee, and from the gross proceeds of the sale of such additional power 2 1/2 per centum shall be paid to the State of Alabama and 2 1/2 per centum to the State of Tennessee. These percentages shall apply to any other dam that may hereafter be constructed and controlled and operated by the board on the Tennessee River or any of its tributaries, the main purpose of which is to control flood waters and where the development of electric power is incidental to the operation of such flood-control dam. In ascertaining the gross proceeds from the sale of such power upon which a percentage is paid to the States of Alabama and Tennessee, the board shall not take into consideration the proceeds of any power sold or delivered to the Government of the United States, or any department or agency of the Government of the United States, used in the operation of any locks on the Tennessee River or for any experimental purpose, or for the manufacture of fertilizer or any of the ingredients thereof, or for any other governmental purpose: *Provided,* That the percentages to be paid to the States of Alabama and Tennessee, as provided in this section, shall be subject to revision and change by the board, and any new percentages established by the board, when approved by the President, shall remain in effect until and unless again changed by the board with the approval of the President. No change of said percentages shall be made more often than once in five years, and no change shall be made without giving to the States of Alabama and Tennessee an opportunity to be heard.

SEC. 14. The board shall make a thorough investigation as to the present value of Dam Numbered 2, and the steam plants at nitrate plant numbered 1, and nitrate plant numbered 2, and as to the cost of Cove Creek Dam, for the purpose of ascertaining how much of the value or the cost of said properties shall be allocated and charged up to (1) flood control, (2) navigation, (3) fertilizer, (4) national defense, and (5) the development of power. The findings thus made by the board, when approved by the President of the United States, shall be final, and such findings shall thereafter be used in all allocation of value for the purpose of keeping the book value of said properties. In like manner, the cost and book value of any dams, steam plants, or other similar improvements hereafter constructed and turned over to said board for the purpose of control and management shall be ascertained and allocated.

SEC. 15. In the construction of any future dam, steam plant, or other facility, to be used in whole or in part for the generation or transmission of electric power the board is hereby authorized and empowered to issue on the credit of the United States and to sell serial bonds not exceeding $50,000,000 in amount, having a maturity not more than fifty years from the date of issue thereof, and bearing interest not exceeding 3 1/2 per centum per annum. Said bonds shall be issued and sold in amounts and prices approved by the Secretary of the Treasury, but all such bonds as may be so issued and sold shall have equal rank. None of said bonds shall be sold below par, and no fee, commission, or compensation whatever shall be paid to any person, firm, or corporation for handling, negotiating the sale, or selling the said bonds. All of such bonds so issued and sold shall have all the rights and privileges accorded by law to Panama Canal bonds, authorized by section 8 of the Act of June 28, 1902, chapter 1302, as amended by the Act of December 21, 1905 (ch. 3, sec. 1, 34 Stat. 5), as not compiled in section 743 of title 31 of the United States Code. All funds derived from the sale of such bonds shall be paid over to the Corporation.

SEC. 16. The board, whenever the President deems it advisable, is hereby empowered and directed to complete Dam Numbered 2 at Muscle Shoals, Alabama, and the steam plant at nitrate plant Numbered 2, in the vicinity of Muscle Shoals, by installing Dam Numbered 2 the additional power united according to the plans and specifications of said dam, and the additional power unit in the steam plant at nitrate plant Numbered 2.

SEC. 17. The Secretary of War, or the Secretary of the Interior, is hereby authorized to construct, either directly or by contract to the lowest responsible bidder, after due advertisement, a dam in and across Clinch River in the State of Tennessee, which has by long custom become known and designated as the Cove Creek Dam, together with the transmission line from Muscle Shoals, according to the latest and most approved designs, including power house and hydroelectric installations and equipment for the generation of power, in order that the waters of the said Clinch River may be impounded and stored above said dam for the purpose of increasing and regulating the flow of the Clinch River and the Tennessee River below, so that the maximum amount of primary power may be development at Dam Numbered 2 and at any and all other dams below the said Cove Creek Dam: *Provided, however,* That the President is hereby authorized by appropriate order to direct the employment by the Secretary of War, or by the Secretary of the Interior, of such engineer or engineers as he may designate, to perform such duties and obligations as he may deem proper, either in the drawing of plans and building or construction of the same. The president may, by such order, place the control of the construction of said dam in the hands of such engineer or engineers taken from private life as he may desire: *And provided further,* That the President is hereby expressly authorized, without regard to the restriction or limitation of any other statute, to select attorneys and assistants for the purpose of making any investigation he may deem proper to ascertain whether, in the control and management of Dam Numbered 2, or any other dam or property owned by the Government in the Tennessee River Basin, or in the authorization of any improvement therein, there has been any undue or unfair advantage given to private persons, partnerships, or corporations, by any officials or employees of the Government, or whether in any such matters the Government has been injured, or unjustly deprived of any of its rights.

SEC. 18. In order and enable and empower the Secretary of War, the Secretary of the Interior, or the Board to carry out the authority hereby conferred, in the most economical and efficient manner, he or it is hereby authorized and empowered in the exercise of the powers of national defense in aid of navigation, and in the control of the flood waters of the Tennessee and Mississippi Rivers, constituting channels of interstate commerce, to exercise the right of eminent domain for all purposes of this Act, and to condemn all lands, easements, rights of way, and other area necessary in order to obtain a site for said Cove Creek Dam, and the flowage rights for the reservoir of water above said dam, and to negotiate and conclude contracts with the States, counties, municipalities, and all State agencies and with railroads, railroad corporations, common carriers, and all public utility commissions and any other person, firm, or corporation, for the relocation of railroad tracks, highways, highway bridges, mills, ferries, electric-light plants, and any and all other properties, enterprises, and projects whose removal may be necessary in order to carry out the provisions of this Act. When said Cove Creek Dam, transmission line, and power house shall have been completed, the possession, use, and control thereof shall be intrusted to the Corporation for use and operation in connection with the general Tennessee Valley project, and to promote flood control and navigation in the Tennessee River.

SEC. 19. The Corporation, as an instrumentality and agency of the Government of the United States for the purpose of executing its constitutional powers, shall have access to the Patent Office of the United States for the purpose of studying, ascertaining, and copying all methods, formulae, and scientific information (not including access to pending applications for patents) necessary to enable the Corporation to use and employ the most efficacious and economical process for the production of fixed nitrogen, or any essential ingredient of fertilizer, or any method of improving and cheapening the production of hydroelectric power, and any owner of a patent whose patent rights may have been thus in any way copied, used, infringed, or employed by the exercise of this authority by the Corporation shall have as the exclusive remedy a cause of action against the Corporation to be instituted and prosecuted on the equity side of the appropriate district court of the United States, for the recovery of reasonable compensation for such infringement. The Commissioner of Patents shall furnish to the Corporation, at its request and without payment of fees, copies of documents on file in his office: *Provided,* That the benefits of this section shall not apply to any art, machine, methods of manufacture, or composition of matter, discovered or invented by such employee during the time of his employment or service with the Corporation or with the Government of the United States.

SEC. 20. The Government of the United States hereby reserves the right, in case of war or national emergency declared by Congress, to take possession of all or any part of the property described or referred to in this Act for the purpose of manufacturing explosives or for other war purposes; but, if this right is exercised by the Government, it shall pay the reasonable and fair damages that may be suffered by any party whose contract for the purchase of electric power or fixed nitrogen or fertilizer ingredients is hereby violated, after the amount of the damages has been fixed by the United States Court of Claims in proceedings instituted and conducted for the purpose under rules prescribed by the court.

SEC. 21. (a) All general penal statutes relating to the larceny, embezzlement, conversion, or to the improper handling, retention, use, or disposal of public moneys or property of the United States, shall apply to the moneys and property of the Corporation and to moneys and properties of the United States intrusted to the Corporation.

(b) Any person who, with intent to defraud the Corporation, or to deceive any director, officer, or employee of the Corporation or any officer or employee of the United States (1) makes any false entry in any book of the Corporation, or (2) makes any false report or statement for the Corporation, shall, upon conviction thereof, be fined not more than $10,000 or imprisoned not more than five years, or both.

(c) Any person who shall receive any compensation, rebate, or reward, or shall enter into any conspiracy, collusion, or agreement, express or implied, with intent to defraud the Corporation or wrongfully and unlawfully to defeat its purposes, shall, on conviction thereof, be fined not more than $5,000 or imprisoned not more than five years, or both.

SEC. 22. To aid further the proper use, conservation, and development of the natural resources of the Tennessee River drainage basin and of such adjoining territory as may be related to or materially affected by the development consequent to this Act, and to provide for the general welfare of the citizens of said areas, the President is hereby authorized, by such means or methods as he may deem proper within the limits of appropriations made therefor by Congress, to make such surveys of and general plans for said Tennessee basin and adjoining territory as may be useful to the Congress and to the several States in guiding and controlling the extent, sequence, and nature of development that may be equitably and economically advanced through the expenditure of public funds, or through the guidance or control of public authority, all for the general purpose of fostering an orderly and proper physical, economic, and social development of said areas; and the President is further authorized in making said surveys and plans to cooperate with the States affected thereby, or subdivisions or agencies of such States, or with cooperative or other organizations, and to make such studies, experiments, or demonstrations as may be necessary and suitable to that end.

SEC. 23. The President shall, from time to time, as the work provided for in the preceding section progresses, recommend to Congress such legislation as he deems proper to carry out the general purposes stated in said section, and for the especial purpose of bringing about in said Tennessee drainage basin and adjoining territory in conformity with said general purposes (1) the maximum amount of flood control; (2) the maximum development of said Tennessee River for navigation purposes; (3) the maximum generation of electric power consistent with flood control and navigation; (4) the proper use of marginal lands; (5) the proper method of reforestation of all lands in said drainage basin suitable for reforestation; and (6) the economic and social well-being of the people living in said river basin.

SEC. 24. For the purpose of securing any rights of flowage, or obtaining title to or possession of any property, real or personal, that may be necessary or may become necessary, in the carrying out of any of the provisions of this Act, the President of the United States for a period of three years from the date of the enactment of this Act, is hereby authorized to acquire title in the name of the United States to such rights or such property, and to provide for the payment for same by directing the board to contract to deliver power generated at any of the plants now owned or hereafter owned or constructed by the Government or by said Corporation, such future delivery of power to continue for a period not exceeding thirty years. Likewise, for one year after the enactment of this Act, the President is further authorized to sell or lease any parcel or part of any vacant real estate now owned by the Government in said Tennessee River Basin, to persons, firms, or corporations who shall contract to erect thereon factories or manufacturing establishments, and who shall contract to purchase of said Corporation electric power for the operation of any such factory or manufacturing establishment. No contract shall be made by the President for the sale of any of such real estate as may be necessary for present or future use on the part of the Government for any of the purposes of this Act. Any such contract made by the President of the United States shall be carried out by the board: *Provided,* That no such contract shall be made that will in any way abridge or take away the preference right to purchase power given in this Act to States, counties, municipalities, or farm organizations: *Provided further,* That no lease shall be for a term to exceed fifty years: *Provided further,* That any sale shall be on condition that said land shall be used for industrial purposes only.

SEC. 25. The Corporation may cause proceedings to be instituted for the acquisition by condemnation of any lands, easements, or rights of way which, in the opinion of the Corporation, are necessary to carry out the provisions of this Act. The proceedings shall be instituted in the United States district court for the district in which the land, easement, right of way, or other interest, or any part thereof, is located, and such court shall have full jurisdiction to divest the complete title to the property sought to be acquired out of all persons or claimants and vest the same of the United States in fee simple, and to enter a decree quieting the title thereto in the United States of America.

Upon the filing of a petition for condemnation and for the purpose of ascertaining the value of the property to be acquired, and assessing the compensation to be paid, the court shall appoint three commissioners who shall be disinterested persons and who shall take and subscribe an oath that they do not own any lands, or interest or easement in any lands, which it may be desirable for the United States to acquire in the furtherance of said project, and such commissioners shall not be selected from the locality wherein the land sought to be condemned lies. Such commissioners shall receive a per diem of not to exceed $15 for their services, together with an additional amount of $5 per day for subsistence for time actually spent in performing their duties as commissioners.

It shall be the duty of such commissioners to examine into the value of the lands sought to be condemned, to conduct hearings and receive evidence, and generally to take such appropriate steps as may be proper for the determination of the value of the said lands sought to be condemned, and for such purpose the commissioners are authorized to administer oaths and subpoena witnesses, which said witnesses shall receive the same fees as are provided for witnesses in the Federal courts. The said commissioners shall thereupon file a report setting forth their conclusions as to the value of the said property sought to be condemned, making a separate award and valuation in the premises with respect to each separate parcel involved. Upon the filing of such award in court the clerk of said court shall give notice of the filing of such award to the parties to said proceeding, in manner and form as directed by the judge of said court.

Either or both parties may file exceptions to the award of said commissioners within twenty days from the date of the filing of said award in court. Exceptions filed to such award shall be heard before three Federal district judges unless the parties, in writing, in person, or by their attorney, stipulate that the exceptions may be heard before a lesser number of judges. On such hearing such judges shall pass *de novo* upon the proceedings had before the commissioners, may view the property, and may take additional evidence. Upon such hearings that said judges shall file their own award, fixing therein the value of the property sought to be condemned, regardless of the award previously made by the said commissioners.

At any time within thirty days from the filing of the decision of the district judges upon the hearing on exceptions to the award made by the commissioners, either party may appeal from such decision of the said judges to the circuit court of appeals, and the said circuit court of appeals shall upon the hearing on said appeal dispose of the same upon the record, without regard to the awards or findings theretofore made by the commissioners of the district judges, and such circuit court of appeals shall thereupon fix the value of the said property sought to be condemned.

Upon acceptance of an award by the owner of any property herein provided to be appropriated, and the payment of the money awarded or upon the failure of either party to file exceptions to the award of the commissioners within the time specified, or upon the award of the commissioners, and the payment of the money by the United States pursuant thereto, or the payment of the money awarded into the registry of the court by the Corporation, the title to said property and the right to possession thereof shall pass to the United States, and the United States shall be entitled to a writ in the same proceeding to dispossess the former owner of said property, and all lessees, agents, and attorneys of such former owner, and to put the United States, by its corporate creature and agent, the Corporation, into possession of said property.

In the event of any property owned in whole or in part by minors, or insane persons, or incompetent persons, or estates of deceased persons, then the legal representatives of such minors, insane persons, incompetent persons, or estates shall have power, by and with the consent and approval of the trial judge in whose court said matter is for determination, to consent to or reject the awards of the commissioners herein provided for, and in the event that there be no legal representatives, or that the legal representatives for such minors, insane persons, or incompetent persons shall fail or decline to act, then such trial judge may, upon motion, appoint a guardian *ad litem* to act for such minors, insane persons, or incompetent persons, and such guardian *ad litem* shall act to the full extent and to the same purpose and effect as his ward could act, if competent, and such guardian *ad litem* shall be deemed to have full power and authority to respond, to conduct, or to maintain any proceeding herein provided for affecting his said ward.

SEC. 26. The net proceeds derived by the board from the sale of power and any of the products manufactured by the Corporation, after deducting the cost of operation, maintenance, depreciation, amortization, and an amount deemed by the board as necessary to withhold as operating capital, or devoted by the board to new construction, shall be paid into the Treasury of the United States at the end of each calendar year.

SEC. 27. All appropriations necessary to carry out the provisions of the Act are hereby authorized.

SEC. 28. That all Acts or parts of Acts in conflict herewith are hereby repealed, so far as they affect the operations contemplated by this Act.

SEC. 29. The right to alter, amend, or repeal this Act is hereby expressly declared and reserved, but no such amendment or repeal shall operate to impair the obligation of any contract made by said Corporation under any power conferred by this Act.

SEC. 30. The sections of this Act are hereby declared to be separable, and in the event any one or more sections of this Act be held to be unconstitutional, the same shall not affect the validity of other sections of this Act.

Approved, May 18, 1933.

From *Congressional Record,* H.R. 5081, 73rd Cong., 1st sess. (May 18, 1933), chap. 21. See also: http://www.tva.gov/abouttva/pdf/TVA_ACT.pdf.

26
The Manhattan Project and the Atomic Energy Act

United States

DID YOU KNOW . . . ?

➤ A single letter from Albert Einstein prompted the American decision to develop an atomic bomb.

➤ The Manhattan Project ran from 1942 to 1945, with an overall cost of $20 billion.

➤ By 1943, 66,000 people were working on the project; later there were 75,000.

➤ The first atomic-bomb test was on July 16, 1945.

➤ The bombs had names: Gadget, Little Boy, and Fat Man. A fourth bomb was never used.

➤ Hitler abandoned the idea of developing an atomic bomb.

➤ Several German scientists escaped Nazi Germany and eventually helped develop the atomic bomb in the United States.

In a supreme irony, Hitler followed the advice of his military men and abandoned the idea of developing atomic weapons; so far as is known, there were none in Germany during World War II. Equally ironic is that several indispensable scientists who made U.S. atomic research and development possible were themselves refugees from Germany. If the same group of scientists and physicists had helped Hitler make an atom bomb, World War II might have had an entirely different outcome. It is true that Germany had lost the war before the atomic bomb was dropped on Hiroshima. However, if Germany had retained

Atomic bomb test, part of Exercise Desert Rock, Yucca Flats, Nevada, 1951. Courtesy of the National Archives and Records Administration.

its scientists, if they had cooperated in a program similar to the Manhattan Project, and if they had successfully developed the atomic bomb, history might have been written far differently.

HISTORY

The Manhattan Project, which created the first atomic bomb, was a massive U.S. research and development program extending from 1942 to 1945.

In 1939, on the eve of World War II, American scientists alerted the White House to a research project they believed possessed enormous military potential. Late the preceding year, German scientists had successfully split (fissioned) uranium atoms. In the United States, scientists Enrico Fermi (1938 Nobel laureate in physics) and Leo Szilard realized that the enormous energy released by fission, if harnessed in a chain reaction, could be used to make "extremely powerful bombs of a new type," as stated in a letter written by physicist Albert Einstein to President Franklin D. Roosevelt on August 2, 1939. That letter, reproduced here, suggests that research should be stepped up, especially since Germany—rumored to be at work on new weapons—controlled a major source of uranium. Alexander Sachs, an economist and friend of Roosevelt's, delivered Einstein's letter, along with an explanatory letter of his own (also reproduced here), and a memo from Szilard summing up the current science on fission and underscoring the need for the United States to secure a supply of uranium.

However, in September 1939 Germany invaded Poland, and Roosevelt was preoccupied with the eruption of war in Europe. Sachs did not see Roosevelt until the following month, when he finally passed along the letters. A few days later, Roosevelt sent a reply to Einstein (also reproduced here), saying he had appointed a board to look into Einstein's proposal.

Despite skeptical scientists who believed production of such weapons would take years, a British report in mid-1941 supported the possibility and urged that it be given high priority. American urgency heightened with the December 1941 entry of the United States into the war. It was clear, however, that a huge number of scientific and engineering issues would need to be resolved to create such a weapon. Aside from the enormous complexity of designing and building an atomic bomb, there was the difficult problem of obtaining a supply of fissionable material.

Roosevelt placed the project under the supervision of the War Department and signaled that time was "very much of the essence" in a memo dated March 11, 1942 (reproduced here), and addressed to Vannevar Bush, head of the Office of Scientific Research and Development. Bush had reported that the probability of a successful weapon was greater than previously believed, and that if the project were fast-tracked, a bomb could be produced as early as 1944.

The project's code name stemmed from the fact that some preliminary research had been done at Columbia University in Manhattan, New York, but now work proceeded in a network of scientific and production facilities. Just as Einstein had warned, Germany blocked off supplies of uranium from Czechoslovakia, so the United States had to obtain it from Belgium. However, in late 1942 in a reactor at the University of Chicago, Fermi achieved the first controlled, self-sustaining fission chain reaction. Soon thereafter plants for

Scientists and workmen rig the world's first atomic bomb, raising it up into a 100-foot tower at the Trinity bomb test site in the desert near Alamogordo, New Mexico, 1945. Courtesy of AP / Wide World Photos.

producing fissionable materials (uranium 235 and plutonium 239) were built in Tennessee (at the Clinton Engineer Works, later called Oak Ridge) and Washington State (at the Hanford Engineering Works). Much of the actual construction of a bomb was done at a new laboratory established in a remote location at Los Alamos, New Mexico.

The first test of the new weapon took place at Alamogordo Air Base in New Mexico in the early morning of July 16, 1945. Upon explosion there was an intense light, followed by a blast of heat, then a powerful shock wave; a growing fireball appeared, giving way to a mushroom-shaped cloud rising approximately 40,000 feet (12,000 meters) into the atmosphere. As he watched, Los Alamos Lab Director J. Robert Oppenheimer said it reminded him of a passage from the *Bhagavad Gita:* "I am become death, / The destroyer of worlds" (http://www. pbs.org/wgbh/amex/bomb/peopleevents/pandeAMEX65.html). The following month a uranium bomb was dropped on Hiroshima, Japan; three days later a plutonium bomb was dropped on Nagasaki, and within days Japan surrendered. A fourth bomb went unused.

CULTURAL CONTEXT

At both the scientific and political levels, many worried about the weapons they feared others were developing. This factor created a culture of fear—and innovation.

Albert Einstein's 1939 letter to President Roosevelt occurred several years prior to the decision to drop the bomb on Japan; the actual bombing occurred in 1945, during the Truman administration.

One of the first considerations was financial. As it turned out, the atomic devices were surprisingly economical. The overall cost of the Manhattan Project was estimated to be $20 billion (1996 dollars) for *Gadget* (dropped on Alamogordo, July 16, 1945), *Little Boy* (Hiroshima, August 6, 1945), and *Fat Man* (Nagasaki, August 9, 1945), and a fourth bomb that was not used. War expenditures for smaller bombs, mines, and grenades cost $31.5 billion. Guns and rifles cost $24 billion, not including ammunition. Tanks were the biggest expense, at $64 billion (http://www.brook.edu/fp/projects/nucwcost/manhattn. htm). So, relatively speaking, $20 billion for four atomic bombs was actually cheaper than traditional equipment.

While December 7, 1941, the bombing of Pearl Harbor by the Japanese, was the date when the United States entered the war, it was not until sometime later that the U.S. government seriously considered using the bomb. The documents featured here predate that decision but give the reader some sense of the planning that went into later decisions. The first document is the letter that forever changed society's concept of war—and brought all its accompanying fears into stark relief. The letter, written August 2, 1939, from his home on Long Island, was signed by Dr. Albert Einstein, who reported, "Some recent work by E. Fermi and L. Szilard, which has been communicated to me in manuscript, leads me to expect that the element uranium may be turned into a new and important source of energy in the immediate future. Certain aspects of this situation which has arisen seem to call for watchfulness and, if necessary, quick action on the part of the Administration."

Such letters are not simply sent through the post; they are presented. Einstein's letter was hand-carried by Alexander Sachs, who wrote to President Roosevelt requesting a personal meeting. Owing to urgent demands of the war, it was not until October 19, 1939, that President Roosevelt finally responded to Einstein's letter, saying, "I found this data of such import that I have convened a Board consisting of the Bureau of Standards and a chosen representative of the Army and Navy to thoroughly investigate the possibilities of your suggestion regarding the element of uranium. I am glad to say that Dr. Sachs will cooperate and work with this committee. . . . Please accept my sincere thanks."

PLANNING

Planning began in 1939. From that time on, much occurred, but most of it was necessarily undocumented because of its top-secret nature. Tensions of the time can be seen in a deal between President Roosevelt and British prime minister Winston Churchill. Churchill had been holding back approval of Operation Overlord, the joint British/American plan for a cross-channel invasion of Europe. But Churchill was pressing hard for information on the atomic bomb. Roosevelt had thus far followed the advice of his security advisors who cautioned against letting the British into American secrets. However, Roosevelt suddenly relaxed his position, and thereafter Churchill sent a British team to Los Alamos to work under General Leslie Groves, military head of the Manhattan Project, and Robert Oppenheimer, the scientific director. At the same time, Churchill withdrew his objections to Operation Overlord, which took place on June 6, 1944, as the invasion of Normandy (Persico, 362).

Planning involved a crucial decision: should the bomb be merely demonstrated but not used? A Roosevelt adviser suggested that the President tell an enemy country, perhaps Japan, that he would be bombing a certain target and that everybody should leave. Then, after conducting the bombing, the inevitable shock and horror would be sufficient to stop the war. Why was this course of action not pursued? Because Roosevelt was concerned that the bomb had never been tested, and if for any reason it simply fell to the ground without exploding, then the enemy would have a full-scale model that could be reverse-engineered and replicated. The eventual decision was made to drop as few bombs as possible in the hope that the enemy would promptly surrender.

At this point, Leslie Groves entered the picture. Born in Albany, New York, Groves went to MIT and then West Point, where he studied engineering, graduating from the General Staff School in 1934 and the Army War College in 1939. The following year he was charged with building the Pentagon. Groves was Roosevelt's choice for director of the Manhattan Project; Groves in turn chose

Oppenheimer to direct Los Alamos. It was Groves who studied and planned where to drop the bombs, if it became necessary to do so. It was Groves's order to General Carl Spaatz, in the Pacific theater of war, to "deliver its first special bomb as soon as weather will permit visual bombing after about 3 August 1945" (http://www.nuclearfiles.org/menu/library/biographies/bio_groves-leslie.htm). Note that although technology was advanced enough to build and deliver a bomb, actual launch was still dependent on visual targeting at this time.

Is it fair to say that between Roosevelt and Churchill, the research and development of atomic power won the war? This has been a matter of much debate. But certainly the bomb would not have been ready if the planning and development work had not been put in motion by President Roosevelt and, of course, Einstein and Sachs.

BUILDING

When Leslie Groves became head of the Manhattan Project, he was empowered to find a location and construct whatever facilities were needed to develop an atomic bomb before the Germans did the same.

There were several reasons why Oak Ridge, Tennessee, was chosen. One was that the activities to be conducted would require enormous amounts of electricity, already available there as a result of the building of the Tennessee Valley Authority (TVA). Clean water was abundant, and there was a rail line, adequate roads, and land that could be acquired at little cost. Many families farmed in the area, but moving was not expected to pose a difficulty for the army, although it was sensitive to the dislocation of families. This time, however, the urgency of the times meant that few niceties were observed: families woke up to find a notice nailed to their fencepost instructing them to move. In just one year, everybody was cleared away from the desired site; eventually 59,000 acres were made ready.

The site was also advantageous for security reasons, as the Clinch River bordered the reservation on three sides; on the exposed side, a patrolled fence was all that was needed to ensure the secrecy of the work going on inside.

Oak Ridge was a natural choice for other reasons. By 1941, the TVA had been completed and was the largest producer of electricity in the United States, ensuring a steady and dependable supply of electricity. Scientific and intellectual knowledge abounded, with the University of Tennessee at Knoxville home to many scientists who remained in the area after the TVA was completed. Topographically, Oak Ridge was ideally situated for the five different plants that would be within easy transport distance but could remain separated by mountain ridges that afforded protection to the whole if one were compromised or damaged.

For security reasons, the innocuous name *Clinton Engineer Works* was given to the overall site that was comprised of five plants: the K-25 Gaseous Diffusion Plant, the Y-12 Electromagnetic Plant, the headquarters of the Clinton

Engineer Works and central utilities, Clinton Laboratories, and the S-50 Thermal Diffusion Plant. The total spent on Oak Ridge was $1,188,352,000. The total spent for the entire Manhattan Project was $1,889,604,000—clearly Oak Ridge represented the biggest investment.

The architectural firm of Skidmore, Owings, and Merrill was chosen to build the town that would house Oak Ridge workers. Supplied only with topographical information and the number of housing units, stores, schools, and places of worship needed, the firm designed the entire town. Once plans were approved, the architectural and building teams were put on a train in Chicago and given a sealed envelope. Only after the train had left the station, the story goes, were the architects permitted to open the envelope to learn they were going to Tennessee.

The town was laid out with avenues named for the U.S. states, arranged by the alphabet. In an era of planned communities, Oak Ridge could be compared to the town of Norris, a planned community for workers of the TVA. Building materials that produced results quickly were used, for time was of the essence; a fast-bonding asbestos/cement, the so-called cemestos, was used so that buildings could be raised quickly.

Unfortunately, African American employees were segregated not only by race but also by gender. The Oak Ridge community claimed (just as had happened with the TVA and the town of Norris) that it was complying with local customs regarding race. At the time, Tennessee was a segregated state; however, local resistance grew, and in 1955 Oak Ridge became one of the first cities in the south to desegregate its public schools. The Oak Ridge school system was excellent. Workers for the project demanded high-quality education for their dislocated families, and teachers were offered higher salaries than those paid in other school districts.

However, building facilities wasn't the only preparation getting underway. A huge workforce was needed; scientists were recruited

General Groves (right) and Dr. Oppenheimer (left) examine the remains of the tower from which a test atomic bomb was detonated near Alamogordo, New Mexico, 1945. Courtesy of the Library of Congress.

and drafted into the army in a special engineering division. Some had no military training but plenty of technical know-how. The army was well trained for fast recruitment. When President Roosevelt started the Civilian Conservation Corps, he gave the job of organization and recruitment to the army. Within one week of his executive order, young men were enrolled and assigned to public-works projects. Women also worked in the scientific environment. In the spring of 1943, 66,000 people were working on the project, a number that later increased to 75,000.

While Oak Ridge was built to house the R&D effort and the production of fissionable U-235, the bombs were actually put together in Los Alamos, New Mexico. After the uranium was ready for use in a bomb, officials had to hand-carry the material to Los Alamos. Only a little more than $74 million was spent on this facility.

IMPORTANCE IN HISTORY

It is noteworthy that the ability to work cooperatively was, of necessity, developing steadily in the United States. An earlier multiparty building consortium that had built the Hoover Dam (Six Companies, Inc.), marked a turning point in U.S. management history when the affiliated companies worked together so well that they finished two years early. The ancient Romans showed a similar capacity to work efficiently in large teams on large projects. Likewise, the essential competency to work cooperatively in a large-scale, multipartner firm was part of the success of the Manhattan Project.

Guns to butter, swords to plowshares—these are stock phrases in English that speak of potentially dangerous things being turned to peaceful and beneficial use. The Manhattan Project opened the door to a new energy source for the world. In 1946, the original intent of Einstein's letter to President Roosevelt was realized: discovery of "a new and important source of energy," and steps were taken, in the form of the Atomic Energy Act of 1946, to provide guidelines for the safe and appropriate use of this potent new form of energy. The Atomic Energy Act of 1946, presented in this chapter, has five goals:

(1) A program to assist and foster private research and development to encourage maximum scientific progress;
(2) A program to control scientific and technical information, which will permit the dissemination of such information to encourage scientific progress, and for the sharing on a reciprocal basis of information concerning the practical industrial application of atomic energy as soon as effective and enforceable safeguards against its use for destructive purposes can be devised;
(3) A program of federally conducted research and development to assure the Government of adequate scientific and technical accomplishment;

(4) A program for Government control of the production, ownership and use of fissionable material to assure the common defense and security and to insure the broadest possible exploitation of the fields; and

(5) A program of administration that will be consistent with the foregoing policies and with international arrangements made by the United States, and which will enable the Congress to be currently informed so as to take further legislative action as may hereafter be appropriate.

Might civilian use of atomic energy be paralleled in the future if a high-level decision is made to develop microwave transmission of electricity?

The world may someday undergo a paradigm shift in its source of electricity. If the world decided to use the sun as its major source of energy, every part of the earth could have as much electric power as it needed and wanted, provided agreement is reached as to appropriate controls. Peter E. Glaser, inventor and original patent holder for solar-power satellites, has stated that the way to proceed might be to have a receiving antenna (rectenna) on an island, possibly near Europe, and then have the energy transmission via tubes under the Atlantic—floating submerged tunnels.

FOR FURTHER REFERENCE

Books and Articles

Dyson, Freeman. *Weapons and Hope*. New York: Harper and Row, 1984.

Nitze, Paul. H. *From Hiroshima to Glasnost: At the Center of Decision; A Memoir*. New York: Grove Weidenfeld, 1989.

Persico, Joseph E. *Roosevelt's Secret War: FDR and World War II Espionage*. New York: Random House, 2001.

Internet

For a comparison of World War II costs of traditional military vs. the Manhattan Project, see http://www.brook.edu/fp/projects/nucwcost/manhattn.htm.

For an overview of the nuclear age in the 1940s, see http://www.nuclearfiles.org/.

For a nuclear time line in the 1940s, including links to video and audio clips from December 7, 1941, from CBS News, see http://www.radiochemistry.org/history/nuclear_timeline/40s.html.

For the organization chart of the Manhattan Engineer District, see the Society for the Historical Preservation of the Manhattan Project's Web site: http://www.childrenofthemanhattanproject.org/COTMP/MED_chart_Printable.htm.

For information about the Clinton Engineer Works (forerunner to Oak Ridge), see http://www.vcdh.virginia.edu/HIUS316/mbase/docs/oak.html.

To learn more about the current Oak Ridge Laboratories, see http://www.ornl.gov/ornlhome/about.shtml.

For Y-12 information, see http://www.y12.doe.gov/.
For a brief bio of Leslie Groves, see http://www.nuclearfiles.org/. Search for Leslie Groves.
For a time line of nuclear treaties and agreements from 1963 to 1966, see http://library.thinkquest.org/17940/texts/timeline/treaties.html.

Film and Television

Casablanca. Directed by Michael Curtiz. Burbank, CA: Warner Studios, 1942. To view the trailer of this film set in the period of the European conflict, see http://us.imdb.com/Trailers?0034583&272&28.

Destination Tokyo. Directed by Delmer Daves. Burbank, CA: Warner Studios, 1943.

Dr. Strangelove, or How I Learned to Stop Worrying and Love the Bomb. Directed by Stanley Kubrick. Culver City, CA: Columbia Tri-Star, 1964.

Enola Gay. Directed by Tim Curren. New York: A&E Entertainment, 1995.

Hiroshima: Decision to Drop the Bomb. New York: A&E Home Video, 2001.

Thirty Seconds Over Tokyo. Directed by Mervyn LeRoy. Burbank, CA: Warner Studios, 1944.

Documents of Authorization—I

On the following pages, the reader will find the text of original letters found in the Roosevelt Library. The Internet address following each letter links to a source where the letter can be viewed online.

1. The two-page letter, dated August 2, 1939, from Albert Einstein to President Franklin D. Roosevelt
2. A copy of the letter in reply, dated October 19, 1939, from President Roosevelt to Einstein
3. The original three-page letter, dated October 11, 1939, from Alexander Sachs to President Roosevelt
4. The original memo, dated March 11, 1942, from President Roosevelt to Vannevar Bush

1. Letter from Albert Einstein to President Franklin D. Roosevelt

Albert Einstein
Old Grove Rd.
Nassau Point
Peconic, Long Island

August 2nd, 1939
F. D. Roosevelt
President of the United States
White House
Washington, D.C.

Sir:

Some recent work by E. Fermi and L. Szilard, which has been communicated to me in manuscript, leads me to expect that the element uranium may be turned into a new and important source of energy in the immediate future. Certain aspects of this situation which has arisen seem to call for watchfulness and, if necessary, quick action on the part of the Administration. I believe therefore that it is my duty to bring to your attention the following facts and recommendations:

In the course of the last four months it has been made probable—through the work of Joliot in France as well as Fermi and Szilard in America—that it may become possible to set up a nuclear chain reaction in a large mass of uranium, by which vast amounts of power and large quantities of new radium-like elements would be generated. Now it appears almost certain that this could be achieved in the immediate future.

This new phenomena would also lead to the construction of bombs, and it is conceivable—though much less certain—that extremely powerful bombs of a new type may thus be constructed. A single bomb of this type, carried by boat and exploded in a port, might very well destroy the whole port together with some of the surrounding territory. However, such bombs might very well prove to be too heavy for transportation by air.

The United States has only very poor ores of uranium in moderate quantities. There is some good ore in Canada and the former Czechoslovakia, while the most important source of uranium is Belgian Congo.

In view of this situation you may think it desirable to have some permanent contact maintained between the administration and the group of physicists working on chain reactions in America. One possible way of achieving this might be for you to entrust with this task a person who has your confidence and who could perhaps serve in an unofficial capacity. His task might comprise the following:

a) to approach Government Departments, keep them informed of the further development, and put forward recommendations for Government action, giving particular attention to the problem of securing a supply of uranium ore for the United States;

b) to speed up the experimental work, which is at present being carried on within the limits of the budgets of University laboratories, by providing funds, if such funds be required, through his contacts with private persons who are willing to make contributions for this cause, and perhaps also by obtaining the co-operation of industrial laboratories which have the necessary equipment.

I understand that Germany has actually stopped the sale of uranium from the Czechoslovakian mines which she has taken over. That she should have taken such an early action might perhaps be understood on the ground that the son of the German

Under-Secretary of State, von Weizsäcker, is attached to the Kaiser Wilhelm Institute in Berlin where some of the American work on uranium is now being repeated.

Yours very truly,

/s/ A. Einstein

From Roosevelt Library: http://www.fdrlibrary.marist.edu/psf/box5/a64a01.html.

2. Letter of Reply from President Franklin D. Roosevelt to Albert Einstein

THE WHITE HOUSE
Washington, DC
October 19, 1939

My dear Professor:

I want to thank you for your recent letter and the most interesting and important enclosure.

I found this data of such import that I have convened a Board consisting of the head of the Bureau of Standards and a chosen representative of the Army and Navy to thoroughly investigate the possibilities of your suggestion regarding the element of uranium.

I am glad to say that Dr. Sachs will cooperate and work with this Committee and I feel this is the most practical and effective method of dealing with the subject.

Please accept my sincere thanks.

Very sincerely yours,

/s/ Franklin D. Roosevelt
    Dr. Albert Einstein,
    Old Grove Road,
    Nassau Point,
    Peconic, Long Island,
    New York.

From Roosevelt Library: http://www.fdrlibrary.marist.edu/psf/box5/a64a01.html.

3. Letter from Alexander Sachs to President Franklin D. Roosevelt

ONE SOUTH WILLIAM STREET
NEW YORK
October 11, 1939

Dear Mr. President:

With approaching fulfillment of your plans in connection with revision of the Neutrality Act, I trust that you may now be able to accord me the opportunity to present a communication from Dr. Albert Einstein to you and other relevant material bearing on experimental work by physicists with far-reaching significance for National Defense.

Briefly, the experimentation that has been going on for half a dozen years on atomic disintegration has culminated this year (a) in the discovery by Dr. Leo Szilard and Professor Fermi that the element, uranium, could be split by neutrons and (b) in the opening up of the probability of chain reactions,—that is, that in this nuclear process uranium itself may emit neutrons. This new development in physics holds out the following prospects:

1. The creation of a new source of energy which might be utilized for purposes of power production;
2. The liberation from such chain reaction of new radio-active elements, so that tons rather than grams of radium could be made available in the medical field;
3. The construction, as an eventual probability, of bombs of hitherto unenvisaged potency and scope: As Dr. Einstein observes, in the letter which I will leave with you, "a single bomb of this type carried by boat and exploded in a port might well destroy the whole port together with some of the surrounding territory!"

In connection, then, with the practical importance of this work—for power, healing and national defense purposes—it needs to be borne in mind that our supplies of uranium are limited and poor in quality as compared with the large sources of excellent uranium in the Belgian Congo and, next in line, Canada and former Czechoslovakia. It has come to the attention of Dr. Einstein and the rest of the group concerned with this problem that Germany has actually stopped the sale of uranium from the Czechoslovakian mines it seized. This action must be related to the fact that the son of the German Under-Secretary of State, Karl von Weizsaecker, had been an assistant at the Kaiser Wilhelm Institute in Berlin to some of the great physicists now resident in this country who are carrying forward these experiments on uranium.

Mindful of the implications of all this for democracy and civilization in the historic struggle against the totalitarianism that has exploited the inventions of the free human spirit, Dr. Szilard, in consultation with Professor E. P. Wigner, head of the physics department of Princeton, and Professor E. Teller of George Washington University, sought to aid this work in the United States through the formation of an association for scientific collaboration, to intensify the cooperation of physicists in the democratic countries—such as Professor Joliot in Paris, Professor Lindemann of Oxford and Dr. Dirac of Cambridge—and to withhold publication of the progress in the work on chain reactions. As the international crisis developed this summer, these refugee scholars and the rest of us in consultation with them unanimously agreed that it was their duty, as well as desire, to apprise you at the earliest opportunity of their work and to enlist your cooperation.

In view of the danger of German invasion of Belgium, it becomes urgent to make arrangements—preferably through diplomatic channels—with the Union Miniere du Haut-Katanga, whose head office is at Brussels, to make available abundant supplies of uranium to the United States. In addition, it is necessary to enlarge and accelerate the experimental work, which can no longer be carried out within the limited budgets of the departments of theoretical physics in our universities. It is believed that public-spirited executives in our leading chemical and electrical companies could be persuaded to make available certain amounts of uranium oxide and quantities of graphite, and to bear the considerable expense of the newer phases of the experimentation. An alternative plan would be the enlistment of one of the foundations to supply the necessary materials and funds. For either plan and for all the purposes, it would seem advisable to adopt the suggestion of Dr. Einstein that you designate an individual and a committee to serve as a liaison between the scientists and the Executive Departments.

In the light of the foregoing, I desire to be able to convey in person, in behalf of these refugee scholars, a sense of their eagerness to serve the nation that has afforded them hospitality, and to present Dr. Einstein's letter together with a memorandum which Dr. Szilard prepared after some discussion with me and copies of some of the articles that have appeared in scientific journals. In addition, I would request in their behalf a conference with you in order to lay down the lines of policy with respect to the Belgian source of supply and to arrange for a continuous liaison with the Administration and the Army and Navy Departments, as well as to solve the immediate problems of necessary materials and funds.

With high regard,

Yours sincerely,
/s/ Alexander Sachs

The President,
The White House,
Washington, D.C.

From Roosevelt Library: http://www.fdrlibrary.marist.edu/psf/box5/a64b01.html.

4. Letter from President Franklin D. Roosevelt to Dr. Vannevar Bush

THE WHITE HOUSE
Washington, DC
March 11, 1942

MEMORANDUM FOR DR. VANNEVAR BUSH:

I am greatly interested in your report of March ninth and I am returning it herewith for your confidential file. I think the whole thinking should be pushed not only in regard to development, but also with due regard to time. This is very much of the essence. I have no objection to turning over future progress to the War Department on condition that you yourself are certain that the War Department has made all adequate provision for absolute secrecy.

F.D.R.

From: Roosevelt Library: http://www.fdrlibrary.marist.edu/psf/box2/a13w01.html.

Documents of Authorization—II

AN ACT

For the development and control of atomic energy.

Be it enacted by the Senate and House of Representatives of the United States of America in congress assembled,

DECLARATION OF POLICY

SECTION 1. (a) FINDINGS AND DECLARATION.— Research and experimentation in the field of nuclear chain reaction have attained the stage at which the release of atomic energy on a large scale is practical. The significance of the atomic bomb for military purposes is evident. The effect of the use of atomic energy for civilian purposes upon the social, economic, and political structures of today cannot now be determined. It is a field in which unknown factors are involved. Therefore, any legislation will necessarily be subject to revision from time to time. It is reasonable to anticipate, however, that tapping this new source of energy will cause profound changes in our present way of life. Accordingly, it is hereby declared to be the policy of the people of the United States that, subject at all times to the paramount objective of assuring the common defense and security, the development and utilization of atomic energy shall, so far as practicable, be directed toward improving the public welfare, increasing the standard of living, strengthening free competition in private enterprise, and promoting world peace.

(b) PURPOSE OF ACT.—it is the purpose of this Act to effectuate the policies set out in section 1 (a) by providing, among others, for the following major programs relating to atomic energy;

(1) A program of assisting and fostering private research and development to encourage maximum scientific progress;

(2) A program for the control of scientific and technical information which will permit the dissemination of such information to encourage scientific progress, and for the sharing on a reciprocal basis of information concerning the practical industrial application of atomic energy as soon as effective and enforceable safeguards against its use for destructive purposes can be devised;

(3) A program of federally conducted research and development to assure the Government of adequate scientific and technical accomplishment;

(4) A program for Government control of the production, ownership, and use of fissionable material to assure the common defense and security and to insure the broadest possible exploitation of the fields; and

(5) A program of administration which will be consistent with the foregoing policies and with international arrangements made by the United States, and which will enable the Congress to be currently informed so as to take further legislative action as may hereafter be appropriate.

ORGANIZATION

SEC. 2. (a) ATOMIC ENERGY COMMISSION.—

(1) There is hereby established an Atomic Energy Commission (herein called the Commission), which shall be composed of five members. Three members shall constitute a quorum of the Commission. The President shall designate one member as Chairman of the Commission.

(2) Members of the Commission shall be appointed by the President, by and with the advice and consent of the Senate. In submitting any nomination to the Senate, the President shall set forth the experience and the qualifications of the nominee. The term of office of each member of the Commission taking office prior to the expiration of two years after the date of enactment of this Act shall expire upon the expiration of such two years. The term of office of each member of the Commission taking office after the expiration of two years from the date of enactment of this Act shall be five years, except that (A) the terms of office of the members first taking office after the expiration of two years from the date of enactment of this Act shall expire, as designated by the President at the time of appointment, one at the end of three years, one at the end of four years, one at the end of five years, one at the end of six years, and one at the end of seven years, after the date of enactment of this Act; and (B) any member appointed to fill a vacancy occurring prior to the expiration of the term for which his predecessor was appointed, shall be appointed for the remainder of such term. Any member of the Commission may be removed by the President for inefficiency, neglect of duty, or malfeasance in office. Each member, except the Chairman, shall receive compensation at the rate of $15,000 per annum; and the Chairman shall receive compensation at the rate of $17,500 per annum. No member of the Commission shall engage in any other business, vocation, or employment than that of serving as a member of the Commission.

(3) The principal office of the Commission shall be in the District of Columbia, but the Commission or any duly authorized representative may exercise any or all of its powers in any place. The Commission shall hold such meetings, conduct such hearings, and receive such reports as may be necessary to enable it to carry out the provisions of this Act.

(4) There are hereby established within the Commission—

(A) a General Manager, who shall discharge such of the administration and executive functions of the Commission as the Commission may direct. The General Manager shall be appointed by the President by and with the advice and consent of the Senate, and shall receive compensation at the rate of $15,000 per annum. The Commission may make recommendations to the President with respect to the appointment or removal of the General manager.

(B) a Division of Research, a Division of Production, a Division of Engineering, and a Division of Military Application. Each division shall be under the direction of a Director who shall be appointed by the Commission, and shall receive compensation at the rate of $14,000 per annum. The Director of the Division of Military Application shall be a member of the armed forces. The Commission shall require each such division to exercise such of the Commission's powers under this Act as the Commission may determine, except that the authority granted under section 3 (a) of this Act shall not be exercised by the Division of Research.

(b) GENERAL ADVISORY COMMITTEE.—There shall be a General Advisory Committee to advise the Commission on scientific and technical matters relating to materials, production, and research and development, to be composed of nine members, who shall be appointed from civilian life by the President. Each member shall hold office for a term of six years, except that (1) any member appointed to fill a vacancy occurring prior to the expiration of the term for which his predecessor was appointed, shall be appointed for the remainder of such term; and (2) the terms of office of the members first taking office after the date of the enactment of this Act shall expire, as designated by the President at the time of appointment, three at the end of two years, three at the end of four years, and three at the end of six years, after the date of the enactment of this Act. The Committee shall designate one of its own members as Chairman. The Committee shall meet at least four times in every calendar year. The members of the Committee shall receive a per diem compensation of $50 for each day spent in meetings or conferences, and all members shall receive their necessary traveling or other expenses while engaged in the work of the Committee.

(c) MILITARY LIAISON COMMITTEE.—There shall be a Military Liaison Committee consisting of representatives of the Departments of War and Navy, detailed or assigned thereto, without additional compensation, by the Secretaries of War and Navy in such number as they may determine. The Commission shall advise and consult with the Committee on all atomic energy matters which the Committees deems to relate to military applications, including the development, manufacture, use, and storage of bombs, the allocation of fissionable material for military research, and the control of information relating to the manufacture or utilization of atomic weapons. The Commission shall keep the Committee fully informed of all such matters before it and the Committee shall keep the Commission fully informed of all atomic energy activities of the War and Navy Departments. The Committee shall have authority to make written recommendations to the Commission on matters relating to military applications from time to time as it may deem appropriate. If the Committee at any time concludes that any action, proposed action, or failure to act of the Commission on such matters is adverse to the responsibilities of the Departments of War or Navy, derived from the Constitution, laws, and treaties, the Committee may refer such action, proposed action, or failure to act to the Secretaries of War and Navy. If either Secretary concurs, he may refer the matter to the President, whose decision shall be final.

(d) APPOINTMENT OF ARMY AND NAVY OFFICERS.—Notwithstanding the provisions of section 1222 of the Revised Statutes (U.S.C., 1940 edition, title 10, sec. 576), section 212 of the Act entitled "An Act making appropriations for the legislative, executive, and judicial expenses of the Government for the fiscal year ending June thirtieth, eighteen hundred and ninety-five, and for other purposes", approved July 31, 1894, as amended (U.S.C., 1940 edition, title 5, sec. 62), or any other law, any active or retired officer of the Army or the Navy may serve as Director of the Division of Military Application established by subsection (a) (4) (B) of this section, without prejudice to his commissioned status as such officer. Any such officer serving as Director of the Division of Military Application shall receive, in addition to his pay from the United States as such officer, an amount equal to the difference between such pay and the compensation prescribed in subsection (a) (4) (B) of this section.

RESEARCH

SEC. 3. (a) RESEARCH ASSISTANCE.—The Commission is directed to exercise its powers in such manner as to insure the continued conduct of research and development activities in the fields specified below by private or public institutions or persons and to assist in the acquisition of an ever-expanding fund of theoretical and practical knowledge in such fields. To this end the Commission is authorized and directed to make arrangements (including contracts, agreements, and loans) for the conduct of research and development activities relating to—

(1) nuclear processes;

(2) the theory and production of atomic energy, including processes, materials, and devices related to such production;

(3) utilization of fissionable and radioactive materials for medical, biological, health, or military purposes;

(4) utilization of fissionable and radioactive materials and processes entailed in the production of such materials for all other purposes, including industrial uses; and

(5) the protection of health during research and production activities.

The Commission may make such arrangements without regard to the provisions of section 3709 of the Revised Statutes (U.S.C., title 41, sec. 5) upon certification by the Commission that such action is necessary in the interest of the common defense and security, or upon a showing that advertising is not reasonably practicable, and may make partial and advance payments under such arrangements, and may make available for use in connection therewith such of its equipment and facilities as it may deem desirable. Such arrangements shall contain such provisions to protect health, to minimize danger from explosion and other hazards to life or property, and to require the reporting and to permit the inspection of work performed thereunder, as the Commission may determine; but shall not contain any provisions or conditions which prevent the dissemination of scientific or technical information, except to the extent such dissemination is prohibited by law.

(b) RESEARCH BY THE COMMISSION.—The Commission is authorized and directed to conduct, through its own facilities, activities and studies of the types specified in subsection (a) above.

PRODUCTION OF FISSIONABLE MATERIAL

SEC. 4. (a) DEFINITION.—As used in this Act, the term "produce", when used in relation to fissionable materials, means to manufacture, produce, or refine fissionable materials, as distinguished from source materials as defined in section 5 (b) (1), or to separate fissionable material from other substances in which such material may be contained or to produce new fissionable material.

(b) PROHIBITION.—It shall be unlawful for any person to own any facilities for the production of fissionable materials or for any person to produce fissionable materials, except to the extent authorized by subsection (c).

(c) OWNERSHIP AND OPERATION OF PRODUCTION FACILITIES.—

(1) OWNERSHIP OF PRODUCTION FACILITIES.—The Commission, as agent of and on behalf of the United States, shall be the exclusive owner of all facilities for the production of fissionable material other than facilities which (A) are useful in the conduct of research and development activities in the field specified in section 3, and (B) do not, in the opinion of the Commission, have a potential production rate adequate to enable the operator of such facilities to produce within a reasonable period of time a sufficient quantity of fissionable material to produce an atomic bomb or any other atomic weapon.

(2) OPERATION OF THE COMMISSION'S PRODUCTION FACILITIES.— The Commission is authorized and directed to produce or to provide for the production of fissionable material in its own facilities. To the extent deemed necessary, the Commission is authorized to make, or to continue in effect, contracts with persons obligating them to produce fissionable material in facilities owned by the Commission. The Commission is also authorized to enter into research and development contracts authorizing the contractor to produce fissionable material in facilities owned by the Commission to the extent that the production of such fissionable material may be incident to the conduct of research and development activities under such contracts. Any contract entered into under this section shall contain provisions (A) prohibiting the contractor with the Commission from subcontracting any part of the work he is obligated to perform under the contract, except as authorized by the Commission, and (B) obligating the contractor to make such reports to the Commission as it may deem appropriate with respect to his activities under the contract, to submit to frequent inspection by employees of the Commission of all such activities, and to comply with all safety and security regulations which may be prescribed by the Commission. Any contract made under the provisions of this

paragraph may be made without regard to the provisions of section 3709 of the Revised Statutes (U.S.C., title 41, sec. 5) upon certification by the Commission that such action is necessary in the interest of the common defense and security, or upon a showing that advertising is not reasonably practicable, and partial and advance payments may be made under such contracts. The President shall determine at least once each year the quantities of fissionable material to be produced under this paragraph.

(3) OPERATION OF OTHER PRODUCTION FACILITIES.—Fissionable material may be produced in the conduct of research and development activities in facilities which, under paragraph (1) above, are not required to be owned by the Commission.

(d) IRRADIATION OF MATERIALS.—For the purpose of increasing the supply of radioactive materials, the Commission and persons lawfully producing or utilizing fissionable material are authorized to expose materials of any kind to the radiation incident to the processes of producing or utilizing fissionable material.

(e) MANUFACTURE OF PRODUCTION FACILITIES.—Unless authorized by a license issued by the Commission, no person may manufacture, produce, transfer, or acquire any facilities for the production of fissionable material. Licenses shall be issued in accordance with such procedures as the Commission may by regulation establish and shall be issued in accordance with such standards and upon such conditions as will restrict the production and distribution of such facilities to effectuate the policies and purposes of this Act. Nothing in this section shall be deemed to require a license for such manufacture, production, transfer, or acquisition incident to or for the conduct of research or development activities in the United States of the types specified in section 3, or to prohibit the Commission from manufacturing or producing such facilities for its own use.

CONTROL OF MATERIALS

SEC. 5. (a) FISSIONABLE MATERIALS.—

(1) DEFINITION.—As used in this Act, the term "fissionable material" means plutonium, uranium enriched in the isotope 235, any other material which the Commission determines to be capable of releasing substantial quantities of energy through nuclear chain reaction of the material, or any material artificially enriched by any of the foregoing; but does not include source materials, as defined in section 5 (b) (1).

(2) GOVERNMENT OWNERSHIP OF ALL FISSIONABLE MATERIAL.—All right, title, and interest within or under the jurisdiction of the United States, in or to any fissionable material, now or hereafter produced, shall be the property of the Commission, and shall be deemed to be vested in the Commission by virtue of this Act. Any person owning any interest in any fissionable material at the time of this enactment of this Act, or owning any interest in any material at the time when such material is hereafter determined to

be a fissionable materials, or who lawfully produces any fissionable material incident to privately financed research or development activities, shall be paid just compensation therefor. The Commission may, by action consistent with the provisions of paragraph (4) below, authorize any such person to retain possession of such fissionable material, but no person shall have any title in or to any fissionable material.

(3) PROHIBITION.—It shall be unlawful for any person, after sixty days from the effective date of this Act to (A) possess or transfer any fissionable material, except as authorized by the Commission, or (B) export from or import into the United States any fissionable material, or (C) directly or indirectly engage in the production of any fissionable material outside of the United States.

(4) DISTRIBUTION OF FISSIONABLE MATERIAL.—Without prejudice to its continued ownership thereof, the Commission is authorized to distribute fissionable material owned by it, with or without charge, to applicants requesting such material (A) for the conduct of research or development activities either independently or under contract or other arrangement with the Commission, (B) for use in medical therapy, or (C) for use pursuant to a license issued under the authority of section 7. Such materials shall be distributed in such quantities and on such terms that no applicant will be enabled to obtain an amount sufficient to construct a bomb or other military weapon. The Commission is directed to distribute sufficient fissionable material to permit the conduct of widespread independent research and development activity, to the maximum extent practicable. In determining the quantities of fissionable material to be distributed, the Commission shall make such provisions for its own needs and for the conservation of fissionable material as it may determine to be necessary in the national interest for the future development of atomic energy. The Commission shall not distribute any materials to any applicant, and shall recall any distributed material from any applicant, who is not equipped to observe or who fails to observe such safety standards to protect health and to minimize danger from explosion or other hazard to life or property as may be established by the Commission, or who uses such material in violation of law or regulation of the Commission or in a manner other than as disclosed in the application therefor.

(5) The Commission is authorized to purchase or otherwise acquire any fissionable material or any interest therein outside the United States, or any interest in facilities for the production of fissionable material, or in real property on which such facilities are located, without regard to the provisions of section 3709 of the Revised Statutes (U.S.C. title 41, sec. 5) upon certification by the Commission that such action is necessary in the interest of the common defense and security, or upon a showing that advertising is not reasonably practicable, and partial and advance payments may be made under contracts for such purposes. The Commission is further authorized to take, requisition, or condemn, or otherwise acquire any interest in such facilities or real property, and just compensation shall be made therefor.

(b) SOURCE MATERIALS.—

(1) DEFINITION.—As used in this Act, the term "source material" means uranium, thorium, or any other material which is determined by the Commission, with the approval of the President, to be peculiarly essential to the production of fissionable materials; but includes ores only if they contain one or more of the foregoing materials in such concentration as the Commission may by regulation determine from time to time.

(2) LICENSE FOR TRANSFERS REQUIRED.—Unless authorized by a license issued by the Commission, no person may transfer or deliver, receive possession of or title to, or export from the United States any source material after removal from its place of deposit in nature, except that licenses shall not be required for quantities of source materials which, in the opinion of the Commission, are unimportant.

(3) ISSUANCE OF LICENSES.—The Commission shall establish such standards for the issuance, refusal, or revocation of licenses as it may deem necessary to assure adequate source materials for production, research, or development activities pursuant to this Act or to prevent the use of such materials in a manner inconsistent with the national welfare. Licenses shall be issued in accordance with such procedures as the Commission may by regulation establish.

(4) REPORTING.—The Commission is authorized to issue such regulations or orders requiring reports of ownership, possession, extraction, refining, shipment, or other handling of source materials as it may deem necessary, except that such reports shall not be required with respect to (A) any source material prior to removal from its place of deposit in nature, or (B) quantities of source materials which in the opinion of the Commission are unimportant or the reporting of which will discourage independent prospecting for new deposits.

(5) ACQUISITION.—The Commission is authorized and directed to purchase, take, requisition, condemn, or otherwise acquire, supplies of source materials or any interest in real property containing deposits of source materials to the extent it deems necessary to effectuate the provisions of this Act. Any purchase made under this paragraph may be made without regard to the provisions of section 3709 of the Revised Statutes (U.S.C., title 41, sec. 5) upon certification by the Commission that such action is necessary in the interest of the common defense and security, or upon a showing that advertising is not reasonably practicable, and partial and advance payments may be made thereunder. The Commission may establish guaranteed prices for all source materials delivered to it within a specified time. Just compensation shall be made for any property taken, requisitioned, or condemned under this paragraph.

(6) EXPLORATION.—The Commission is authorized to conduct and enter into contracts for the conduct of exploratory operations, investigations, and inspections to determine the location, extent, mode of occurrence, use, or conditions of deposits or supplies of source materials, making just compensation for any damage or injury occasioned thereby. Such exploratory operations may

be conducted only with the consent of the owner, but such investigations and inspections may be conducted with or without such consent.

(7) PUBLIC LANDS.—All uranium, thorium, and all other materials determined pursuant to paragraph (1) of this subsection to be peculiarly essential to the production of fissionable material, contained, in whatever concentration, in deposits in the public lands are hereby reserved for the use of the United States subject to valid claims, rights or privileges existing on the date of the enactment of this Act: *Provided, however,* That no individual, corporation, partnership, or association, which had any part, directly or indirectly, in the development of the atomic bomb project, may benefit by any location, entry, or settlement upon the public domain made after such individual, corporation, partnership, or association took part in such project, if such individual, corporation, partnership, or association, by reason of having had such part in the development of the atomic bomb project, acquired confidential official information as to the existence of deposits of such uranium, thorium, or other materials in the specific lands upon which such location, entry, or settlement is made, and subsequent to the date of the enactment of this Act made such location, entry, or settlement or caused the same to be made for his, its, or their benefit. The Secretary of the Interior shall cause to be inserted in every patent, conveyance, lease, permit, or other authorization hereafter granted to sue the public lands or their mineral resources, under any of which there might result the extraction of any materials so reserved, a reservation to the United States of all such materials, whether or not of commercial value, together with the right of the United States through its authorized agents or representatives at any time to enter upon the land and prospect for, mine, and remove the same, making just compensation for any damage or injury occasioned thereby. Any lands so patented, conveyed, leased, or otherwise disposed of may be used, and any rights under any such permit or authorization may be exercised, as if no reservation of such materials had been made under this subsection; except that, when such use results in the extraction of any such material from the land in quantities which may not be transferred or delivered without a license under this subjection, such material shall be the property of the Commission and the Commission may require delivery of such material to it by any possessor thereof after such material has been separated as such from the ores in which it was contained. If the Commission requires the delivery of such material to it, it shall pay the person mining or extracting the same, or to such other person as the Commission determines to be entitled thereto, such sums, including profits, as the Commission deems fair and reasonable for the discovery, mining, development, production, extraction, and other services performed with respect to such material prior to such delivery, but such payment shall not include any amount on account of the value of such material before removal from its place of deposit in nature. If the Commission does not require delivery of such material to it, the reservation made pursuant to this paragraph shall be of no further force or effect.

(c) BYPRODUCT MATERIALS.—

(1) DEFINITION.—As used in this Act, the term "byproduct material" means any radioactive material (except fissionable material) yielded in or made radioactive by exposure to the radiation incident to the processes of producing or utilizing fissionable material.

(2) DISTRIBUTION.—The Commission is authorized to distribute, with or without charge, byproduct materials to applicants seeking such materials for research or development activity, medical therapy, industrial uses, or such other useful applications as may be developed. In distributing such materials, the Commission shall give preference to applicants proposing to use such materials in the conduct of research and development activity or medical therapy. The Commission shall not distribute any byproduct materials to any applicant, and shall recall any distributed materials from any applicant, who is not equipped to observe or who fails to observe such safety standards to protect health as may be established by the Commission or who uses such materials in violation of law or regulation of the Commission or in a manner other than as disclosed in the application therefor.

(d) GENERAL PROVISIONS.—The Commission shall not—

(1) distribute any fissionable material to (A) any person for a use which is not under or within the jurisdiction of the United States, (B) any foreign government, or (C) any person within the United States if, in the opinion of the Commission, the distribution of such fissionable material to such person would be inimical to the common defense and security.

(2) license any person to transfer or deliver, receive possession of or title to, or export from the United States any source material if, in the opinion of the Commission, the issuance of a license to such person for such purpose would be inimical to the common defense and security.

MILITARY APPLICATIONS OF ATOMIC ENERGY

Sec. 6 (a) AUTHORITY.—The Commission is authorized to—

(1) conduct experiments and do research and development work in the military application of atomic energy; and

(2) engage in the production of atomic bombs, atomic bomb parts, or other military weapons utilizing fissionable materials; except that such activities shall be carried on only to the extent that the express consent and direction of the President of the United States has been obtained, which consent and direction shall be obtained at least once each year.

The President from time to time may direct the Commission (1) to deliver such quantities of fissionable materials or weapons to the armed forces for such use as he deems necessary in the interest of national defense or (2) to authorize the armed forces to manufacture, produce, or acquire any equipment or device utilizing fissionable material or atomic energy as a military weapon.

(b) PROHIBITION.—It shall be unlawful for any person to manufacture, produce, transfer, or acquire any equipment or device utilizing fissionable material or atomic energy as a military weapon, except as may be authorized by the Commission. Nothing in this subsection shall be deemed to modify the provisions of section 4 of this Act, or to prohibit research activities in respect of military weapons, or to permit the export of any such equipment or device.

UTILIZATION OF ATOMIC ENERGY

SEC. 7. (a) LICENSE REQUIRED.—It shall be unlawful, except as provided in sections 5 (a) (4) (A) or (B) or 6 (a), for any person to manufacture, produce, or export any equipment or device utilizing fissionable material or atomic energy or to utilize fissionable material or atomic energy with or without such equipment or device, except under and in accordance with a license issued by the Commission authorizing such manufacture, production, export, or utilization. No license may permit any such activity if fissionable materials are produced incident to such activity, except as provided in sections 3 and 4. Nothing in this section shall be deemed to require a license for the conduct of research or development activities relating to the manufacture of such equipment or devices or the utilization of fissionable material or atomic energy, or for the manufacture or use of equipment or devices for medical therapy.

(b) REPORT TO CONGRESS.—Whenever in its opinion any industrial, commercial, or other nonmilitary use of fissionable material or atomic energy has been sufficiently developed to be of practical value, the Commission shall prepare a report to the President stating all the facts with respect to such use, the Commission estimate of the social, political, economic, and international effects of such use and the Commission's recommendations for necessary or desirable supplemental legislation. The President shall then transmit this report to the Congress together with his recommendation. No license for any manufacture, production, export, or use shall be issued by the Commission under this section until after (1) a report with respect to such manufacture, production, export, or use has been filed with the Congress; and (2) a period of ninety days in which the Congress was in session has elapsed after the report has been so filed. In computing such period of ninety days, there shall be excluded the days on which either House is not in session because of an adjournment of more than three days.

(c) ISSUANCE OF LICENSES.—After such ninety-day period, unless hereafter prohibited by law, the Commission may license such manufacture, production, export, or use in accordance with such procedures and subject to such conditions as it may by regulation establish to effectuate the provisions of this Act. The Commission is authorized and directed to issue licenses on a nonexclusive basis and to supply to the extent available appropriate quantities of fissionable material to licensees (1) whose proposed activities will serve some useful purpose proportionate to the quantities of fissionable material to be consumed; (2) who are equipped to observe such safety standards to protect health and to minimize danger from explosion or other hazard to life or property as the Commission may establish; and (3) who agree to make available to the Commission such technical information and data concerning their activities pursuant to such licenses as the Commission may determine necessary to encourage similar activities by as many licensees as possible. Each such license shall be issued for a specified period, shall be revocable at any time by the Commission in accordance with such procedures as the Commission may establish, and may be renewed upon the expiration of such period. Where activities under any license might serve to maintain or to foster the growth of monopoly, restraint of trade, unlawful competition, or other trade position inimical to the entry of new, freely competitive enterprises in the field, the Commission is authorized and directed to refuse to issue such license or to establish such conditions to prevent these results as the Commission, in consultation with the Attorney General may determine. The Commission shall report promptly to the Attorney General any information it may have with respect to any utilization of fissionable material or atomic energy which appears to have these results. No license may be given to any person for activities which are not under or within the jurisdiction of the United States, to any foreign government, or to any person within the United States if, in the opinion of the Commission, the issuance of a license to such person would be inimical to the common defense and security.

(d) BYPRODUCT POWER.—If energy which may be utilized is produced in the production of fissionable material, such energy may be used by the Commission, transferred to other Governmental agencies, or sold to public or private utilities under contracts providing for reasonable resale prices.

INTERNATIONAL ARRANGEMENTS

SEC. 8. (a) DEFINITION.—As used in this Act, the term "international arrangement" shall mean any treaty approved by the Senate or international agreement hereafter approved by the Congress, during the time such treaty is approved by the Senate or international agreement hereafter approved by the Congress, during the time such treaty or agreement is in full force and effect.

(b) EFFECT OF INTERNATIONAL ARRANGEMENTS.—Any provision of this Act or any action of the Commission to the extent that it conflicts with the provisions of any international arrangement made after the date of enactment of this Act shall be deemed to be of no further force or effect.

(c) POLICIES CONTAINED IN INTERNATIONAL ARRANGEMENT.—In the performance of its functions under this Act, the Commission shall give maximum effect to the policies contained in any such international arrangement.

PROPERTY OF THE COMMISSION

SEC. 9. (a) The President shall direct the transfer to the Commission of all interests owned by the United States or any Government agency in the following property:

(1) All fissionable material; all atomic weapons and parts thereof; all facilities, equipment, and materials for the processing, production, or utilization of fissionable material or atomic energy; all processes and technical information of any kind, and the source thereof (including data, drawings, specifications, patents, patent applications, and other sources) relating to the processing, production, or utilization of fissionable material or atomic energy; and all contracts, agreements, leases, patents, applications for patents, inventions and discoveries (whether patented or unpatented), and other rights of any kind concerning any such items;

(2) All facilities, equipment, and materials, devoted primarily to atomic energy research and development; and

(3) Such other property owned by or in the custody or control of the Manhattan Engineer District or other Government agencies as the President may determine.

(b) In order to render financial assistance to those States and localities in which the activities of the Commission are carried on and in which the Commission has acquired property previously subject to States and local taxation, the Commission is authorized to make payments to State and local governments in lieu of property taxes. Such payments may be in the amounts, at the times, and upon the terms the Commission deems appropriate, but the Commission shall be guided by the policy of not making payments in excess of the taxes which would have been payable for such property in the condition in which it was acquired, except in cases where special burdens have been cast upon the State or local government by activities of the Commission, the Manhattan Engineer District or their agents. In any such case, any benefit accruing to the State or local governments by reason of such activities shall be considered in determining the amount of the payment. The Commission, and the property, activities, and income of the Commission, are hereby expressly exempted from taxation in any manner or form by any State, county, municipality, or any subdivision thereof.

CONTROL OF INFORMATION

Sec. 10. (a) POLICY.—It shall be the policy of the Commission to control the dissemination of restricted data in such a manner as to assure the common defense and security. Consistent with such policy, the Commission shall be guided by the following principles:

(1) That until Congress declares by joint resolution that effective and enforceable international safeguards against the use of atomic energy for

destructive purposes have been established, there shall be no exchange of information with other nations with respect to the use of atomic energy for industrial purposes; and

(2) That the dissemination of scientific and technical information relating to atomic energy should be permitted and encouraged so as to provide that free interchange of ideas and criticisms which is essential to scientific progress.

(b) RESTRICTIONS.—

(1) The term "restricted data" as used in this section means all data concerning the manufacture or utilization of atomic weapons, the production of fissionable material, or the use of fissionable material in the production of power, but shall not include any data which the Commission from time to time determines may be published without adversely affecting the common defense and security.

(2) Whoever, lawfully or unlawfully, having possession of, access to, control over, or being entrusted with, any document, writing, sketch, photograph, plan, model, instrument, appliance, note or information involving or incorporating restricted data—

(A) communicates, transmits, or discloses the same to any individual or person, or attempts or conspires to do any of the foregoing, with intent to injure the United States or with intent to secure an advantage to any foreign nation, upon conviction thereof, shall be punished by death or imprisonment for life (but the penalty of death or imprisonment for life imposed only upon recommendation of the jury and only in cases where the offense was committed with intent to injure the United States); or by a fine of not more than $20,000 or imprisonment for not more than twenty years, or both;

(B) communicates, transmits, or discloses the same to any individual or person, or attempts or conspires to do any of the foregoing, with reason to believe such data will be utilized to injure the United States or to secure an advantage to any foreign nation, shall, upon conviction, be punished by a fine of not more than $10,000 or imprisonment for not more than ten years, or both.

(3) Whoever, with intent to injure the United States or with intent to secure an advantage to any foreign action, acquires or attempts or conspires to acquire any document, writing, sketch, photograph, plan, model, instrument, appliance, note or information involving or incorporating restricted data shall, upon conviction thereof, be punished by death or imprisonment for life (but the penalty of death or imprisonment for life may be imposed only upon recommendation of the jury and only in cases where the offense was committed with intent to injure the United States); or by a fine of not more than $20,000 or imprisonment for not more than twenty years, or both.

(4) Whoever, with intent to injure the United States or with intent to secure an advantage to any foreign nation, removes, conceals, tampers with, alters, mutilates, or destroys any document, writing, sketch, photograph, plan, model, instrument, appliance, or note involving or incorporating restricted data and

used by any individual or person in connection with the production of fissionable material, or research or development relating to atomic energy, conducted by the United States, or financed in whole or in part by Federal funds, or conducted with the aid of fissionable material, shall be punished by death or imprisonment for life (but the penalty of death or imprisonment for life may be imposed only upon recommendation of the jury and only in cases where the offense was committed with intent to injure the United States); or by a fine of not more than $20,000 or imprisonment for not more than twenty years or both.

(5) (A) No person shall be prosecuted for any violation under this section unless and until the Attorney General of the United States has advised the Commission with respect to such prosecution and no such prosecution shall be commenced except upon the express direction of the Attorney General of the United States.

(B) (i) No arrangement shall be made under section 3, no contract shall be made or continued in effect under section 4, and no license shall be issued under section 4 (e) or 7, unless the person with whom such arrangement is made, the contractor or prospective contractor, or the prospective licensee agrees in writing not to permit any individual to have access to restricted data until the Federal Bureau of Investigation shall have made an investigation and report to the Commission on the character, associations, and loyalty of such individual and the Commission shall have determined that permitting such person to have access to restricted data will not endanger the common defense or security.

(ii) Except as authorized by the Commission in case of emergency, no individual shall be employed by the Commission until the Federal Bureau of Investigation shall have made an investigation and report to the Commission on the character, associations, and loyalty of such individual.

(iii) Notwithstanding the provisions of subparagraphs (i) and (ii), during such period of time after the enactment of this Act as may be necessary to make the investigation, report, and determination required by such paragraphs, (a) any individual who was permitted access to restricted data by the Manhattan Engineer District may be permitted access to restricted data and (b) the Commission may employ any individual who was employed by the Manhattan Engineer District.

(iv) To protect against the unlawful dissemination of restricted data and to safeguard facilities, equipment, materials, an other property of the Commission, the President shall have authority to utilize the services of any Government agency to the extent he may deem necessary or desirable.

(C) All violations of this Act shall be investigated by the Federal Bureau of Investigation and by the Department of Justice.

(6) This section shall not exclude the applicable provisions of any other laws, except that no Government agency shall take any action under such other laws inconsistent with the provisions of this section.

(c) INSPECTIONS, RECORDS, AND REPORTS.—The Commission is—

(1) authorized by regulation or order to require such reports and the keeping of such records with respect to, and to provide for such inspections of, activities and studies of types specified in section 3 and of activities under licenses issued pursuant to section 7 as may be necessary to effectuate the purposes of this Act;

(2) authorized and directed by regulation or order to require regular reports and records with respect to, and to provide for frequent inspections of, the production of fissionable material in the conduct of research and development activities.

PATENTS AND INVENTIONS

SEC. 11. (a) PRODUCTION AND MILITARY UTILIZATION.

(1) No patent shall hereafter be granted for any invention or discovery which is useful solely in the production of fissionable material or in the utilization of fissionable material or atomic energy for a military weapon. Any patent granted for any such invention or discovery is hereby revoked, and just compensation shall be made therefor.

(2) No patent hereafter granted shall confer any rights with respect to any invention or discovery to the extent that such invention or discovery is used in the production of fissionable material or in the utilization of fissionable material or atomic energy for a military weapon. Any rights conferred by any patent heretofore granted for any invention or discovery are hereby revoked to the extent that such invention or discovery is so used, and just compensation shall be made therefor.

(3) Any person who has made or hereafter makes any invention or discovery useful in the production of fissionable material or in the utilization of fissionable material or atomic energy for a military weapon shall file with the Commission a report containing a complete description thereof, unless such invention or discovery is described in an application for a patent filed in the Patient Office by such person within the time required for the filing of such report. The report covering any such invention or discovery shall be filed on or before whichever of the following is the latest: (A) The sixtieth day after the date of enactment of this Act; (B) the sixtieth day after the completion of such invention or discover; or (C) the sixtieth day after such person first discovers or first has reason to believe that such invention or discovery is useful in such production or utilization.

(b) USE OF INVENTIONS FOR RESEARCH.—No patent hereafter granted shall confer any rights with respect to any invention or discovery to the extent that such invention or discovery is used in the conduct of research or development activities in the fields specified in section 3. Any rights conferred by any patent heretofore granted for any invention or discovery are hereby revoked to the extent that such invention or discovery is so used, and just compensation shall be made therefor.

(c) NONMILITARY UTILIZATION.—

(1) It shall be the duty of the Commission to declare any patent to be affected with the public interest if (A) the invention or discovery covered by the patent utilizes or is essential in the utilization of fissionable material or atomic energy; and (B) the licensing of such invention or discovery under this subsection is necessary to effectuate the policies and purposes of this Act.

(2) Whenever any patent has been declared, pursuant to paragraph (1), to be affected with the public interest—

(A) The Commission is hereby licensed to use the invention or discovery covered by such patent in performing any of its powers under this Act; and

(B) Any person to whom a license has been issued under section 7 is hereby licensed to use the invention or discovery covered by such patent to the extent such invention or discovery is used by him in carrying on the activities authorized by his license under section 7.

The owner of the patent shall be entitled to a reasonable royalty fee for any use of an invention or discovery licensed by this subsection. Such royalty fee may be agreed upon by such owner and the licensee, or in the absence of such agreement shall be determined by the Commission.

(3) No court shall have jurisdiction or power to stay, restrain, or otherwise enjoin the use of any invention or discovery by a licensee, to the extent that such use is licensed by paragraph (2) above, on the ground of infringement of any patent. If in any action for infringement against such licensee the court shall determine that the defendant is exercising such license, the measure of damages shall be the royalty fee determined pursuant to this section, together with such costs, interest, and reasonable attorney's fees as may be fixed by the court. If no royalty fee has been determined, the court shall stay the proceeding until the royalty fee is determined pursuant to this section. If any such licensee shall fail to pay such royalty fee, the patentee may bring an action in any court of competent jurisdiction for such royalty fee, together with such costs, interest, and reasonable attorney's fees as may be fixed by the court.

(d) ACQUISITION OF PATENTS.—The Commission is authorized to purchase, or to take, requisition, or condemn, and make just compensation for, (1) any invention or discovery which is useful in the production of fissionable material or in the utilization of fissionable material or atomic energy for a military weapon, or which utilizes or is essential in the utilization of fissionable material or atomic energy, or (2) any patent or patent application covering any such invention or discovery. The Commissioner of Patents shall notify the Commission of all applications for patents heretofore and hereafter filed which in his opinion disclose such inventions or discoveries and shall provide the Commission access to all such applications.

(e) COMPENSATION AWARDS, AND ROYALTIES.—

(1) PATENT COMPENSATION BOARD.—The Commission shall designate a Patent Compensation Board, consisting of two or more employees of the Commission, to consider applications under this subsection.

(2) ELIGIBILITY.—

(A) Any owner of a patent licensed under subsection (c) (2) or any license thereunder may make application to the Commission for the determination of a reasonable royalty fee in accordance with such procedures as it by regulation may establish.

(B) Any person seeking to obtain the just compensation provided in subsections (a), (b), or (d) shall make application therefor to the Commission in accordance with such procedures as it may by regulation establish.

(C) Any person making any invention or discovery useful in the production of fissionable material or in the utilization of fissionable material or atomic energy for a military weapon who is not entitled to compensation therefore under subsection (a) and who has complied with subsection (a) (3) above may make application to the Commission for, and the Commission may grant, an award.

(D) Any person making application under this subsection shall have the right to be represented by counsel.

(3) STANDARDS.—

(A) In determining such reasonable royalty fee, the Commission shall take into consideration any defense, general or special, that might be pleaded by a defendant in an action for infringement, the extent to which, if any, such patent was developed through federally financed research, the degree of utility, novelty, and importance of the invention or discovery, and may consider the cost to the owner of the patent of developing such invention or discovery or acquiring such patent.

(B) In determining what constitutes just compensation under subsection (a), (b), or (d) above, the Commission shall take into account the considerations set forth in paragraph (A) above, and the actual use of such invention or discovery, and may determine that such compensation be paid in periodic payments or in a lump sum.

(C) In determining the amount of any award under paragraph (2) (C) of this subsection, the Commission shall take into account the considerations set forth in paragraph (A) above, and the actual use of such invention or discovery. Awards so made may be paid by the Commission in periodic payments or in a lump sum.

(4) JUDICIAL REVIEW.—Any person aggrieved by any determination of the Commission of an award or of a reasonable royalty fee may obtain a review of such determination in the Court of Appeals for the District of Columbia by filing in such court, within thirty days after notice of such determination, a written petition praying that such determination be set aside. A copy of such petition shall be forthwith served upon the Commission and thereupon the Commission shall file with the court a certified transcript of the entire record in the proceeding, including the findings and conclusions upon which the determination was based. Upon the filing of such transcript the court shall have exclusive jurisdiction upon the record certified to it to affirm the determination in its entirety or

set it aside and remand it to the Commission for further proceedings. The findings of the Commission as to the facts, if supported by substantial evidence, shall be conclusive. The court's judgment shall be final, subject, however, to review by the Supreme Court of the United States upon writ of *certiorari* on petition therefor under section 240 of the Judicial Code (U.S.C., title 28, sec. 347), by the Commission or any party to the court proceeding.

GENERAL AUTHORITY

Sec. 12. (a) In the performance of its functions the Commission is authorized to—

(1) establish advisory boards to advise with and make recommendations to the Commission on legislation, policies, administration, research, and other matters;

(2) establish by regulation or order such standards and instructions to govern the possession and use of fissionable and byproduct materials as the Commission may deem necessary or desirable to protect health or to minimize danger from explosions and other hazards to life or property;

(3) make such studies and investigations, obtain such information, and hold such hearings as the Commission may deem necessary or proper to assist it in exercising any authority provided in this Act, or in the administration or enforcement of this Act, or any regulations or orders issued thereunder. For such purposes the Commission is authorized to administer oaths and affirmations, and by subpoena to require any person to appear and testify, or to appear and produce documents, or both, at any designated place. No person shall be excused from complying with any requirements under this paragraph because of his privilege against self-incrimination, but the immunity provisions of the Compulsory Testimony Act of February 11, 1893 (U.S.C., title 49, sec. 46), shall apply with respect to any individual who specifically claims such privilege. Witnesses subpoenaed under this subsection shall be paid the same fees and mileage as are paid witnesses in the district courts of the United States;

(4) appoint and fix the compensation of such officers and employees as may be necessary to carry out the functions of the Commission. Such officers and employees shall be appointed in accordance with the civil-service laws and their compensation fixed in accordance with the Classification Act of 1923, as amended, except that to the extent the Commission deems such action necessary to the discharge of its responsibilities, personnel may be employed and their compensation fixed without regard to such laws. The Commission shall make adequate provision for administrative review of any determination to dismiss any employee;

(5) acquire such materials, property, equipment, and facilities, establish or construct such buildings and facilities, and modify such buildings and facilities from time to time as it may deem necessary, and construct, acquire, provide, or arrange for such facilities and services (at project sites where such facilities and services are not available) for the housing, health, safety, welfare, and recreation of personnel employed by the Commission as it may deem necessary;

(6) with the consent of the agency concerned, utilize or employ the services or personnel of any Government agency or any State or local government, or voluntary or uncompensated personnel, to perform such functions on its behalf as may appear desirable;

(7) acquire, purchase, lease, and hold real and personal property as agent of any on behalf of the United States and to sell, lease, grant, and dispose of such real and personal property as provided in this Act; and

(8) without regard to the provisions of the Surplus Property Act of 1944 or any other law, make such disposition as it may deem desirable of (A) radioactive materials, and (B) any other property the special disposition of which is, in the opinion of the Commission, in the interest of the national security.

(b) SECURITY.—The President may, in advance, exempt any specific action of the Commission in a particular matter from the provisions of law relating to contracts whenever he determines that such action is essential in the interest of the common defense and security.

(c) ADVISORY COMMITTEES.—The members of the General Advisory Committee established pursuant to section 2 (b) and the members of advisory boards established pursuant to subsection (a) (1) of this section may serve as such without regard to the provisions of sections 109 and 113 of the Criminal Code (18 U.S.C., secs. 198 and 203) or section 19 (e) of the Contract Settlement Act of 1944, except insofar as such sections may prohibit any such member from receiving compensation in respect of any particular matter which directly involved the Commission or in which the Commission is directly interested.

COMPENSATION FOR PRIVATE PROPERTY ACQUIRED

SEC. 13. (a) The United States shall make just compensation for any property or interests therein taken or requisitioned pursuant to sections 5 and 11. The Commission shall determine such compensation. If the compensation so determined is unsatisfactory to the person entitled thereto, such person shall be paid 50 per centum of the amount so determined, and shall be entitled to sue the United States in the Court of Claims or in any district court of the United States in the manner provided by sections 24 (20) and 145 of the Judicial Code to recover such further sum as added to said 50 per centum will make up such amount as will be just compensation.

(b) In the exercise of the rights of eminent domain and condemnation, proceedings may be instituted under the Act of August 1, 1888 (U.S.C., title 40, sec. 257), or any other applicable Federal statute. Upon or after the filing of the condemnation petition, immediate possession may be taken and the property may be occupied, used, and improved for the purposes of this Act, notwithstanding any other law. Real property acquired by purchase, donation, or other means of transfer may also be occupied, used and improved for the purposes of this Act, prior to approval of title by the Attorney General.

JUDICIAL REVIEW AND ADMINISTRATIVE PROCEDURE

SEC. 14. (a) Notwithstanding the provisions of section 12 of the Administrative Procedure Act (Public Law 404, Seventy-ninth Congress, approved June 11, 1946) which provide when such Act shall take effect, section 10 of such Act (relating to judicial review) shall be applicable, upon the enactment of this Act, to any agency action under the authority of this Act or by any agency created by or under the provisions of this Act.

(b) Except as provided in subsection (a), no provision of this Act shall be held to supersede or modify the provisions of the Administrative Procedure Act.

(c) As used in this section the terms "agency action" and "agency" shall have the same meaning as is assigned to such terms in the Administrative Procedure Act.

JOINT COMMITTEE ON ATOMIC ENERGY

SEC. 15. (a) There is hereby established a Joint Committee on Atomic Energy to be composed of nine Members of the Senate to be appointed by the President of the Senate, and nine Members of the House of Representatives to be appointed by the Speaker of the House of Representatives. In each instance not more than five members shall be members of the same political party.

(b) The joint committee shall make continuing studies of the activities of the Atomic Energy Commission and of problems relating to the development, use, and control of atomic energy. The Commission shall keep the joint committee fully and currently informed with respect to the Commission's activities. All bills, resolutions, and other matters in the Senate or the House of Representatives relating primarily to the Commission or to the development, use, or control of atomic energy shall be referred to the joint committee. The members of the joint committee who are Members of the Senate shall from time to time report to the Senate, and the members of the joint committee who are Members of the House of Representatives shall from time to time report to the House, by bill or otherwise, their recommendations with respect to matters within the jurisdiction of their respective Houses which are (1) referred to the joint committee or (2) otherwise within the jurisdiction of the joint committee.

(c) Vacancies in the membership of the joint committee shall not affect the power of the remaining members to execute the functions of the joint committee, and shall be filled in the same manner as in the case of the original selection. The joint committee shall select a chairman and a vice chairman from among its members.

(d) The joint committee, or any duly authorized subcommittee thereof, is authorized to hold such hearings, to sit and act at such places and times, to require, by subpoena or otherwise, the attendance of such witnesses and the production of such books, papers, and documents, to administer such oaths, to take such testimony, to procure such printing and binding, and to make such expenditures as it deems advisable. The cost of stenographic services to report such hearings shall not be in excess of 25 cents per hundred words. The provisions of sections 102 to 104, inclusive, of the Revised Statutes shall apply in case of any failure of any witness to comply with a subpoena or to testify when summoned under authority of this section.

(e) The joint committee is empowered to appoint and fix the compensation of such experts, consultants, technicians, and clerical and stenographic assistants as it deems necessary and advisable, but the compensation so fixed shall not exceed the compensation prescribed under the Classification Act of 1923, as amended, for comparable duties. The committee is authorized to utilize the services, information, facilities, and personnel of the departments and establishments of the Government.

ENFORCEMENT

SEC. 16. (a) Whoever willfully violates, attempts to violate, or conspires to violate, any provision of sections 4 (b), 4 (e), 5 (a) (3), or 6 (b) shall, upon conviction thereof, be punished by a fine of not more than $10,000 or by imprisonment for not more than five years, or both, except that whoever commits such an offense with intent to injure the United States or with intent to secure an advantage to any foreign nation shall, upon conviction thereof, be punished by death or imprisonment for life (but the penalty of death or imprisonment for life may be imposed only upon recommendation of the jury and only in cases where the offense was committed with intent to injure the United States); or by a fine of not more than $20,000 or by imprisonment for not more than twenty years, or both.

(b) Whoever willfully violates, attempts to violate, or conspires to violate, any provision of this Act other than those specified in subsection (a) and other than section 10 (b), or of any regulation or order prescribed or issued under sections 5 (b) (4), 10 (c), or 12 (a) (2), shall, upon conviction thereof, be punished by a fine of not more than $5,000 or by imprisonment for not more than two years, or both, except that whoever commits such an offense with intent to injure the United States or with intent to secure an advantage to any foreign nation shall, upon conviction thereof, be punished by a fine of not more than $20,000 or by imprisonment for not more than twenty years, or both.

(c) Whenever in the judgment of the Commission any person has engaged or is about to engage in any acts or practices which constitute or will constitute a violation of any provision of this Act, or any regulation or order issued thereunder, it may make application to the appropriate court for an order enjoining such acts or practices, or for an order enforcing compliance with such provision, and upon a showing by the Commission that such person has engaged or is about to engage in any such acts or practices a permanent or temporary injunction, restraining order, or other order may be granted.

(d) In case of failure of refusal to obey a subpoena served upon any person pursuant to section 12 (a) (3), the district court for any district in which such person is found or resides or transacts business, upon application by the Commission, shall have jurisdiction to issue an order requiring such person to appear and give testimony or to appear and produce documents, or both, in accordance with the subpoena; and any failure to obey such order of the court may be punished by such court as a contempt thereof.

REPORTS

SEC. 17. The Commission shall submit to the Congress, in January and July of each year, a report concerning the activities of the Commission. The Commission shall include in such report, and shall at such other times as it deems desirable submit to the Congress, such recommendations for additional legislation as the Commission deems necessary or desirable.

DEFINITIONS

SEC. 18. As used in this Act—

(a) The term "atomic energy" shall be construed to mean all forms of energy released in the course of or as a result of nuclear fission or nuclear transformation.

(b) The term "Government agency" means any executive department, commission, independent establishment, corporation wholly or partly owned by the United States which is an instrument of the United States, board, bureau, division, service, office, officer, authority, administration, or other establishment, in the executive branch of the Government.

(c) The term "person" means any individual, corporation, partnership, firm, association, trust, estate, public or private institution, group, the United States or any agency thereof, any government other than the United States, any political subdivision of any such government, and any legal successor, representative, agent, or agency of the foregoing, or other entity, but shall not include the Commission or officers or employees of the Commission in the exercise of duly authorized functions.

(d) The term "United States" when used in a geographical sense, includes all Territories and possessions of the United States and the Canal Zone.

(e) The term "research and development" means theoretical analysis, exploration, and experimentation, and the extension of investigative findings and theories of a scientific or technical nature into practical application for experimental and demonstration purposes, including the experimental production and testing of models, devices, equipment, materials, and processes.

(f) The term "equipment or device utilizing fissionable material or atomic energy" shall be construed to mean any equipment or device capable of making use of fissionable material or peculiarly adapted for making use of atomic energy and any important component part especially designed for such equipment or devices, as determined by the Commission.

(g) The term "facilities for the production of fissionable material" shall be construed to mean any equipment or device capable of such production and any important component part especially designed for such equipment or devices, as determined by the Commission.

APPROPRIATIONS

SEC. 19. There are hereby authorized to be appropriated such sums as may be necessary and appropriate to carry out the provisions and purposes of this Act. The Acts appropriating such sums may appropriate specified portions thereof to be accounted for upon the certification of the Commission only. Funds appropriated to the Commission shall, if obligated by contract during the fiscal year for which appropriated, remain available for expenditure for four years following the expiration of the fiscal year for which appropriated. After such four-year period, the unexpended balances of appropriations shall be carried to the surplus fund and covered into the Treasury.

SEPARABILITY OF PROVISIONS

SEC. 20. If any provision of this Act, or the application of such provision to any person or circumstances, is held invalid, the remainder of this Act or the application of such provision to persons or circumstances other than those as to which it is held invalid, shall not be affected thereby.

SHORT TITLE

SEC. 21. This Act may be cited as the "Atomic Energy Act of 1946".
Approved August 1, 1946.

From U.S. Code, Title 42, Ch. 23, "Atomic Energy Act of 1946." A version can be found on the Web at: http://0-www.osti.gov.library.unl.edu/atomicenergyact.pdf.

27
The Alaska Highway

United States and Canada

DID YOU KNOW . . . ?

- ➤ In 1725, Czar Peter the Great ordered Captain Vitus Bering to explore the Alaskan seacoast. He discovered what would later be called the Bering Strait.
- ➤ Russia sold the land that would become Alaska to the United States in 1867.
- ➤ The Alaska Highway was built in 1942, in just over six months, across 1,645 miles (2,647 kilometers) of rugged terrain.
- ➤ 10,000 soldiers and several thousand civilians did the work.
- ➤ The highway now extends 1,500 miles (2,400 kilometers).
- ➤ The Alaska Highway is the northernmost section of the Pan-American Highway.

The initial Alaska Highway was built in just over six months—amazingly fast for a road that was blasted through craggy mountains and bulldozed through dense forests. It was not easy, as the poem "Mile 392," written by retired sergeant Troy Hise, suggests:

> The Alaska Highway
> winding in and winding out
> fills my mind with serious doubt
> as to whether the lout
> who planned this route
> was going to hell or coming out. (http://www.themilepost.com/history.html)

HISTORY

The Japanese attack on Pearl Harbor, Hawaii, in December 1941 drew the United States into World War II, and with the sudden focus on the Pacific

A scenic stretch of the Alaska Highway in 1942. Courtesy of the Library of Congress.

theater of war, there was immediate need for an overland route to supply U.S. forces based in Alaska. Sharpened by Japan's occupation of three Aleutian islands in mid-1942, a road was built at breakneck speed through sub-Arctic territory, a largely uncharted wilderness between Fairbanks, Alaska, and the railroad terminus at Dawson Creek, British Columbia.

The highway was a colossal undertaking, and some voices in the U.S. military initially argued against it, questioning the wisdom of allocating to such a grandiose scheme a quantity of supplies and men that were surely needed elsewhere; meanwhile, the Navy believed it could hold the sea-lanes open. But the window of opportunity was limited: troops and supplies had to be delivered to the area while the ground was still frozen, so the decision was made to go ahead. By early February 1942, a route was chosen, and in mid-February a directive to start work was issued. The diplomatic notes from mid-March 1942, reproduced here, officially signaled the agreement of the U.S. and Canadian governments to terms under which the project would be carried out.

It was a prodigious endeavor. The route selected had military advantages (planners wanted to link up with existing airstrips along the way; an inland route was less exposed to possible Japanese air attack), but the overland route was far from the easiest or most direct option; a coastal route would likely have presented fewer construction difficulties.

About 10,000 soldiers were brought in for the project. The first trainload arrived at Dawson Creek in early March, even before the exchange of diplomatic notes. The troops, plus several thousand civilians working under the U.S. Public Roads Administration, bulldozed out a rough "pioneer road" 1,645 miles (2,647 kilometers) long, in a little over half a year. It was a heroic feat. The workers traversed mountains, forests, soft bog lands (muskeg), ice, and scores of rivers, not to mention coping with swarms of mosquitoes and blackflies; those who stayed on until the following year had to endure bitter winter cold. The muddy pioneer road became the basis for construction of an improved, all-weather military road the next year. This construction was handled by an even larger work crew consisting of thousands of U.S. and Canadian private contract workers. The temperature at the dedication ceremony, held on November 20, 1942, at Soldiers' Summit, a point about 1,000 miles up the road from Dawson Creek, was about −35°F (−37°C).

The portion of the Alaska Highway that lies in Canada, approximately 1,200 miles (1,900 kilometers), was transferred to Canadian control in 1946, and the entire road opened to civilian traffic two years later. Today, the Alaska Highway—its surface improved, its path smoothed out, and now extending about 1,500 miles (2,400 kilometers)—remains in use by tourists and truckers.

CULTURAL CONTEXT

Real development of Alaska began when Czar Peter the Great, in 1725, hired a seafaring Dane to explore the waters and land of the Northwest Coast. His name was Vitus Bering, and in 1728 he sailed through the strait that now bears his name: the Bering Strait.

Alaska represents a story of three different treasures that sparked three waves of engineering. The first treasure was fur: the Russians nearly wiped out the entire species of sea otter in their greedy desire for the soft and waterproof fur. James A. Michener, in his novel *Alaska*, describes a poignant scene of a mother sea otter floating on her back with her baby cub nestled on her belly, until both mother and cub are slain. It would not be until 1911 that sea otters were given status as a protected species. But the quantity and quality of fur to be found in Alaska attracted Russian ships and settlers, who took advantage of the native people. Unfortunately for Russia, dire economic straits prevailed, and in October 1867 Russia sold Alaska to the United States.

The next treasure that caused a wave of interest was gold, which brought the railroads to Alaska. Although early findings of the precious ore were discovered in small deposits as early as 1832, shortly after the United States bought Alaska, gold began to appear more frequently. Then larger amounts were found at Cook Inlet and Mastadon Creek in 1894. In 1896 Bonanza Creek lived up to its name: the Klondike River yielded its hitherto unknown store of gold. By 1897, the river had sent $1 million worth of gold into the port of Seattle on the *Portland*. The next year, the Klondike yielded $200 million. With such activity, the U.S. Congress authorized construction of a railroad in 1898. By 1905, the railroad and telegraph linked Fairbanks to Valdez.

As gold began to run out, another kind of gold was discovered—liquid gold. In 1944 the Alaska-Juneau Gold Mine shut down, but at the same time oil and gas exploration began. The possibility of oil was first hinted in 1853 when a modest slick of oil appeared at Cook Inlet; it was not until 1891 that the first oil claims were staked in the area. When gold was discovered there in 1894, oil was completely forgotten. The year 1944 brought another wave of engineering, and by 1968 oil and gas had been found on the North Slope; the sale of that lease for $900 million attracted considerable attention the next year. Authorized on November 16, 1973, and built by 1977, the Alaska pipeline was the symbol of the third wave of engineering.

Following the Japanese attack on Pearl Harbor, Alaska in particular and the entire West Coast of the United States suddenly became more vulnerable to the possibility of attack from Japan. Without a highway to bring supplies to Alaska, access to the airports and airbases that had been built in Alaska by Canada would be unacceptably slow. It was decided and agreed between the United States and Canada that such a road had to be built.

Engineers are shown pounding bridge supports into a riverbed that will become a part of the Alaska Highway, 1942. Courtesy of the Library of Congress.

The competencies required for large-scale engineering had become ingrained in the American character. Technology for road building was available; skilled management of work teams had been evolving ever since the Erie Canal, the transcontinental railroad, and the Tennessee Valley Authority. The Army Corps of Engineers knew how to build public projects quickly with hired laborers and outside experts, as they had done in many projects. With a war in progress, the corps was even more motivated and supported. Indeed, the military origins of the road are clear from its first title: "Military Highway to Alaska."

PLANNING

Because its border was so long and difficult to secure, and because it was so far removed from the lower 48 states, Alaska was disturbingly vulnerable. When the possibility of an invasion by Japan seemed distinctly real, there was a rush for greater protection. The U.S. War Department planned the Alaska Highway to link airfields from Edmonton, Alberta, to Fairbanks, Alaska. The entire project was planned from the air, partly because its *raison d'être* was to link up existing airfields, and partly because time was deemed too short for the customary topographical surveys. It seemed a sensible way to lay out the route, but back on the ground there were many obstacles. Like the Trans-Siberian Railway, the Alaska Highway ended up following old native trails and winter roads originally worn into the landscape by sleds.

Canada and the United States planned jointly for the Military Highway to Alaska. The documents presented here portray the numerous cooperative exchanges between Ottawa and Washington. The first set of documents, from March 17 and 18, 1942, lays out the basic agreement: "the construction of a highway along the route that follows the general line of airports . . . the respective termini connecting with existing roads in Canada and Alaska." The first section of the executive agreement states that the decision is based "on military considerations and military considerations only." America offered to build and maintain the road, especially since Canada had already built the airfields. Cooperation is the theme of the agreement, and speed is the goal, as seen in section 2(b), "arranging for the highway's completion under contracts made by the United States Public Roads Administration and awarded with a view to insuring the execution of all contracts in the shortest possible time without regard to whether the contractors are Canadian or American." Section 2(d) stipulates that after the war, the section of the highway running through Canada will become "an integral part of the Canadian highway system." For this reason, the highway was initially referred to as the Alcan Highway. An official name is reflected in the second set of documents, from 1943, in which the two nations agree to name the road the "Alaska Highway."

Tax incentives sweetened the deal, as described in section 3(c), with no import duties on equipment and no taxes for workers. Even timber from Canadian land was available to use, as long as the cutting met Canadian environmental standards (section 3[f]).

The spirit of cooperation between the two contiguous countries—and war allies—was clear in the conclusion of the agreement, which states (in paraphrase) that the two countries would work out the details as those details arose. Cooperation and expediency were in the air when the Canadians responded positively to the agreement the next day, March 18, 1942.

BUILDING

With agreement in place and a plan ready to be implemented, construction on the Alaska Highway began on March 8, 1942—not a moment too soon, for as feared, Japan invaded the Aleutian Islands on June 7 of that same year. With so much at stake, the United States acted quickly. A decision was made to work from both ends toward the middle; the Army Corps of Engineers sent soldiers from the 36th Regiment to work from the southern end, while the 340th Regiment tackled the northern end.

A monument marks the end of the Alaska Highway, Delta Junction, Alaska. Courtesy of Shutterstock.

Skilled workers and engineers overcame extraordinarily difficult odds: 90-degree turns and 25 percent grades! Bulldozers and blasting could be heard at every work site. The urgent need for speedy completion meant that construction took place during all seasons and in all conditions; rain drenched sections of the road, turning it into a quagmire; bitter cold hampered every effort and made building extremely difficult. After the highway opened, it needed improvement almost immediately because it had been built so quickly.

There was much celebrating when the two work crews met on September 24, 1942, at Contact Creek. In October, a vehicle traversed the new highway, and on November 20, 1942, the official dedication took place with a ceremony at Soldiers' Summit.

IMPORTANCE IN HISTORY

The Alaska Highway resulted from a military emergency, which brought on cordial and cooperative teamwork for common defense. All the normal logjams were removed, rights were acquired, and Canada smoothed out potential barriers, making it possible to use their land and timber, and even remitted taxes. The result was a key improvement in the basic infrastructure for the region, and a long-term boost for the economies of Canada and the United States. The completion of the Alaska Highway added the northernmost section of the Pan-American Highway.

The agreements presented here, especially those from 1942, which established and authorized the building of the Alaska Highway, represent the essence of trust and cooperation, and offer a model that is relevant today.

FOR FURTHER REFERENCE

Books and Articles

Christy, Jim. *Rough Road to the North: Travels along the Alaska Highway*. Toronto: Doubleday Canada, 1980.

Coates, Kenneth, and William R. Morrison. *The Alaska Highway in World War II: The U.S. Army of Occupation in Canada's Northwest*. Toronto: University of Toronto Press, 1992.

Huber, Thomas Patrick, and Carole J. Huber. *The Alaska Highway: A Geographical Discovery*. Boulder: University Press of Colorado, 2000.

Michener, James A. *Alaska*. New York: Fawcett Crest, 1988.

Twitchell, Heath. *Northwest Epic: The Building of the Alaska Highway*. New York: St. Martin's Press, 1992.

Internet

For a brief time line, see http://www.themilepost.com/timeline.html.

For a history, with pictures, and the text of Troy Hise's lament, see http://www.themilepost.com/history.html. (© Morris Communications, 1998.)

For information about the highway as a part of the U.S. program, see http://www.fhwa.dot.gov/infrastructure/blazer01.htm.

For music of the native people of Alaska, including CDs by Sylvestor Ayek, John Pingayak, Evelyn Alexander, Shxat'kwaan Daners, and Ethan Petticrew and the Atka Dancers, including sound clips, see http://www.oyate.com/.. (The CDs are produced in partnership with American Indian Music, Lee Productions, and Oyate Records, who seek to preserve traditional native music, ensure musicians are paid for their lifework, and educate others on the diversity of native culture. Tel: 1-800-486-8940.)

Film and Television

Building the Alaska Highway. Produced and directed by Tracy Heather Strain. Boston: WGBH Educational Foundation, PBS Television, 2005, http://www.pbs.org/wgbh/amex/alaska/.

Documents of Authorization

MILITARY HIGHWAY TO ALASKA

Exchange of notes at Ottawa March 17 and 18, 1942
   Entered into force March 18, 1942
   Supplemented by agreement of May 4 and 9, 1942, and April 10, 1943
   56 Stat. 1458; Executive Agreement Series 246
   From the American Minister to the Secretary of State for External Affairs
   Legation of the United States of America
Ottawa, Canada
March 17, 1942
No. 626

Sir:

1. As you are aware, on February 26th, 1942, the Permanent Joint Board on Defence approved a recommendation as a result of which the two Sections proposed to their respective Governments:

"the construction of a highway along the route that follows the general line of airports, Fort St. John-Fort Nelson-Watson Lake-Whitehorse-Boundary-Big Delta, the respective termini connecting with existing roads in Canada and Alaska."

This recommendation, based as it was on military considerations and military considerations only, and having the endorsement of the Service Departments of the two countries, has been approved by both Governments.

2. My Government, being convinced of the urgent necessity for the construction of this highway and appreciating the burden of war expenditure already incurred by Canada, in particular on the construction of the air route to Alaska, is prepared to undertake the building and wartime maintenance of the highway. Subject to the provision by Canada of the facilities set forth in paragraph three of this Note, the Government of the United States is prepared to:

(a) Carry out the necessary surveys for which preliminary arrangements have already been made, and construct a Pioneer Road by the use of United States Engineer troops for surveys and initial construction;

(b) Arrange for the highway's completion under contracts made by the United States Public Roads Administration and awarded with a view to insuring the execution of all contracts in the shortest possible time without regard to whether the contractors are Canadian or American;

(c) Maintain the highway until the termination of the present war and for six months thereafter unless the Government of Canada prefers to assume responsibility at an earlier date for the maintenance of so much of it as lies in Canada;

(d) Agree that at the conclusion of the war that part of the highway which lies in Canada shall become in all respects an integral part of the Canadian highway system, subject to the understanding that there shall at no time be imposed any discriminatory conditions in relation to the use of the road as between Canadian and United States civilian traffic.

3. For its part, my Government will ask the Canadian Government to agree:

(a) To acquire rights of way for the road in Canada (including the settlement of all local claims in this connection), the title to remain in the Crown in the right of Canada or of the Province of British Columbia as appears more convenient;

(b) To waive import duties, transit or similar charges on shipments originating in the United States and to be transported over the highway to Alaska, or originating in Alaska and to be transported over the highway to the United States;

(c) To waive import duties, sales taxes, license fees or other similar charges on all equipment and supplies to be used in the construction or maintenance of the road by the United States and on personal effects of the construction personnel;

(d) To remit income tax on the income of persons (including corporations) resident in the United States who are employed on the construction or maintenance of the highway;

(e) To take the necessary steps to facilitate the admission into Canada of such United States citizens as may be employed on the construction or maintenance of the highway, it being understood that the United States will undertake to repatriate at its expense any such persons if the contractors fail to do so;

(f) To permit those in charge of the construction of the road to obtain timber, gravel and rock where such occurs on Crown lands in the neighborhood of the right of way, providing that the timber required shall be cut in accordance with the directions of the appropriate Department of the Government of the province in which it is located, or, in the case of Dominion lands, in accordance with the directions of the appropriate Department of the Canadian Government.

4. If the Government of Canada agrees to this proposal, it is suggested that the practical details involved in its execution be arranged directly between the appropriate governmental agencies subject, when desirable, to confirmation by subsequent exchange of notes.

Accept, Sir, the renewed assurances of my highest consideration.

/s/ Pierrepont Moffat
American Minister
The Secretary of State for External Affairs to the American Minister

The Right Honorable
THE SECRETARY OF STATE
FOR EXTERNAL AFFAIRS,
Ottawa.

DEPARTMENT OF EXTERNAL AFFAIRS
CANADA
OTTAWA
March 18, 1942
No. 29

Sir,

I have the honour to acknowledge receipt of your Note of March 17, 1942, No. 626, in which you referred to the recommendation approved by the Permanent Joint Board on Defence, as a result of which the two Sections of the Board proposed to their respective Governments:

"the construction of a highway along the route that follows the general line of airports, Fort St. John-Fort Nelson-Watson Lake-Whitehorse-Boundary-Big Delta, the respective termini connecting with existing roads in Canada and Alaska."

2. As announced on March 6, 1942, the Canadian Government has approved this recommendation and has accepted the offer of the United States Government to undertake the building and wartime maintenance of the highway which will connect the airports already constructed by Canada.

3. It is understood that the United States Government will:

(a) Carry out the necessary surveys for which preliminary arrangements have already been made, and construct a Pioneer Road by the use of United States Engineer troops for surveys and initial construction;

(b) Arrange for the highway's completion under contracts made by the United States Public Roads Administration and awarded with a view to insuring the execution of all contracts in the shortest possible time without regard to whether the contractors are Canadian or American;

(c) Maintain the highway until the termination of the present war and for six months thereafter unless the Government of Canada prefers to assume responsibility at an earlier date for the maintenance of so much of it as lies in Canada;

(d) Agree that at the conclusion of the war that part of the highway which lies in Canada shall become in all respects an integral part of the Canadian highway system, subject to the understanding that there shall at no time be imposed any discriminatory conditions in relation to the use of the road as between Canadian and United States civilian traffic.

4. The Canadian Government agrees:

(a) To acquire rights of way for the road in Canada (including the settlement of all local claims in this connection), the title to remain in the Crown in the right of Canada or of the Province of British Columbia as appears more convenient;

(b) To waive import duties, transit or similar charges on shipments originating in the United States and to be transported over the highway to Alaska, or originating in Alaska and to be transported over the highway to the United States;

(c) To waive import duties, sales taxes, license fees or other similar charges on all equipment and supplies to be used in the construction or maintenance of the road by the United States and on personal effects of the construction personnel;

(d) To remit income tax on the income of persons (including corporations) resident in the United States who are employed on the construction or maintenance of the highway;

(e) To take the necessary steps to facilitate the admission into Canada of such United States citizens as may be employed on the construction or maintenance of the highway, it being understood that the United States will undertake to repatriate at its expense any such persons if the contractors fail to do so;

(f) To permit those in charge of the construction of the road to obtain timber, gravel and rock where such occurs on Crown lands in the neighborhood of the right of way, providing that the timber required shall be cut in accordance with the directions of the appropriate Department of the Province in which it is located, or, in the case of Dominion lands, in accordance with the directions of the appropriate Department of the Canadian Government.

5. The Canadian Government agrees to the suggestion that the practical details of the arrangement be worked out by direct contact between the appropriate governmental agencies subject, when desirable, to confirmation by subsequent exchange of notes.

Accept, Sir, the renewed assurance of my highest consideration.

/s/ W.L. Mackenzie King
Secretary of State
for External Affairs
    Exchange of notes at Washington July 19, 1943
    Entered into force July 19, 1943
    57 State. 1023; Executive Agreement Series 331
    The Secretary of State to the Canadian Minister
Department of State
Washington
July 19, 1943

Sir:

I have the honor to inform you that the Honorable Anthony J. Dimond, Delegate of Alaska, United States House of Representatives, has proposed that the highways from Dawson Creek, British Columbia, to Fairbanks, Alaska, be given the official name "Alaska Highway".

The Government of the United States believes that the name suggested by Mr. Dimond is suitable and in harmony with popular usage. It is of the further opinion that the highway should be jointly named by the Governments of the United States and Canada in view of the location of the greater part of the highway within Canada and in view of the friendly cooperation which has made possible its construction.

In accordance with the foregoing, I have the honor to propose that the highways from Dawson Creed, British Columbia, to Fairbanks, Alaska, be designated the "Alaska Highway". If the Canadian Government is agreeable to this proposal, it is suggested that this note and your reply in that sense shall be considered as placing on record the agreement of the two Governments in this matter.

Accept, Sir, the renewed assurances of my highest consideration.

/s/ CORDELL HULL
The Honorable Leighton McCarthy, J.C.,

Minister of Canada
    The Canadian Minister to the Secretary of State
Canadian Legation
Washington
July 19, 1943
No. 377
I have the honour to inform you that the Government of Canada concurs in the proposal, containing in your note of July 19, 1943, that the highway from Dawson Creek, British Columbia to Fairbanks, Alaska be given the official name "Alaska Highway".

Accept, Sir, the renewed assurances of my highest consideration.

/s/ Leighton McCarthy
The Honourable Cordell Hull,
Secretary of State of the United States,
Washington, D.C.

28

The Snowy Mountains Hydroelectric Power Project

Australia

DID YOU KNOW . . . ?

➤ Australia is the most arid country on earth.
➤ The Snowy Mountains project was one of the twentieth century's major achievements in civil engineering.
➤ It was one of the first projects to use computers for engineering design.
➤ Begun in 1949, it took 25 years to complete.
➤ It includes 16 dams, seven power stations, and 140 miles (225 kilometers) of tunnels, pipelines, and aqueducts.
➤ The project was completed on time and under budget ($1 billion Aus).
➤ The labor force of 100,000 included thousands from other countries.
➤ Snowy's hydroelectric power plants take less than 90 seconds to generate additional electricity.

The American Society of Civil Engineers nominated the Snowy Mountains project as "one of the great engineering achievements of the century" (http://www.users.bigpond.net.au/snowy/bgrnd.html). Situated in southern New South Wales, Australia, the project is one of the most complex water and electricity utilities in the world and the greatest engineering project undertaken in Australia. It has the additional distinction of being one of the most complex, multipurpose, multi-reservoir hydroelectric schemes in the world. It took 25 years to build and consists of 16 major dams, seven power stations, a pumping station, and 140 miles (225 kilometers) of interconnected tunnels, pipelines, and aqueducts. This project was not only completed on time but also under budget (the cost was $1 billion Aus, including interest).

Guthega Power House, part of the Snowy Mountains Hydro-electric scheme, circa 1940. Courtesy of the National Library of Australia.

HISTORY

The Snowy Mountains are located in southwestern Australia, several dozen miles from Canberra, the country's capital. The mountains are part of Australia's Great Dividing Range and are in the only region of the country with elevations of more than 7,000 feet (2,100 meters). Precipitation totals about 120 inches (300 centimeters) a year, feeding rivers that flow to the west or to the east, depending on which side of the Great Dividing Range they drain.

Drought periods plagued the early chapters of Australia's history. From 1813 to 1815 drought plagued the land, and from 1824 to 1829 there was a drought so bad that even the native aboriginal people, who had learned to cope with the problem, were suffering from a lack of water. When a third drought lasted from 1837 to 1840, livestock died, bushes withered, crops failed, and the populace despaired. Even Australia's inland rivers dried up. In a grim kind of amusement, settlers organized horse races on the Murrumbidgee River, using its dry bed as a dusty track.

Since the nineteenth century, there were numerous discussions about constructing hydroengineering works in the Snowy Mountains to take advantage of the snow melt and use the water for irrigation. Interest grew following World War II because of the need for greater generating capacity. In 1949, Australia's parliament passed the Snowy Mountains Hydro-electric Power Act. The measure (reproduced here), created the Snowy Mountains Hydro-electric Authority to carry out and operate the project for the purpose of generating electricity. Also in 1949, the federal government and affected states drew up preliminary agreements—officially legalized a decade or so later after resolving certain constitutional issues—regarding the sharing of the water and its use for power and irrigation.

Australia carried out the project to divert water westward for the purpose of generating electricity and irrigating the country's arid interior. Work began in 1949 and continued for 25 years. Upon completion, the complex project covered an area of about 3,000 square miles (8,000 square kilometers), with a pumping station and seven generating stations that produce a combined capacity of 3.74 million kilowatts. The chief water-storage reservoir, Lake Eucumbene, has a capacity of 3,890,000 acre-feet (4,798,000,000 cubic meters). The scheme's longest tunnel, between the Snowy River and Lake Eucumbene, stretches 14.6 miles (23.5 kilometers). The American Society of Civil Engineers has called the project, which was inspired by the Tennessee Valley Authority in the United States, one of the twentieth century's major achievements in civil engineering.

The labor force was bolstered by thousands of people from other countries. Many analysts believe the influx of foreign labor helps explain Australia's remarkable postwar industrial expansion.

CULTURAL CONTEXT

Although immense in area, Australia in the late 1940s had fewer than 10 million people, most of British or Irish ancestry. When the Snowy Mountains project started, an intense recruitment campaign began. Two-thirds of the workforce of 100,000 people came from over two dozen different countries. Many workers were Europeans who had become homeless following World War II and were recruited from displaced-persons camps. Eager for an opportunity to start a new life, former enemies worked side by side and grew to be friends, validating the promise of Sir William Hudson, first commissioner of the Snowy Mountains Hydro-electric Plant, who had recruited in the European camps: "You won't be Balts or Slavs . . . you will be men of the Snowy!" Although there was cooperation, there was also friendly rivalry. Workers formed competing teams that raced good-naturedly to complete jobs first.

The project was of such colossal magnitude that help from overseas was needed. Technical assistance was provided by the U.S. Bureau of Reclamation, and foreign firms were contracted to handle much of the work.

PLANNING

The year was 1884, and the surveyor-general of New South Wales, P. F. Adams, had an idea: build a dam just above the confluence point of two rivers, the Snowy and the Eucumbene, and then channel the water through a canal that could cut across a small gap he had spotted in the Great Dividing Range, where it would rejoin the water in the Murrumbidgee River.

Once the idea caught on at the highest level, the colonies that would become Australian states began to see the possibilities of Adams's idea. There were many discussions, and two very different uses of the water were debated: irrigation or hydroelectricity. It took a war to decide the official answer. World War II, especially the raging war in the Pacific region, made Australia nervous about the vulnerability of its coastal power plants. It was decided that another, more secure source of electricity should became a national strategic and military priority.

This decision did not, however, quell the hotly debated question about the two uses of water. New South Wales badly needed water for irrigation and wanted the decision to favor diverting water to the Murrumbidgee River, which could then be used for irrigation. Victoria, on the other side of the contentious issue, insisted that the water flow come into the Murray River—also for irrigation but, in addition, for transport. When transport turned to rails and later roads, Victoria still needed water for irrigation, but now it also wanted electricity.

World War II was looming over the horizon, making everyone nervous, especially the government. It decided that hydroelectricity had to be the primary goal for any large-scale effort. The government also wanted to place the project so that its benefits could easily reach the two fastest-growing cities: Melbourne and Sydney.

In 1948, the first plan called for the northern part to generate electricity by taking water from the Eucumbene, Murrumbidgee, and Tooma Rivers and running it into the Tumut River, where the steep fall of the water would be ideal for power generation. The water would then be used for irrigation purposes. The southern part would take water from the Snowy River, channel it into the Murray River, and use the fall rate to produce electricity. Both projects would end in dams that would store the water.

It was decided that electricity and defense were first priorities, so irrigation would become a by-product that could be distributed at no charge. The decision about irrigation was two-edged. It would enable a fundamental change in the agricultural economy—moving from livestock, which require less water, to crops, which require much more. And making the irrigation water free would facilitate added development. The consequences of this decision would take years to emerge.

In May 1949, the nation, desperate for a solution to drought and eager for a decision, gathered around their radios to hear Prime Minister Benjamin Chifley: "The Snowy Mountains plan is the greatest single project in our history. It is a plan for the whole nation, belonging to no one state nor to any group or section. This is a plan for the nation, and it needs the nation to back it." (http://www.users.bigpond.net.au/snowy/politics.html). On July 7, 1949, the Snowy Mountains Hydro-electric Power Act authorized construction of the scheme and created an administrative entity to oversee the work, the Snowy Mountains Hydro-electric Authority.

Although the Australian parliament had enacted the basic law, the proposal languished for almost 10 years because it was not validated in the state parliaments of New South Wales, Victoria, and South Australia. The states eventually relented because it was promised that outside labor would be recruited, and so would not deplete their own scarce human resources, which were urgently needed to run operations in their own regions.

BUILDING

How is it possible to create such vast amounts of water from snow on mountains? Here is a simple explanation. Rain showers fall, and snow melts. That moisture normally forms lakes. Tunnels with pressurized pipelines are built to carry water from the lakes (or reservoirs) to a power station. At the power station, turbines drive electrical generators. Electricity is generated by the energy of the rate of falling water (for steep falls, medium, and mild, there are different sizes of turbines; the Snowy mainly uses a medium-size turbine called a Francis). After it has been used to generate electricity, the water can be stored, used for irrigation, and/or returned to the rivers and lakes. This basic process can have many stages.

It is important to know two things about hydroelectricity: (1) no greenhouse gases are produced that can harm the environment, and (2) it is instantly available, whereas other forms, like coal-based power plants, can take hours to emerge. Therefore, hydroelectricity is an excellent system for coping with fluctuating demand. It takes less than 90 seconds to make extra power available.

Construction on the Snowy Mountains scheme began on October 17, 1949, and was completed in 1974. The first project undertaken was Guthega Dam and Power Station because they were relatively easy and needed nothing unusual in the way of equipment or technology. A Norwegian firm was hired to handle the construction, which was completed in three years, from 1951 to 1954. Australia opened the Guthega Power Station in 1955, and it soon began making money. Many things were learned that prepared the builders for the larger, less certain projects that lay ahead where innovation would certainly be required.

Sydney Harbor, on which the famed Sydney Opera House faces, is big, but Lake Eucumbene is nine times bigger, and it would be the Snowy's biggest reservoir. Tumut, another section of the project, was so enormous that it took from 1954 to 1973 to build, including its underground tunnels and two underground power stations: Tumut 1 and Tumut 2. Tumut 3, even larger than its two predecessors, has pipes so large it is possible to drive a double-decker bus through them—a feat actually demonstrated upon completion of the huge effort. This was only the beginning, however; the complete project includes Blowering Dam and Power Station, Murray River Work, Jindabyne, Island Bend, Geehi Dam, Murray 1 and Murray 2 Dams and Power Stations, and the Khancoban Dam. Consider the amazing facts that the Snowy project:

• comprises 50 miles (80 kilometers) of aqueducts; 90.6 miles (145 kilometers) of tunnels, some going through mountains; 16 dams; and a 4,798-gigaliter water reservoir

- produces $3 billion (Aus) in agricultural crops, products, and services per year
- replaces 5 million tons of carbon dioxide every year
- produces 70 percent of all renewable energy on Australia's eastern mainland grid
- was one of the first projects to use computers for engineering design
- finished on time and under budget
- had 121 people die in its construction, whom the government dedicated a monument to in the 1980s

Laborers on the project who came from other countries were trained. Free English classes were provided to all who needed to learn the language. In the town of Cooma, previously a small settlement, the population swelled from 2,000 to 10,000; later it would become the headquarters of the Snowy Mountains Authority. "The Scheme was the first meal ticket for thousands of new Australians who entered the isolated realm of the Australian high country stockman, the reticent, resolute character embodied in Australia's most cherished legends. It was the strangest mélange of humanity flung together since the construction of the pyramids, with the Snowy people demonstrating to a troubled world the capacity that exists for tolerance and harmony" (http://www. users.bigpond.net.au/snowy/bgrnd.html).

The Snowy people were just what the country needed. During the years 1947–55, the Australian population grew dramatically because of its immigrants; the government mounted a major public-relations campaign to invite new settlers. With mounting concern over the rise of communism in nearby Asia, it was feared there might not be enough people to defend the homeland. So the Snowy people were most welcome, and they helped build the country. The children went to school together, learned English, and came home with playmates of different origins, all of whom were now Australians, thus creating a new culture.

Two existing towns had to be relocated. The towns of Adaminaby and Jindabyne were eventually flooded to become reservoirs. Their residents were dispersed, some joining the new communities being built for workers and immigrants.

IMPORTANCE IN HISTORY

One consequence of the Snowy Mountains project was the devastating effect that free irrigation water had on the crops it was intended to nurture. For 30 years, the results were satisfactory to all. But during the 1980s, the salinity level increased, caused by salts from the weathering of rocks, naturally saline groundwater, and salts deposited over thousands of years by precipitation. This problem is not small; 40 percent of the world's irrigated area is affected by salinity. The problem is especially acute in the Murray Darling Basin, where an initiative was created to promote and coordinate planning for the efficient and sustainable use of water. Indeed, the problem of salinity represents a worldwide threat to irrigation schemes. The Imperial Valley in California still copes with this problem, as do many of the *polders* of the Netherlands.

By 1997, an environmental-impact statement was required for the Snowy Mountains project, and three areas were identified as needing closer inspection: the Snowy River, the Snowy Mountains Hydro-electric Plant, and the Murray and Murrumbidgee irrigation project. The public became involved as never before, which raised the nation's environmental awareness and fostered research into new technologies to deal with the issues. Final disposition was reached in October 1998, when it was determined that open-flow irrigation would be changed to a system of directed pipelines.

Although the workers on the Snowy project came from diverse cultures, the project not only created a monumental hydroengineering complex, but also contributed to transforming Australia into a more heterogeneous society.

FOR FURTHER REFERENCE

Books and Articles

Collis, Brad. *Snowy: The Making of Modern Australia*. Palmerston, Australia: Tabletop Press, 2002. Sections of this book are available on the web at: http://www.users.bigpond.net.au/snowy.

Neal, Laura, ed. *It Doesn't Snow Like It Used To: Memories of Monaro and the Snowy Mountains*. Ultimo, Australia: Stateprint, 1988.

Unger, Margaret. *Voices from the Snowy: The Personal Experiences of the Men and Women Who Worked on One of the World's Great Engineering Feats; The Snowy Mountains Scheme*. Kensington, Australia: NSWU Press, 1989.

Internet

For Snowy's own Web site, see http://www.snowyhydro.com.au/.

For information on the diversity of the Snowy workforce, including people from more than 30 lands, see http://www.users.bigpond.net.au/snowy/bgrnd.html.

For "The Politics of Water," see http://www.users.bigpond.net.au/snowy/politics.html.

For information about the Snowy transition to present times, see http://www.users.bigpond.net.au/snowy/overview.html.

For information about the problem of salinity and irrigation, see http://web.bryant.edu/~langlois/ecology/pollution.html.

Documents of Authorization

SNOWY MOUNTAINS HYDRO-ELECTRIC POWER

No. 25 of 1949

An Act relating to the Construction and Operation of the Works, for the Generation of Hydro-electric Power in the Snowy Mountains Area.

[Assented to 7th July, 1949.]

WHEREAS additional supplies of electricity are required for the purposes of defence works and undertakings:

AND WHEREAS the construction of further defence works and the establishment of further defence undertakings will require additional supplies of electricity:

AND WHEREAS it is desirable that provision should be made now to enable increased supplies of electricity to be immediately available in time of war:

AND WHEREAS the consumption of electricity in the Australian Capital Territory and, in particular, at the Seat of Government within that Territory, is increasing and is likely to continue to increase:

AND WHEREAS it is desirable that the generation of electricity for the purposes referred to in this preamble should be undertaken in such an area and in such a manner as to be least likely to suffer interruption in time of war:

AND WHEREAS, by reason of the foregoing, it is desirable that provision should be made now for the generation of electricity by means of hydro-electric works in the Snowy Mountains Area:

BE it therefore enacted by the King's Most Excellent Majesty, the Senate, and the House of Representatives of the Commonwealth of Australia, as follows:—

PART I.—PRELIMINARY.

1. This Act may be cited as the *Snowy Mountains Hydro-electric Power Act 1949*.

2. This Act shall come into operation on the day on which it receives the Royal Assent.

3. This Act is divided into Parts, as follows:—

Part I.—Preliminary.
Part II.—The Snowy Mountains Hydro-electric Authority.
Part III.—Functions and Powers of the Authority.
Part IV.—Officers and Employees of the Authority.
Part V.—Finances of the Authority.
Part VI.—Miscellaneous.

4. In this Act, unless the contrary intention appears—

"Associate Commissioner" means an Associate Commissioner holding office under this Act;

"easement" includes a license or a right in the nature of an easement;

"officer" means an officer of the Authority;

"owner," in relation to any land, includes any person having an estate or interest in that land;

"the Authority" means the Snowy Mountains Hydro-electric Authority;

"the Commissioner" means the Commissioner constituting the Authority.

5. This Act shall bind the Crown in right of a State.

6.—(1.) For the purposes of this Act, the Snowy Mountains Area shall be an area of land in the south-eastern portion of the State of New South Wales and the north-eastern portion of the State of Victoria defined in accordance with this section.

(2.) The Governor-General may, by Proclamation, define the boundaries of the Snowy Mountains Area and may, from time to time, by Proclamation, vary the boundaries as so defined.

PART II.—THE SNOWY MOUNTAINS HYDRO-ELECTRIC AUTHORITY.

7.—(1.) For the purposes of this Act, there shall be an Authority to be known as the Snowy Mountains Hydro-electric Authority.

(2.) The Authority shall be constituted by a Commissioner, shall be a corporation sole with perpetual succession and an official seal, may acquire, hold and dispose of real and personal property and shall be capable of suing and being sued in its corporate name.

(3.) All courts, judges and persons acting judicially shall take judicial notice of the seal of the Authority affixed to any document and shall presume that it was duly affixed.

(4.) The Commissioner shall be appointed by the Governor-General.

8.—(1.) The Commissioner shall be assisted by two Associate Commissioners, each of whom shall be appointed by the Governor-General.

(2.) An Associate Commissioner shall give such advice and assistance to the Commissioner as the Commissioner requires and shall perform such duties as the Commissioner directs.

9.—(1.) The Commissioner and the Associate Commissioners first appointed under this Act shall be appointed—

(a) in the case of the Commissioner—for a period of seven years; and

(b) in the case of the Associate Commissioners—for period of six years and five years, respectively.

(2.) After the appointment of the Commissioner and the Associate Commissioners first appointed under this Act, each further appointment shall be for a period of seven years.

(3.) Where the period of appointment of the Commissioner or an Associate Commissioner has expired he shall be eligible for re-appointment.

(4.) Notwithstanding the preceding provisions of this section, where the period of appointment of the Commissioner or of an Associate Commissioner would, but for this sub-section, expire after he attains the age of sixty-five years, his appointment shall be for a period ending upon the date upon which he attains that age.

10. The Commissioner and each Associate Commissioner shall be paid salary and allowances at such rates as the Governor-General determines.

11. The Minister may grant leave of absence to the Commissioner or an Associate Commissioner upon such terms and conditions as to payment of salary or otherwise as the Minister determines.

12. The Governor-General may terminate the appointment of the Commissioner or of an Associate Commissioner for inability, inefficiency or misbehaviour.

13. The office of the Commissioner or of an Associate Commissioner shall be vacated—

(a) if he engages in any paid employment outside the duties of his office;

(b) if he becomes bankrupt, applies to take the benefit of any law for the relief of bankrupt or insolvent debtors, compounds with his creditors or makes an assignment of his salary for their benefit;

(c) if he resigns his office by writing under his hand addressed to the Governor-General and the resignation has been accepted;

(d) if he is absent from duty, except on leave granted by the Minister, for fourteen consecutive days or for twenty-eight days in any twelve months; or

(e) if he, in any way, otherwise than as a member, and in common with the other members, of an incorporated company consisting of not less than twenty-five persons—

(i) becomes concerned or interested in any contract or agreement entered into by or on behalf of the Authority; or

(ii) participates or claims to participate in the profit of any such contract or agreement or any benefit or emolument arising from any such contract or agreement.

14.—(1.) In the event of the office of Commissioner becoming vacant at any time, or in the event of the illness or absence of the Commissioner, the Governor-General may appoint a person to be Acting Commissioner.

(2.) An Acting Commissioner appointed in the event of the office of Commissioner becoming vacant shall hold office during the pleasure of the Governor-General but shall not in any event continue in office after the expiration of twelve months from the occurrence of the vacancy in the office of Commissioner.

(3.) An Acting Commissioner appointed in the event of the illness or absence of the Commissioner shall hold office during that illness or absence but his appointment may at any time be terminated by the Governor-General.

(4.) An Acting Commissioner shall have all the powers and perform all the duties of the Commissioner.

(5.) An Acting Commissioner shall be paid salary and allowances at such rates (if any) as the Governor-General determines.

15.—(1.) The Authority may, in relation to any particular matter or class of matters, or to any particular place, by writing under its seal, delegate to an Associate commissioner or an office all or any of its powers under this Act (except this power of delegation), so that the delegated powers may be exercised by the delegate with respect to the matter or class of matters, or to the place, specified in the instrument of delegation.

(2.) Every delegation under this section shall be revocable at will and no delegation shall prevent the exercise of any power by the Authority.

PART III.—FUNCTIONS AND POWERS OF THE AUTHORITY.

16. The functions of the Authority shall be—

(a) to generate electricity by means of hydro-electric works in the Snowy Mountains Area; and

(b) to supply electricity so generated to the Commonwealth—

(i) for defence purposes; and

(ii) for consumption in the Australian Capital Territory.

17.—(1.) For the purpose of performing its functions under the last preceding section, the Authority shall have power to construct, maintain, operate, protect, manage, and control works—

(a) for the collection, diversion and storage of water in the Snowy Mountains area;

(b) for the generation of electricity in that area;

(c) for the transmission of electricity generated by the Authority; and

(d) incidental or related to the construction, maintenance, operation, protection, management or control of any works specified in the preceding paragraphs of this sub-section.

(2.) The Authority shall have power to construct, maintain, operate, protect, manage and control works which, in the opinion of the Authority, are necessary or desirable for the purpose of preventing or mitigating injurious effects of any works referred to in the last preceding sub-section.

18. The Authority shall have, in addition to the powers specifically conferred upon it by this Act, such other powers as are necessary or convenient for the performance of its functions under this Act, and, in particular, and without limiting the generality of the foregoing, shall have power—

(a) to purchase land;

(b) to take land on lease;

(c) to take easements over land;

(d) to sell or otherwise dispose of land vested in the Authority but not required for the purposes of the Authority;

(e) to lease land vested in the Authority the use of which is not for the time being required by the Authority;

(f) to release any easement over land;

(g) to purchase or take on hire plant, machinery, equipment or other goods;

(h) to dispose of plant, machinery, equipment or other goods owned by the Authority but not required by the Authority;

(i) to provide transport, accommodation, provisions, medical treatment, hospital facilities and amenities for officers and employees of the Authority and their families; and

(j) to do anything incidental to any of its powers.

19. The Commissioner, an Associate Commissioner, an officer or employee of the Authority, or any other person authorized by the Authority so to do, may, for the purposes of this Act, without any previous notice—

(a) enter upon land (including land owned or occupied by the Crown in right of a State) for the purpose of inspecting the land;

(b) make surveys, take levels, sink bores, dig pits and examine the soil; and

(c) do any other thing necessary for ascertaining the suitability of the land for the purposes of the Authority.

20. The Authority, or any person authorized by the Authority so to do, may, for the purposes of this Act—

(a) after giving not less than seven days' notice in writing to the occupier of land (including land owned or occupied by the Crown in right of a State), enter upon and occupy that land;

(b) on or from land so occupied—

(i) construct, build or place any plant, machinery, equipment or goods;

(ii) take sand, clay, stone, earth, gravel, timber, wood or other materials or things;

(iii) make cuttings or excavations;

(iv) deposit sand, clay, stone, earth, gravel, timber, wood or other materials or things;

(v) erect workshops, sheds and other buildings;

(vi) make roads; and

(vii) manufacture and work materials of any kind; and

(c) demolish, destroy or remove, on or from land so occupied, any plant, machinery, equipment, goods, workshop, shed, building or road.

21. The Authority may raise or lower the level of a lake, river or stream in the Snowy Mountains Area and impound, divert and use the waters of a lake, river or stream in that area.

PART IV.—OFFICERS AND EMPLOYEES OF THE AUTHORITY.

22.—(1.) The authority may appoint such officers as it thinks necessary for the purposes of this Act.

(2.) The selection of persons for appointment as officers under this section shall be made in accordance with such requirements as the Public Service Board determines.

(3.) Subject to the next succeeding sub-section, a person shall not be appointed as an officer of the Authority unless—

(a) he is a British subject;

(b) the Commissioner or an Associate Commissioner is satisfied, upon medical examination, as to his health and physical fitness; and

(c) he makes and subscribes an oath or affirmation of allegiance in accordance with the form in the Schedule to the Constitution.

(4.) The Authority may, with the approval of the Minister, appoint a person who is not a British subject and has not made and subscribed the oath or affirmation of allegiance.

(5.) The appointment, transfer, or promotion of a person to a position the salary, or the maximum salary, of which exceeds One thousand five hundred pounds, or such higher amount as is prescribed, per annum, shall be subject to the approval of the Minister.

(6.) For the purposes of the last preceding sub-section, the salary of a position shall not be deemed to be affected by variations made in accordance with variations in the cost of living and shall not include any allowance.

(7.) Officers shall not be subject to the *Commonwealth Public Service Act 1922 1948* but shall hold office under such terms and conditions as are, subject to the approval of the Public Service Board, determined by the Authority.

23.—(1.) The Authority may employ such temporary or casual employees as it thinks fit, on such terms and conditions as the Authority thinks necessary for the purposes of this Act.

(2.) A person shall not be employed under this section unless, when required by the Authority so to do, he makes and subscribes an oath or affirmation of allegiance in accordance with the form in the Schedule of the Constitution.

24. Nothing in this Act shall prevent the making of an industrial award, order, determination or agreement under any Act in relation to officers or employees appointed or employed under this Act or affect the operation of any such award, order, determination or agreement in relation to any such officer or employee.

PART V.—FINANCES OF THE AUTHORITY.

25.—(1.) The Authority shall have power to borrow money on overdraft from the Commonwealth Bank of Australia upon the guarantee of the Treasurer.

(2.) The Treasurer may, out of moneys appropriated by this Parliament for the purposes of this Act, make advances to the Authority of such amounts and upon such terms as he thinks fit.

(3.) Except with the consent of the Treasurer, the Authority shall not have power to borrow money otherwise than in accordance with this section.

26. The Authority shall open and maintain an account or accounts with the Commonwealth Bank of Australia and may open and maintain an account or accounts with such other bank or banks as the Treasurer approves.

27. Subject to this Act, the moneys of the Authority—

(a) shall be applied by the Authority—

(i) in payment or discharge of the expenses, charges, and other obligations incurred or undertaken by the Authority under this Act;

(ii) in payment of the salaries and allowances of the Commissioner and of the Associate Commissioners and of any Acting Commissioner; and

(iii) in re-payment of advances made to the Authority by the Treasurer under this Act, in accordance with the terms upon which those advances were made; and

(b) may be invested on fixed deposit with the Commonwealth Bank of Australia or in securities of, or guaranteed by, the Government of the Commonwealth.

28.—(1.) The Authority shall keep accounts in such form as the Treasurer approves.

(2.) The accounts of the Authority shall be subject to inspection and audit, at least once yearly, by the Auditor-General for the Commonwealth.

(3.) The Auditor-General shall report to the Minister the result of each inspection and audit.

29. The income, property and operations of the Authority shall be subject to taxation (other than income tax) under the laws of the Commonwealth but shall not be subject to taxation under any law of a State to which the Commonwealth is not subject.

30. The Authority may, with the approval of the Treasurer, set aside, out of its revenue, such sums as it thinks proper for depreciation of assets, insurance or other purposes.

31. The Authority shall not, except with the approval of the Minister, enter into any contract involving the payment or receipt of an amount exceeding One hundred thousand pounds.

32. The price at which electricity is supplied or sold by the Authority shall be such as the Treasurer, after receipt of a recommendation by the Minister, determines.

PART VI.—MISCELLANEOUS.

33.—(1.) In the exercise of its powers under this Act, the Authority shall cause as little detriment and inconvenience and do as little damage as possible.

(2.) Where the owner of land in the Snowy Mountains Area is injuriously affected by the exercise, in relation to that land, of any of the powers conferred by this Act, compensation shall be paid by the Authority.

(3.) Where land (whether within or without the Snowy Mountains Area) is entered or occupied in pursuance of section twenty of this Act, the Authority shall be liable to pay compensation to the owner or occupier of the land, or to both, as the case requires, and the compensation so payable shall include compensation in respect of—

(a) damage of a temporary character as well as of a permanent character; and

(b) the taking of sand, clay, stone, earth, gravel, timber, wood, materials or things by the Authority.

(4.) The provisions of Divisions 2, 3 and 5 of Part IV., and of section sixty, of the *Lands Acquisition Act* 1906–1936 shall, so far as they are applicable, and subject to the next succeeding sub-section, be applicable in relation to claims for compensation against the Authority.

(5.) In the application of those provisions—

(a) any reference therein to the Commonwealth or to the Minister shall be read as a reference to the Authority; and

(b) the reference in paragraph (b) of sub-section (1.) of section thirty-three of the *Lands Acquisition Act* 1906–1936 to damage suffered by reason of the exercise of any powers under Part III. of that Act shall be read as a reference to damage suffered by reason of the exercise of any powers under this Act.

34. Where the Commissioner, an Acting Commissioner, an Associate Commissioner or an officer was, immediately prior to his appointment under this Act, an officer of the Public Service of the Commonwealth, his service as the Commissioner, as Acting Commissioner, as an Associate Commissioner or as an officer of the Authority shall, for the purpose of determining his existing and accruing rights, be taken into account as if it were services in the Public Service of the Commonwealth and the *Officers' Rights Declaration Act* 1928–1940 shall apply as if this Act and this section had been specified in the Schedule to that Act.

35.—(1.) The *Commonwealth Employees' Compensation Act* 1930–1948 shall apply to the Commissioner, to an Acting Commissioner, to the Associate Commissioners and to be officers and employees of the Authority as if they were employees within the meaning of that Act.

(2.) Any liability to pay compensation under that Act as applied by this section shall be borne by the Authority.

36. The Authority may arrange with a Minister of State or authority of the Commonwealth or of a State for the performance by that Minister or authority of any work on behalf of the Authority.

37. Except as prescribed, a person shall not, in the Snowy Mountains Area, carry out any work, or make any use of the water in a lake, river or stream, whereby any works, or purposes works, of the Authority, or the use, or proposed use, of water by the Authority, is or may be injuriously affected or interfered with.

38. Any water used by the Authority for the generation of electricity shall be discharged into a lake, river or stream in the Snowy Mountains Area.

39. The Authority may sell to a State, or to an authority of a State, electricity generated by the Authority which is not immediately required by the Commonwealth for defence purposes or for consumption in the Australian Capital Territory.

40.—(1) The Authority shall, as soon as practicable after the thirtieth day of June in each year, prepare and furnish to the Minister a report on the operations of the Authority during the year ended on that date, together with financial accounts in respect of that year in such form as the Treasurer approves.

(2.) Before submitting the financial accounts to the Minister, the Authority shall submit them to the Auditor-General for the Commonwealth for report as to their correctness or otherwise.

(3.) The report and financial accounts of the Authority, together with the report of the Auditor-General as to those accounts, shall be laid before each House of the Parliament within fifteen sitting days of that House after their receipt by the Minister.

(4.) The Authority shall furnish to the Minister such other reports, and such documents and information, relating to the operations of the Authority, as the Minister requires.

41. The Governor-General may make regulations, not inconsistent with this Act, prescribing all matters which, by this Act, are required or permitted to be prescribed, or which are necessary or convenient to be prescribed, for carrying out or giving effect to this Act and, in particular, for prescribing penalties not exceeding a fine of Fifty pounds or imprisonment for a period not exceeding three months, or both, for offences against the regulations.

From Act No. 25 of 1949, 109–18. Federal Register of Legislative Instruments, No. 25, 1949, 109-118. For further information on the Web, see http://www.frli.gov.au.

29
The Mont Blanc Tunnel

Italy, France, and Switzerland

DID YOU KNOW . . . ?

- ► The first Alpine tunnel was built in 1475–1480 by Ludovic II, Marquis of Saluzzo, in Piedmont, Italy.
- ► Construction of the Mont Blanc Tunnel began in 1953 and was finished in 1965.
- ► The tunnel is 7.2 miles (11.6 kilometers) long, and 32 feet (10 meters) wide. It is one of the world's longest vehicle tunnels.
- ► Built to accommodate 350,000 vehicles, the tunnel now handles 2 million vehicles annually.
- ► More than 40 lives were lost in a catastrophic fire in the tunnel in 1999.
- ► Road accidents now cause more injuries than war.
- ► The city of Geneva (Switzerland) is a leading shareholder of this tunnel between Italy and France.

The Mont Blanc Tunnel is a dramatic illustration of a worldwide failure that has not yet been sufficiently addressed. Most traffic accidents are caused by collisions, and collisions inside tunnels are particularly dangerous. While we have had roads since the Romans and cars since Henry Ford, we have not yet solved the problem of traffic injuries and fatalities. Several tunnels—for instance, the 1920s tunnel across the St. Gotthard—have been equipped to carry cars and trucks on specially designed railway flatcars. Nevertheless, road tunnels have continued to proliferate, but without the safety features that could be provided by rail or pallet transport.

Construction of the Mont Blanc tunnel between Italy and France. © Keystone / Getty Images.

HISTORY

Nature gave central Europe several topographical features that hindered economic development routes and the expansion of trade. For instance, standing between France and Italy and extending into Switzerland and beyond are the Alps, a chain of mountains whose highest point is known as Mont Blanc, elevation 15,771 feet (4,807 meters).

Transport routes are connective, and a route such as the Gotthard Pass undoubtedly played a part in establishing the Swiss Confederation in the thirteenth century. The pass included a suspended roadway along Schöllenen Gorge as well as a bridge, which together created one of the first shortened north–south routes (Löw and Einstein).

The first recorded Alpine tunnel was built in 1475–1480 by miners working around Piedmont. This innovation was one of convenience—the Marquis of Saluzzo, Ludovic II, disliked paying tolls to use passes, especially when he had to leave the territory of his marquisate. So he hired miners to dig a route 247.5 feet (75 meters) long and 8.25 feet (2.5 meters) wide. The Pertus du Viso (Hole of Viso) was completed in 1480.

When the first rail routes were established through the mountains, commerce and communication increased rapidly. Soon, more tunnels were built. The most famous was the Mont Cenis–Frejus in the French Alps. The French began the project in 1857 and it was completed in 1871.

In 1953, France and Italy agreed to jointly build and operate a motor-vehicle tunnel between Chamonix–Mont Blanc in France and Courmayeur in Italy. The resulting Mont Blanc Vehicular Tunnel opened in 1965. It measures 7.2 miles (11.6 kilometers) in length, two-thirds of which lies within France, and 32 feet (10 meters) wide. A single-bore, two-way tunnel, it is one of the longest of the world's vehicular tunnels.

The Mont Blanc tunnel was of great benefit to the regional economy, but it also generated a traffic boom that transformed the tunnel into an accident waiting to happen. Originally designed for 350,000 vehicles a year, by the late 1990s the tunnel was handling 2 million, many of them trucks.

A number of truck fires have occurred in the tunnel, but in March 1999 a Belgian truck carrying flour and margarine caught fire in the middle of the tunnel. Temperatures exceeded 1,800°F (1,000°C); asphalt melted and tires exploded. More than 40 people died, many of them killed by smoke and fumes blown downwind through the tunnel. The tunnel, which carried nearly half of the truck traffic between France and Italy, was closed and traffic was diverted to a few alternative crossings, which quickly became congested. The tunnel finally reopened in early 2002, although some communities vehemently opposed its use for truck traffic.

CULTURAL CONTEXT

Julius Caesar's chief engineer was a tunneling expert, and there has been speculation about whether he might have built a tunnel across the English Channel. Sir Pierson Dixon, onetime former British Ambassador in Paris, wrote a novel that mentions Julius Caesar's chief engineer, who built tunnels through the mountains around Rome. After sessions of the Roman Senate, when the senators wanted to get to their estates quickly, instead of going around the mountains—a tough trip by any measure, but especially so by chariot—the senators zipped through mountain tunnels.

The investment capacity of the private sector offers access to financial resources in the trillions of dollars, far beyond the reach of funds available through direct government budgets. Pioneer research and development can, however, be undertaken by the public sector, and participation through guarantees or other insurance arrangements can release private funds.

The palleted automated transit system designed by MIT's David Gordon Wilson resulted from a General Motors grant to identify feasible improvements in road safety. The challenge to the public and the public-relations industry is to generate the political support needed if the problem of highway safety is to find commensurate redesign and reconstruction incentives. The Channel Tunnel, although privately financed, offers the safety of transport on railway flatcars. The accident rates of the Mont Blanc and other tunnels will persist until the problem is viewed statistically and solutions are devised and adopted.

PLANNING

The Mont Blanc Convention (reproduced here as Document I) specified that responsibility for building and maintaining the tunnel would be shared equally between France and Italy; a French company would build half, and an Italian company the other half. It was subsequently agreed that Switzerland, which stood to benefit economically from the project, would help finance and administer the tunnel, as spelled out in a separate *procès-verbal* (official report of the meeting) of May 16, 1953 (reproduced here as Document II).

The convention addresses numerous issues, among them ownership of any discoveries found during the digging (article 4) and membership of a supervisory commission and provisions for the chairmanship (article 5). Article 8 covers responsibility for tunnel operations, which takes on added weight since the 1999 accident and fire. Article 9 anticipates possible hazards and notes that any specialized personnel responsible for ventilation are an exception to the rule that the project must have an equal balance of French and Italian personnel. Subsequent articles cover procedures for handling crimes (article 13) and resolving disputes (articles 15–17).

The second authorizing document, quite different from the first, was signed in May 1953. Titled "Final *Procès-Verbal:* Signed at Rome, on 16 May, 1953," this document replaces the *procès-verbal* that was adopted when the formal authorization agreement was signed in March 1953. The second document exemplifies the difference between planning and building, between theory and practice.

In this document, a new player appears: the state and city of Geneva. At some point, the planners must have realized that Switzerland would benefit greatly from the building of the Mont Blanc Tunnel. Hence, one of the most significant clauses of the *Procès-Verbal* is article 12, which states, "The French and Italian construction companies shall obtain material, equipment and other supplies in Switzerland equivalent to approximately half the amount of the subsidy of the State and City of Geneva insofar as price and exchange conditions permit." The *procès-verbal* also mentions private investors for the first time.

In the end, the Mont Blanc Tunnel project experienced a cost overrun of at least 100 percent because the costs of construction could not be accurately specified and anticipated. Tunneling through the Alps required innovations that had to be tested on-site. It is inherently difficult to gauge accurately the costs of such innovations.

BUILDING

The Mont Blanc tunnel was constructed by a syndicate of five companies: Bouygues Construction, Dumez-GTM, Freyssinet International, Impreglio S.p.A., and Vinci Construction Grands Projets. The Italians started work on their side of the Alps in 1958, and the French on their side six months later; the two met on August 14, 1962.

Since the rock within the immense mountain was under extreme stress, a preparatory pilot bore was driven to cut the risk of rock collapsing from the advancing tunnel walls. Predicting the location and properties of fault zones 1,000 to 2,500 meters below ground surface is an extremely difficult, if not impossible, task (Löw and Einstein).

The tunnel itself was excavated at its full width from the outset, one of the first major rock tunnels to be constructed in this "full-face" manner.

Adequate ventilation in the tunnel was of particular concern, as evidenced in the Mont Blanc Convention's proviso that the tunnel's ventilation specialists did not need to conform to the general requirement that operating and maintenance staff would "consist of an equal number of Italian and French nationals who shall be of equal rank."

Among the causes of fire in vehicular tunnels are accumulation of combustible debris along the roadway and within vent shafts, fuel spills and oils on the road surface, flammable materials carried as cargo, short circuits or electrical malfunctions in control or power cables, malfunctions in ventilation and air-handling equipment, and human-caused hazards such as motor-vehicle accidents.

IMPORTANCE IN HISTORY

From the 1960s to the late 1990s, European trade grew dramatically, as freight tripled both by rail and truck throughout Europe. Vehicular accidents were common; when they occurred in a tunnel, they were often disastrous. In the Gotthard Tunnel on October 24, 2001, two trucks smashed, and the ensuing fire closed the tunnel for days.

Simon Löw and Herbert Einstein have written a seminal article about advances in Alpine tunneling. The Gotthard Tunnel (35.4 miles, 57 kilometers) and the Lötschberg Tunnel (21.5 miles, 34.6 kilometers) will be built lower in the mountain than previous tunnels and will therefore be able to bear heavier transport weight. Switzerland has long had a limit of 28 tons for trucks, in effect forcing the use of rail transport for much cargo. The European Union is adopting a standardized 40-ton limit, forcing Switzerland to amend its long-standing limit.

The real importance of the Mont Blanc Tunnel may turn out to be its demonstration of the shortcomings of traditional designs for road tunnels. Future tunnel construction may benefit from improved designs by systems-oriented engineers of the future. Such designs could reduce the statistical risks of collisions and their resultant out-of-control fires. In 1998 worldwide road-traffic fatalities totaled 1,170,694; worldwide road-traffic injuries for that year totaled 38,848,625. In general, the automobile has been more lethal to humanity than warfare and terrorism. Eventually it may become illegal to build road tunnels that have inherently unacceptable risks of disaster. The European public suffers from the absence of a Europe-wide procedure for reviewing the safety of road-tunnel designs.

In pursuit of specialization and expertise, many degree programs in engineering concentrate on a single discipline, such as roads, railways, shipping, and aerospace. However, the unsatisfactory safety record of road tunnels is one illustration of the need for a degree program offering interdisciplinary training and certification of *transport* engineers and managers. An acknowledged certification process would reduce insurance rates and have the effect of reassuring and attracting additional investors.

FOR FURTHER REFERENCE

Books and Articles

Davidson, Frank P. *Macro: A Clear Vision of How Science and Technology Will Shape Our Future*. New York: William Morrow, 1983.

Dixon, Pierson. *Farewell, Catullus*. London: Hollis and Carter, 1953.

Levy, Bertrand. "Le tunnel du Mont-Blanc." *Travaux*, November 2001. No. 780. A magazine published by Fédération Nationale des Travaux Publics, http://www.fntp.fr.

Louveau, L. "Le tunnel du Mont-Blanc. 'Déjà dix-huit années de bons et loyaux services.'" *Travaux*, February 1984, .No. 585. For location, see http://www.fntp.fr.

Wilson, David Gordon. "Palleted Automated Transportation (PAT)—A View of Developments at the Massachusetts Institute of Technology." Tokyo: *IATSS Research* 13, no. 1, (1989): 53–60 . For further information on IATSS, see http://www.iatss.or.jp/english.

Internet

For a chronology of the Mont Blanc Tunnel, with bibliography before and after the 1999 fire, see the International Database and Gallery of Structures at http://www.structurae.net/.

For Simon Löw's and Herbert Einstein's work on rail tunnels in Switzerland, and background on Alpine tunnels, see http://www.geotimes.org/feb03/feature_tunnel.html.

For provisions against tunnel fires in New York, see http://www/panynj.gov/pr/116–99.html.

For information about a leading tunneling company, see http://www.bouygues-construction.com/en/.

For a news story about the Mont Blanc Tunnel fire, see http://news.bbc.co.uk/1/hi/world/europe/1856504.stm.

For a list of notable tunnels, both rail and vehicular, see the useful summary at the Pearson Education Web site: http://www.infoplease.com/ipa/A0001340.html.

For the first Alpine tunnel, 1475–1480, see http://www.ltf-sas.com/pages/articles.php?art_id=265.

For statistics relating to international injury and fatality statistics, see http://www.safecarguide.com/exp/statistics/statistics.htm and http://www.scienceserving-society.com/p/141.htm.

For more about the concept of a trans-Atlantic tunnel with interactive 3-D options, see http://dsc.discovery.com/convergence/engineering/transatlantictunnel/interactive.html.

Film and Television

Extreme Engineering: Transatlantic Tunnel. DVD. Discovery Channel. Powderhouse Productions, Somerville, MA, 2003.

Documents of Authorization—I

THE MONT-BLANC TUNNEL CONVENTION OF 1953

CONVENTION BETWEEN ITALY AND FRANCE CONCERNING THE CONSTRUCTION AND OPERATION OF A TUNNEL UNDER MONT-BLANC, SIGNED AT PARIS ON 14 MARCH 1953.

The President of the Italian Republic and the President of the French Republic, recognizing that the establishment of permanent road communication between the two countries by means of a tunnel connecting the valleys of Chamonix and Courmayeur through Mont-Blanc would be in their common interest, have decided to conclude a convention to this end and have accordingly designated as their Plenipotentiaries:

The President of the Italian Republic: Mr. Eugenio Prato, *Assistant Director General for Economic Affairs* in the Ministry of Foreign Affairs:

The President of the French Republic: Mr. Francois de Panafieu, Minister Plenipotentiary; who, having communicated their full powers, have agreed upon the following provisions.

ARTICLE I

The Italian Government and the French Government undertake to share equally the task of tunneling through Mont-Blanc on the basis of the technical plan submitted by the Mont-Blanc Tunnel Syndicate.

ARTICLE 2

Construction of the tunnel shall be assigned to a French company and an Italian company, each of which shall complete half of the total length.

The articles of association of these companies shall be approved by their respective Governments.

The companies shall receive concessions for the exclusive purpose of constructing and operating the tunnel and its annexes; these concessions shall be granted by their respective Governments in accordance with the conditions fixed by the present Convention.

ARTICLE 3

The Italian Government and the French Government shall come to an agreement respecting the terms of the concession granted by each of them and of the specifications attached thereto.

The two Governments shall endeavour to fix terms which shall be as similar as possible and shall not subsequently modify these terms except subject to a previous agreement. Any change in the tolls, as well as any repurchase or cancellation of the concessions, shall take place only by agreement between the two Governments.

ARTICLE 4

The waters and useful minerals found in the course of building the tunnel shall be allocated under the laws of the State on whose territory the discovery is made, irrespective of which company makes the discovery.

ARTICLE 5

The Italian Government and the French Government shall each appoint one-half of the members of a Supervisory Commission which shall consist of six members.

The Chairman, who shall have a casting vote in the event of a tie, shall be of Italian and French nationality alternately. He shall be appointed from among the members of the Commission, by agreement between the two Governments, for a period of one year.

The Supervisory Commission shall oversee the progress of work and submit its observations in the form of reports addressed simultaneously to the Italian and French Governments.

In case of emergency, it shall have the power to order the execution or cessation of specified work for reasons of safety.

In extreme emergency, the Chairman shall have the power to act in place and on behalf of the Commission.

ARTICLE 6

The two Governments undertake to construct, in due course, the sections which are to link the tunnel entrances with the Italian and French highway systems.

ARTICLE 7

The concessionary companies shall assign the operation of the tunnel to an incorporated company and each shall subscribe one-half of the capital of such incorporated company, the Board of Directors of which shall be composed of an equal number of representatives of the concessionary companies.

The Chairman of the Board of Directors, who shall be appointed for five years, shall be of Italian and French nationality alternately.

The Managing Director shall not be of the same nationality as the Chairman.

The company shall distribute one-half of the receipts to each of the two concessionary companies, after the deduction of the sums necessary for the operation, maintenance and upkeep of the tunnel.

ARTICLE 8

The concessionary companies shall operate the tunnel at their own risk. A mixed Franco-Italian commission shall be responsible for the supervision of the operation, maintenance and upkeep of the tunnel.

ARTICLE 9

The operating and maintenance staff shall, as a rule, consist of an equal number of Italian and French nationals who shall be of equal rank.

By way of exception the staff responsible for the ventilation may consist of specialised persons approved by the Supervisory Commission without reference to the rule in the preceding paragraph.

ARTICLE 10

Every year the Board of Directors of the Operating Company shall submit a documented report on its work to the Italian and French Governments.

ARTICLE 11

The concessions provided for in article 2 shall terminate seventy years from the date which shall be fixed by agreement between the Italian and French Governments upon delivery of the works.

ARTICLE 12

Monetary, fiscal, customs, and social questions arising out of the construction and operation of the tunnel shall be covered by special agreements between the Italian Government and the French Government.

ARTICLE 13

Each Government shall assume the cost of its own customs, police and medical services.

The operating Company shall be responsible for regulating traffic in the tunnel and shall require the police officers entrusted with this duty to be sworn in, in conformity with the laws of both countries.

ARTICLE 14

A vertical line from the frontier in the open air shall constitute the Franco-Italian frontier inside the tunnel.

Qualified and authorised representatives of both sides shall have the right to cross the frontier freely for any customs, or police inquiry within the limits of the concessions.

ARTICLE 15

Immediately upon termination of the two concessions for any reason whatsoever, the tunnel shall become the common and indivisible property of the Italian and French States and shall be operated jointly on the basis of equal rights and responsibilities.

The terms of the joint administration shall be governed by an agreement previously entered into between the two Governments.

ARTICLE 16

The Italian Government and the French Government shall by common agreement appoint a single arbitrator to settle any disputes between the two concessionary companies.

Any difficulty arising in connection with carrying out the arbitrator's decision shall be settled in accordance with article 17.

ARTICLE 17

Any dispute between the two Governments concerning the interpretation or application of the present Convention or concerning one of the concessions granted under the terms of article 2 shall, if not settled within a reasonable period by diplomatic or other amicable means, be submitted to an arbitral tribunal whose decision shall be binding.

The arbitral tribunal shall consist of two members and a referee. Each of the two Governments shall designate one member. The referee, who shall not be a national of either of the two countries, shall be appointed by agreement between the two Governments.

If the referee has not been jointly appointed within a period of six months from the time the dispute is submitted for arbitration by one of the two Governments, the President of the International Court of Justice shall make the appointment at the request of the party which first makes an application.

ARTICLE 18

The present Convention shall enter into force upon the exchange of the instruments of ratification.

IN WITNESS WHEREOF, the respective Plenipotentiaries have signed the present Convention and have thereto affixed their seals.

DONE at Paris, 14 March, 1953, in duplicate.

For the President of the Italian Republic	For the President of the French Republic
E. PRATO	F. DE PANAFIEU

Documents of Authorization—II

FINAL PROCÈS-VERBAL

SIGNED AT ROME, ON 16 MAY, 1953

The Inter-Governmental Commission for the Mont-Blanc Tunnel met at Rome from 12 to 16 May, 1953, and adopted the provisions of the present financial *Procès-Verbal,* which replace those of the financial *Procès-Verbal* adopted at the Paris session of 10 to 14 March 1953.

1. The French and Italian concessionary companies established by the Convention signed on 14 March 1953, shall be constituted in accordance with the conditions laid down below.

2. The capital of the French concessionary company shall be distributed as follows:

French State	210 million francs
French municipal bodies	40 " "
State and City of Geneva	30 " "
Private individuals	120 " "
	400 million francs

The subscribers shall, moreover, pay to the French concessionary company the following amounts by way of subsidy:

French State	1,790 million francs
French municipal bodies	10 " "
State and City of Geneva	220 " "
	2,020 million francs

The French Government shall extend its guarantee to medium-term credits to an amount of 2,000 million francs, which shall be replaced subsequently by the issue of bonds also guaranteed by the French State.

3. The capital of the Italian concessionary company shall be distributed as follows:

Italian State and public bodies	342 million lire
State and City of Geneva	50 " "
Private subscribers	408 " "
	800 million lire

The subscribers shall, moreover, pay to the Italian concessionary company the following amounts by way of subsidy:

Italian Government	2,743 million lire
Italian public bodies	915 " "
State and City of Geneva	379 " "
	4,037 million lire

The Italian concessionary company shall obtain the remaining funds required to carry out its contractual obligations, including the 1,200 million lire referred to in Article 5 below, by means of loans placed in Italy.

4. In case any variation should occur in the rate of exchange of the French franc and the lira with the Swiss franc, it is agreed that the contributions of the State and City of Geneva shall remain fixed at three million Swiss francs each.

The subsidies contributed by the State and City of Geneva shall be paid to each of the concessionary companies in the following manner, in proportion as the work of each progresses:

—one-quarter on completion of the installation of the equipment at the start of the work;

—a second quarter on completion of the first third of the tunneling operation;

—a third quarter on completion of the second third of the tunneling operation;

—the final quarter on completion of the tunneling and of the lining of the tunnel.

5. If the concessionary companies agree that one of them should carry out a part of the work for which the other is responsible, the company performing the work shall be reimbursed for such work together with interest at the rate of 5 per cent to be charged against the profits earned by the other company before payment of the statutory interest on A and B shares and up to one-third of said amount annually.

The Italian delegation agrees forthwith, on behalf of the Italian concessionary company, to carry out at the request of the French concessionary company a quantity of work which shall be assessed on the basis of the prices of the contracts entered into by the Italian concessionary, up to a limit of 1,200 million lire.

6.(a) The capital of the French company shall consist of A and B shares. The B shares shall constitute the capital subscribed by the French Government; the A shares shall constitute the capital subscribed by the other shareholders.

(b) The articles of association of the French concessionary company shall fix the method of determining profits.

(c) A and B shares shall receive statutory interest at the rate of 6 per cent of their nominal value, to be paid by priority out of profits.

(d) A shares shall be redeemable from the proceeds of a special fund maintained by drawing against profits and shall be replaced by jouissance shares which shall participate only in the distribution of excess profits.

(e) Excess profits shall be distributed between shareholders and the subsidising municipal bodies.

In the French company, excess profits shall be distributed in the following manner:

—one-quarter to the shareholders;

—three-quarters to the subsidising public bodies.

7. The Commission recommends that the articles of association adopted by the Italian concessionary company shall be as similar as possible.

8. The Board of Directors of the French concessionary company shall consist of the following:

a Chairman appointed by the French Government;

six Directors appointed by the French Government;

three Directors representing the private group;

two Directors representing the State and City of Geneva;

one Director representing the French public bodies.

9. Switzerland shall be represented in a similar manner on the Board of Directors of the Italian concessionary company.

10. The articles of Association of the Operating Company shall be submitted for the approval of the French and Italian Governments.

11. The Board of Directors of the Operating Company shall include two non-voting advisory Directorships, which shall be reserved for representatives of the State and City of Geneva.

12. The French and Italian construction companies shall obtain material, equipment and other supplies in Switzerland equivalent to approximately half the amount of the subsidy of the State and City of Geneva, insofar as price and exchange conditions permit.

Rome, 16 May, 1953.

/s/ E. PRATO C. SAUSER-HALL F. DE PANAFIEU

30
The Founding of Brasília

Brazil

DID YOU KNOW . . . ?

➤ Brasília is shaped like an airplane.
➤ It may be the first city designed using aerial photography; it was built to be viewed from the air.
➤ Legend says the city was first dreamed of by Dom Bosco, the Italian saint who founded the Roman Catholic Salesian Order.
➤ Construction began in 1957, and the city was dedicated in 1960.
➤ Chosen for its supply of water, the location is now experiencing water shortages.
➤ In 1987, Brasília became a UNESCO World Heritage site.

Founded in the second half of the twentieth century as Brazil's capital, Brasília was not the first world city created specifically to serve as the seat of a nation's government. The chosen location was, however, unusually remote from the country's existing large metropolitan areas. It is roughly 540 miles (870 kilometers) from São Paulo (even farther from Rio de Janeiro), a far greater distance than from Washington, DC, to New York; from Canberra to Sydney; or even from St. Petersburg to Moscow.

HISTORY

As early as the late eighteenth century—shortly after Rio de Janeiro, with its famous beaches, became the seat of the Brazilian government—officials

An aerial view of Brasília, 1957. © Bettmann / Corbis.

began to discuss the advisability of moving the capital into the interior to eliminate the danger of naval attack.

In 1823, after the founding of Brazil as an independent political entity in 1822, José Bonifacio announced the decision to build a new capital named Brasília. Fifty years later, Dom Bosco (an Italian saint and founder of the Roman Catholic Salesian Order) had a prophetic dream that predicted a town between parallels 15 and 20. Even before the new city was built, a shrine was erected, exactly on the 15th parallel, dedicated to Dom Bosco in memory of his mystical prediction.

Despite the huge expenditures of such a move, the idea remained alive throughout the ensuing decades, during which time Brazil proclaimed itself an independent empire (1822) and abolished the monarchy (1889). The constitution of 1891 mandated the removal of the capital to the Plateau of Brazil (Planalto Central), the area where Brasília was eventually built. Momentum for relocating the capital accelerated after World War II.

National security was not the only reason to move the capital. Rio de Janeiro was becoming crowded, and its far southern location meant it was too distant from the rest of the country. A new capital farther inland would promote development of Brazil's relatively underpopulated interior. During the 1955 presidential campaign, candidate Juscelino Kubitschek de Oliveira pledged to build the new capital on the plateau. Kubitschek won the election, and in April 1956 he proposed a bill establishing an agency to carry out the creation of the new federal capital, the name of which would be Brasília, in a new Federal District to be cut out of Goiás State. The measure, reproduced here, was unanimously approved by the Brazilian Congress in September 1956.

The creation of the new city's infrastructure and buildings required moving huge numbers of workers and construction machines to the plateau. Transportation links were built to Brazil's major cities. In April 1960 the Plaza of the Three Powers was dedicated, and the federal government began a slow process of relocating from Rio de Janeiro—a process that continued even during years when the country was caught up in political crisis. (Some analysts say the immense cost of building the new city contributed to the economic and governmental instability that helped fuel a 1964 military coup.)

By 2000, some elements of architect and urban planner Lúcio Costa's master plan still remained to be built, but the Federal District's population had reached two million.

CULTURAL CONTEXT

In a small office at Cornell University, in New York State, a doctoral student looked at photographs taken from an airplane. Was that limestone in one area? What would be the implications of building there? What does that shadow mean? Perhaps the boundary between two types of soil? How deep is the soil atop the bedrock? What are those white dots all over the field? Why are they shaped like cones?

The student was Hollister Kent, who was working with his partner on the ground, a Brazilian who was exploring for possible locations to build Brazil's new capital of Brasília. During World War II, Kent had been part of the 10th Mountain Division of the U.S. Army, where he learned how to read aerial photography. After the war, Kent became an architecture student at Cornell; his professor was a consultant to a company that was engaged in the project. When his professor became suddenly ill, young Kent was sent to Brazil.

For some critics, the city that took shape was grand but austere. French writer Simone de Beauvoir, along with Jean-Paul Sartre, visited the city, which she described as having an air of "elegant monotony" (de Beauvoir, 576 78). Many bureaucrats were dismayed at the prospect of moving to an outpost in the hinterland that lacked the amenities of Rio de Janeiro. Gradually, however, Brasília and its satellite towns grew in population, as well as cultural and entertainment opportunities. The city's university opened in 1962, and the cathedral in 1970. In 1987 UNESCO named Brasília a World Heritage site.

PLANNING

The site chosen for Brasília had three strong points. First, it was less vulnerable than Rio de Janeiro to attack by sea. If Rio had remained the capital, a coastal attack could have devastated the economy of Brazil and its political administration. Moving the capital to an inland site therefore made military sense. Second, a growing capital like Brasília needed plenty of clean water, and this inland site had ample supplies. Third, it was strategically and economically sound to bring attention to the largely undeveloped interior of Brazil.

The site for Brasília was a sparsely inhabited, savanna-like area with a sub-tropical climate, situated at an elevation of approximately 3,500 feet (1,070 meters) at the headwaters of several rivers. Brazilian architect and urban planner Lúcio Costa, an advocate of architectural modernism, was selected to plan the city. He designed the city in the shape of an airplane. A curving "Highway Axis" extends from the north to the southwest, tying together the chief residential neighborhoods; embassies and commercial areas are also located along this axis. Government and civic buildings lie along a straight northwest–southeast "Monumental Axis." Municipal buildings cluster at the northwestern end of the Monumental Axis; at the other end, on an artificial lake, is the Plaza of the Three Powers, with buildings for the executive, judicial, and legislative branches of government. Because of its unique airplane shape, the city can never grow beyond the boundaries mandated by the designer. So growth is managed by adding small towns around the periphery of the airplane-shaped city.

The document presented here is significant because it not only authorizes the building of the new city but also gives its exact location in longitude and latitude (chapter 1, article I). A construction company was commissioned, the Companhia Urbanizadora da Nova Capital do Brasil (referred to in the document as UC), and the project was funded.

Also included in the documents is the contract with the Belcher firm, an aerial photography firm employed to perform air photo analysis; the firm employed young Kent, the aerial architect. The contract is an interesting mix of law and business, which empowers the private contractor to participate in this public macro project.

Named as chief architect for most of the major public buildings in the city was Brazilian Oscar Niemeyer, whose style was influenced both by Costa (once Niemeyer's teacher) and by the Swiss-French architect Le Corbusier. Niemeyer produced monumental sculpture-like designs that would catch the eye from afar with their interplay of straight and curved shapes and, in combination with the open vistas of Costa's plan, serve as emblems of Brazil's cultural and technological modernity.

BUILDING

Even before Lúcio Costa began constructing the capital, an airport and a temporary residence for the president were built, for how could a new capital function if there was no access for the country's head of state?

Costa began construction of the Plano Piloto in 1957, and within one year the first streets were paved. Costa's proposal, far from a dry technical document with blueprints that only an expert could read, is a short, poetic vision of the new city: a cross in the shape of outstretched arms appearing to embrace the future, "born of a primary gesture of one who marks or takes possession of a place, two axes crossing at right angles, the very sign

The University of Brasília's science building. Courtesy of Shutterstock.

of the Cross" (http://www.infobrasilia.com.br/pilot_plan.htm). Costa drew a triangle of land on which would be built the three seats of power, with two giant green open spaces that would enable all of Brasília to breathe. Every aspect of the city was scripted: the taxis were to be all the same dark gray color; buses were to take a final sweeping turn out of the bus station on a ramp that afforded an inspiring view before bidding farewell to the beautiful city. Even bus drivers' uniforms were designed to include a dark gray cap that conveyed a sense of respect.

While Costa won the contest and was the chief builder, Oscar Niemeyer actually designed many of the buildings, including the Congress, the University of Brasília, the Cathedral and Chapel of our Lady of Fatima, and the Palace of Justice. The very name *Palace of Justice* bespeaks the grandeur of the building.

Niemeyer was one of Costa's oldest students. He graduated from the Ecole Nacional de Belas Artes in 1934 and joined an architectural team that would build the Ministry of Education and Health in Rio. Costa directed the job, and they both worked under the direction of Le Corbusier. The building was completed in 1936, and it quickly became a famed international structure, catapulting the careers of Costa and Niemeyer into the public eye until they were signed to work together on the city of Brasília.

Architecture and city planning are not enough, however, to make a new capital. The president began the process, but Costa and the president both credited Israel Pinheiro with the perseverance that finally brought Brasília to completion. Flying over the city, Kubitschek said of the man he had met long before he became president, "My God, without Israel I'd never have been able to build Brasília" (http://www.infobrasilia.com.br/bsb_h3i.htm). Pinheiro became president of the New Capital Urbanization Agency (NOVACAP), and later became the first mayor of Brasília in May 1960.

No discussion of the building of Brasília would be complete without tribute to the Candangos, the indigenous people who played such an important role in building the city. It is a testimony to their hard work and vision that success was achieved at 9:30 A.M. on April 21, 1960, when Brasília was officially opened and the seat of government switched from Rio to Brasília. A sculpture in the center of Brasília, designed by Bruno Georgio, pays tribute to the Candangos.

IMPORTANCE IN HISTORY

Water, once the chief reason for selecting Brasília's location, now poses a problem. As the population exploded, people began to tap into the public water supply. Illegal punctures and perforations had disastrous effects if they occurred near a septic system. In 2001, the Environmental Secretariat attempted to regulate all underground water utilization, at which time it discovered 2,300 secret or illegal wells; some say the real number is closer to 6,000. Sanitation is not the only concern. Underground water affects cracks in the tectonic plates of the earth. Once the plates are disturbed, they may shift and cause earthquakes that are typically felt immediately. Seismologists caution that water and tectonic plates continue to interact, and as Brasília continues to develop its water systems, there is concern over the unknown long-term effects. For example, an earthquake of 3.7 on the Richter scale hit Brasília in 2001; there was a troubling question as to whether water might have been part of the cause.

The more general danger is a growing shortage of water. In 2001, Secretary of Environment and Hydraulic Resources for the Federal District Antonio Barbosa warned that Brasília's water supply was merely 10 percent above demand. A plan to construct the Corumbá IV Dam, which would extend the water supply into the next century, met with protests from the Brazilian Movement of Dam-Affected People (MAB). This project is only one of many potential dams in Brazil, all of which seek both water and power for the quickly growing population. São Paulo has become one of the largest cities in the world, reaching 10.4 million people in 2000.

A capital located far from the populous urban centers has the advantage of being protected from the undue influences of powerful economic interests that are entrenched in major cities. Therefore, while Brasília contributes to the greater cohesiveness of Brazil, it is already developing the cultural breadth that will in time enhance its stature as a symbol of national unity.

FOR FURTHER REFERENCE

Books and Articles

De Beauvoir, Simone. *La Force des Choses*. Paris: Gallimard, 1963.

Kent, Hollister. "Vera Cruz: Brazil's New Federal Capital." PhD diss., Cornell University, 1956. Copy can be obtained from University Microfilms International, Ann Arbor, Michigan.

Little, P. E. *Abundance Is Not Enough: Water-Related Conflicts in the Amazon River Basin*. Brasília: University of Brasília, 2002.

Niemeyer, Oscar. *Oscar Niemeyer: Notebooks of the Architect*. Brussels: Edition CIVA, 2002.

Ramos, A. R. *The Predicament of Brazil's Pluralism*. Brasília: University of Brasília, 2002.

Internet

For a history of Brasília, see http://www.infobrasilia.com.br/.

To see the complete document of Costa's pilot plan for Brasília, see http://www.info-brasilia.com.br/pilot_plan.htm.

For photos of Brasília and the public archive of the Federal District, see http://www.geocities.com/thetropics/3416/minis_ic.htm.

For information about Brasília's water problems, see http://www.tierramerica.net/2001/0603/iacentos2.shtml.

For information about São Paulo's population, see http://www.demographia.com/db-mumsao.htm.

Music

McPartland, Marian. *Chick Corea: Marian McPartland's Piano Jazz*. Jazz Alliance, TJA120402, 2002. . Includes "Brasilia," from the *Lyric Suite for Sextet*.

Tiso, Wagner. *Memorial*. Biscoito Fino, BF-519, 2002. Popular music of Brazil, sung in Portuguese, performed by Wagner Tiso and Zé Renato.

Documents of Authorization—I

OFFICIAL JOURNAL

Year XCV (95)—No. 217
Federal Capital
Thursday, 20th September 1956
Law No. 2,874 of 19th September 1956

SECTION I

Regulates the move of the Federal Capital and other measures

The President of the Republic:

I announce that the National Congress decrees and I sanction the following law:

CHAPTER I

Art. 1. The Federal Capital of Brazil, which refers to Article 4 of the Act of the Transitionary Dispositions of the Constitution of 18th September 1946, will be located in the region of the Central High Plateau, which has been chosen for this purpose, in an area which will be the Federal District circumscribed by the following lines:

Beginning at the point of lat. 15° 30' S. and long. 48° 12' W. Greenwich. From this point it continues toward the east along the parallel of 15° 30' S. until meeting the meridian of 47° and 25' W. Greenwich. From this point it continues on the same meridian of 47° and 25' W. Greenwich toward the South until the bed of the river "Santa Rita", the effluent at the right bank of the river "Rio Preto". From there, along the bed of the aforementioned river Santa Rita until its confluence with the Rio Preto, right by the mouth of the lagoon "Lagoa Feia." From the confluence of the rivers Santa Rita and Rio Preto, it continues along the bed of the latter in a southerly direction, until crossing the parallel of 16° 03' S. From there, along the parallel of 16° 03' in westerly direction, until meeting the bed of the river "Rio Descoberto." From there to the North, along the bed of the river Rio Descoberto, until meeting the meridian of 48° 12' W. Greenwich. From there to the North along the meridian of 48° 12' W. until crossing the parallel of 15° 30' S., closing the perimeter.

Art. 2. For the execution of the constitutional disposition mentioned in the previous article, the Executive (executive power) is authorized to undertake the following actions:

(a)   To constitute, in the terms of this law, a corporation that will be named "Urbanizing Company of the New Capital of Brazil" ("Companhia Urbanizadora da Nova Capital do Brasil") (UC), with the objectives being established in Article 3;

(b)   To establish and to construct, through the proper organs of the federal administration and with the cooperation of the organs of the administration of the states, a system of transports and communications of the new Federal District with the Federal Units, coordinating this system with the National Traffic Plan;

(c)   To give the guarantee of the National Treasury to the operations of credit negotiated by the UC within Brazil or abroad, for the financing of the services and constructions of the future Capital, or with it related;

(d)   To attribute to the UC, through contracts or concessions, the execution of the construction and services of interest to the new Federal District, that are not included in the specific attributions of the company;

(e)   To sign contracts and agreements with the State of Goiás aiming at the expropriation of the real estates situated in the area of the new Federal District and at their subsequent dismemberment from the territory of the State and incorporation in the domain of the Union;

(f)   To establish norms and conditions for the approval of construction projects in the area of the Federal District, until the local administration is organized;

(g)   To install in the future Federal District, or in the neighboring towns, services of the civil and military organs of the federal government to which public servants are to be allocated, in order to create better conditions for the development of the construction work of the new City.

Single paragraph. At the appropriate time, the National Congress will deliberate regarding the day of the move of the Capital, thus repealing Article 6 of Law No. 1,803 of 5th January 1953.

CHAPTER II

THE "URBANIZING COMPANY OF THE NEW CAPITAL OF BRAZIL"
Section I

The constitution and aims of the Company

Art. 3. The "Urbanizing Company of the New Capital of Brazil" will have as its aim:

1. the planning and execution of the services of localization, urbanization, and construction of the future Capital, directly or through the organs of the administrations at federal, state, or municipal level, or through contracts established with appropriate companies;

2. the acquisition, exchange, alienation, location or rent of real estates in the area of the new Federal District or in any other part of the national territory, within the aims envisioned by this law;

3. the execution, through concessions, of construction and services related to the new Capital, which fall under the authority of the administrations at federal, state, or municipal levels;

4. the practice of all acts concerning the Company's objectives, established in its statutes or authorized by its Board of Administration.

The Company may accept outright donations of movable and immovable rights and properties, as well as conditioned ones, through authorization by decree of the President of the Republic.

Art. 4. The President of the Republic will appoint by decree the representative of the Union who takes part in the constitutional acts of the Company and in the activities referred to in Article 24, Paragraph 2 of this law.

Art. 5. The constitutional acts of the Company will include the approval:

(a) of the evaluation of the rights and properties drawn up to integrate the capital of the Union;

(b) of social statutes;

(c) of the transfer plan of any public services that may pass to the Company.

Art. 6. The constitution of the corporation and any modifications of its statutes will be approved by decree of the President of the Republic.

Any alteration that aims at modifying the system of administration of the Company, as it is established in this law, will depend on the express authorization of the Legislature.

Art. 7. In the organization of the Company, wherever applicable, the norms of the legislation of joint-stock companies will be observed. The disposition of capital in a bank, however, will be dispensed with.

Art. 8. The Company will have its seat in the region defined in Article 1; the time period for its duration is not determined.

SECTION II

On the Company's Capital

Art. 9. The "Urbanizing Company of the New Capital of Brazil" will have as its capital Cr$500,000,000.00 (five hundred million cruzeiros) divided into 500,000 (five hundred thousand) ordinary shares with a nominal value of Cr$1,000.00 (one thousand cruzeiros) each.

Art. 10. The Union will issue the totality of the company's capital, incorporating it in the following manner:

I. the incorporation of the studies, goods, and rights that belong to the assets of the "Commission for the Exploration of the Central High Plateau of Brasília" of 1892, of the "Commission for the Studies of the New Capital of Brazil" of 1946, of the "Commission for the Planning of the Construction and Move of the Federal Capital", created by decree No. 32,976 of 8th June 1953 and altered by decree No. 38,281 of 9th December 1955;

II. the transfer of the entire area of the future Federal District, at cost price and with the addition of the cost of expropriation during the course of the acquisition, except for the areas reserved for public use or for the special use of the Union;

III. the incorporation of other moveable or immovable goods or of rights belonging to the Union, which may result or not in expropriation;

IV. the entry of funds in the amount of Cr$125,000,000.00 (one hundred and twenty five million cruzeiros) necessary for the costs of organization, installation, and the beginning of the services of the Company;

V. the entry of funds in the amount of Cr$195,000,000.00 (one hundred and ninety five million cruzeiros) at a later time when it should be considered necessary.

Par. 1: The capital of the Company can be increased with new funds for that purpose or with the incorporation of goods mentioned in Point III. of this article.

Par. 2: The shares of the Company can be acquired with the authorization of the President of the Republic by a juristic person of national public law, which however, cannot sell them except to the Union itself, to which it is guaranteed in any case to hold at least 51% (fifty one percent) of the share capital.

Art. 11. The corporation may remit, in addition to holders' promissory notes, special titles from which it will receive a premium of 10% (ten percent) for the payment of the urban lands of the new Capital, being due moreover, an interest of 8% (eight percent) per year, regardless of the limit established by law.

SECTION III

On the Administration and Financial Control of the Company

Art. 12. The administration and financial control of the Company will be carried out by a Board of Administration, a Board of Directors and a Board of Treasurers, each with a term of office of 5 (five) years. The respective positions will be filled by nomination of the President of the Republic, and observe the following paragraphs:

Par.1: The Board of Administration will consist of 6 (six) members with equal right to vote, and its decisions are obligatory for the Board of Directors, with only the President being entitled to raise an objection.

Par.2: The Board of Directors will be composed of 1 (one) chairman and 3 (three) directors.

Par.3: The meetings of the Board of Administration will be presided over by the chairman of the Board of Directors, who however only has a *deciding vote* in these meetings.

Par.4: The Board of Administration will meet at least once a week and detailed minutes will be elaborated on its discussions, which, properly authenticated, will be supplied to each of its members.

Par.5: The Board of Treasurers will be composed of 3 (three) effective members and 3 (three) substitutes and will carry out the functions established in the legislation on joint-stock companies, without the restrictions of the decree-law No. 2,928, of 31 of December 1940.

Par.6: One-third of the members of the Board of Administration, the Board of Directors, and the Board of Treasurers will be chosen from three lists of names supplied by the National Executive of the biggest political party that forms part of the opposition in the National Congress.

Par.7: Replacements of members of the Board of Administration, the Board of Directors, and the Board of Treasurers, be they permanent or temporary due to inability exceeding 30 (thirty) days, will be accomplished by the same process as that of the constitution of these organs, as designated in the previous paragraph.

Par.8: It remains with the Board of Administration to decide exclusively, upon suggestions from the Board of Directors, any plans for buying, selling, leasing, or renting of real estate of the property of the Company, as well as on the operations of credit negotiated by same.

Par.9: After attending to the dispositions of this law, the statutes of the Company will regulate the attributions and functions of the Board of Administration and the Board of Directors.

Par.10: The members of the Board of Administration and the Board of Directors are obliged to have their residence in the area mentioned in Article 1.

SECTION IV

On the Acts and Obligations of the Company

Art. 13. The acts of the constitution of the Company, the incorporation of its capital, the properties it may possess as well as the acquisition of rights, moveable and immovable goods which it may carry out, and moreover the instruments it may form part of, will be free from any kind of taxes or fees of a fiscal nature that fall under the competence of the Union, which will negotiate with the other entities of public law, requesting from them the same favors for the Company within the respective areas of their fiscal competence.

Art.14. The Company will benefit from exemption of taxes on the importation of consumer goods and of additional taxes with respect to machinery, its spare parts and accessories, appliances, tools, instruments and materials, which are destined for its works and services. It will pay these taxes however, when these goods are sold to third parties.

Single paragraph: All the materials and goods referred to in this Article, with restriction for articles that are produced similarly on a national level, will be cleared by edicts of the customs inspectors.

Art.15. The Company has the guaranteed right to promote expropriations according to the law in force and within the modifications mentioned in this law.

Art.16. The Company will submit its accounts by the 30th of April each year to the audit division of the Union, which will accept them by passing them on to the National Congress, to whom it corresponds to adopt the measures, which its fiscal action considers to be convenient.

Art.17. The services, works, and construction necessary for the installation of the Government of the Republic in the future Capital will be carried out by the Company independent of any indemnification, it being understood that the payment of any expenses of the rights, goods, favors, and concessions have been granted by virtue of this law.

Art.18. The Executive furthermore will guarantee the right to the Company to utilize all the equipment, services and installations of the organs of the federal administration, whenever these should become necessary for the activities of the Company.

Art.19. The administrative acts and contracts entered into by the Company will be contained in a monthly bulletin edited by the Company, and from which copies will be distributed to the members of the National Congress, the authorities of the ministries, interested offices, class units, and organs of publicity.

Art. 20. The directors of the Company are obliged to present any kind of information that is requested by the National Congress with respect to its acts and decisions.

Art. 21. In the contracts of works and services or in the acquisition of materials from natural or juristic persons the Company must:

a) Establish invitation of administrative tenders for contracts with a value greater than Cr$1,000,000.00 (one million cruzeiros) and up to Cr$10,000,000.00 (ten million cruzeiros), while it is the prerogative of the Board of Administration, upon proposal of the Board of Directors, to dispense with this requirement, based on a substantiated decision that must be demonstrated in the minutes.

b) Establish invitation of public tenders for contracts with a value greater than Cr$10,000,000.00 (ten million cruzeiros), in which case the Board of Administration is allowed to dispense with this formality, within the cautions of the above section. This decision must be communicated within 5 (five) days to the President of the Republic who still can order a call for the tender.

SECTION V

On the Personnel of the Company

Art. 22. The employees of the "Urbanizing Company," in their relations to the Company, are exclusively subject to the norms of labor legislation, being classified in the different institutions for retirement and pension funds, in the context of social security, according to the nature of their functions.

Art. 23. Military personnel, civil public servants of the Union, of the administrative units, and of entities of the mixed economy, can serve in the Company according to the decree-law No. 6,877 of 16th September 1944.

CHAPTER II

GENERAL AND FINAL DISPOSITIONS

Art. 24. For all legal purposes, decree No. 480, of 30th April 1955, issued by the Governor of the State of Goías, is ratified, who declared the area referred to in Article 1 as a public utility and a necessity in accord with the social interest, for the purpose of expropriation.

Par.1: The expropriations already underway can continue to be delegated to the government of the state, or they can pass directly to the Union for realization.

Par.2: In the transfer to the domain of the Union of real estate acquired by the Government of Goías and in the acts of direct expropriation in which it may come to interfere, and moreover in the acts of their incorporation into the "Urbanizing Company of the Federal Capital", the Union will be represented by the person who is referred to in Article 4 of this law.

Par.3: Whenever expropriations are realized in friendly mutual agreement, the expropriated former owners will benefit from exemption of taxes on income and former profits for the transfer to the expropriator of the respective immovable properties.

Par.4: The immovables that have been expropriated in the area of the new Federal District and the ones referred to in Article 5 can be freely alienated by the expropriating power and by the subsequent proprietors, but without granting them any preferences like the ones given to the expropriated former owners.

Art. 25. The lots of urban land of the future Federal District will become indivisible, if alienated by the "Urbanizing Company of the New Capital of Brazil." The alienation of other areas of land of the mentioned district to natural or juristic persons of civil law is expressively prohibited.

The "Urbanizing Company of the New Capital of Brazil" will organize plans that will guarantee the full economic use of the rural properties, only carrying them out directly or by renting.

Art. 26. The Institute of Social Security, the Society of Mixed Economy and the administrative units of the Union are authorized to acquire titles and promissory notes of the "Urbanizing Company of the New Capital of Brazil," referred to in Article 11 of this law.

These titles can also be sold to military personnel, federal servants, servants of administrative units, and of the societies of mixed economy of the Union, if they authorize the deduction of the due installments, payable in 60 (sixty) months, in their respective payrolls.

Art. 27. In order to guarantee the necessary supplies for the constructions of the new Capital, the roads projected to link the new Federal District to the industrial centers of São Paulo and Belo Horizonte and to the river harbor of Pirapora in the State of Minas Gerais, are included in the category of first priority.

Art. 28. Once this law is in effect, the lots of land in which the rural properties will be divided, which exist up to a distance of 30 (thirty) kilometers from the outer perimeter of the new Federal District in areas of less then 20 (twenty) hectares, can only be inscribed in the Register of Real Estate and put up for sale, after the public areas of those lots have been provided with running water, electrical light, sewage pipes, curbs and asphalt roads.

Art. 29. The legislation regarding joint-stock companies will be applied as a subsidiary of this law to the "Urbanizing Company of the New Capital of Brazil."

Art. 30. The balance of funds from Fund 4, title 4.3.00, subtitle 4.3.01, item 1: "Expenditures with the expropriation of the totality of the areas of the new Federal District, including compensation to the state of Goías," shall be transferred to the Ministry of Finance and attributed to the Ministry of Justice in the current budget.

Art. 31. A special credit of Cr$125,000,000.00 (one hundred and twenty five million cruzeiros) shall be opened to attend to the dispositions of Article 10, Item IV of this law.

Art. 32. The Executive will establish a method for dissolving the "Planning Commission for the Construction and Move of the Federal Capital," after the contracts that have been closed with third parties are transferred to the "Urbanizing Company of the New Capital of Brazil."

Art. 33. The name "Brasília" is given to the new capital.

Art. 34. This law goes into effect on the date of its publication, any contrary dispositions being revoked.

Rio de Janeiro, 19th September 1956,
135th year of the Independence and 68th year of the Republic.

JUSCELINO KUBITSCHEK

Nereu Ramos	Ernesto Dornello
Antônio Alves	Cámara Clovis Salgado
Henrique Lott	Parsifal Barroso
José Carlos de Macedo Soares	Henrique Fiehus
S. Paes de Almeida	Mauricio de Medeiros
Lúcio Meira	

From *Diário Oficial*, Year 95, No. 217, "Federal Capital," September 20, 1956. Translated from Portuguese by Christine Boedler.

Documents of Authorization—II

Terms of the contract made up by mutual assent of the Comissão do Vale do São Francisco (São Francisco Valley Commission) and Donald J. Belcher & Associates, Incorporated, to perform the studies of airphoto-analysis and interpretation indispensable to the selection of the most suitable sites for the location of the New Federal Capital.

At 25th day of February of 1954, the Civil Engineer Paulo Paltier de Queiroz, Superintendent Director of the Comissão do Vale do São Francisco, in view of the arrangements made for this purpose with the Comissão de Localização da Nova Capital Federal and by delegation of this, according to what is determined by item 2 of the Decree no. 32,976 of June 8, 1953, and using the legal and governmental attributions, conferred upon him by the Law no. 541 of December 15, 1948 and rulership approved by the Decree no. 29,807 of June 25, 1951, has agreed and contracted, in the head office of the Comissão do Vale do São Francisco at 210, President Wilson Avenue, under the terms of Letter "b" of item 51 of the Public Accounting code and duly authorized by the President of Brasil, with Donald J. Belcher & Associates,

Incorporated, of Ithaca, New York–U.S.A., the achievement of specialized technical work connected with the airphoto-analysis and photo interpretation necessary in the selection of the most suitable sites for the location of the New Federal capital, under the following clauses and conditions:

FIRST CLAUSE—Donald J. Belcher & Associates, Incorporated are bound, by the contract, to accomplish the airphoto-analysis and photo interpretation, necessary to the selection of the most suitable sites for the location of the new Federal Capital, involving the whole area limited, by the parallels 15 ° 30' S and 17° 00' S and meridians 46° 30' and 49° 30' W.G. under the terms of item 1 of Law no. 1803 of January 5, 1953.

SECOND CLAUSE—Besides the referred studies in the aforesaid clause of this contract, Donald J. Belcher & Associates, Inc. bind themselves to perform the surveys and mapping eventually necessary to the completion of the referred studies.

THIRD CLAUSE—The Comissão do Vale do São Francisco will furnish, free of charge to Donald J. Belcher & Associates, Inc. the following items, within the time stated below:

a. 4 complete sets of all the airphotographs covering the described area in the first clause of this contract, within 5 days after the validity of this contract.
b. 1 uncontrolled mosaic compiled in the same scale of the aerial photographs and photographically reproduced in order to obtain the necessary copies, within 30 days after the validity of this contract.
c. 4 photo-indexes corresponding to the aerial photographs within 5 days after the validity of this contract.
d. The necessary permission for the material mentioned in the preceding items to go out and return to this country under the responsibility of Donald J. Belcher & Associates, Incorporated, without any troubles for the latter.

FOURTH CLAUSE—The studies of airphoto-analysis and photo-interpretation which constitute the subject of this contract, aiming specifically the section of the most suitable sites under the technical view point for the location of the New Federal Capital will obey the requirements established in the SS1 and 2 of Law 1, 803 of January 5, 1953, as follows:

a. Favorable climate and sanitary conditions.
b. Water resources and potential hydroelectric sites.
c. Access by roads, railroads and airways.
d. Suitable topographic conditions.
e. Engineering characteristics of soils and facilities of construction materials.
f. Neighboring agricultural soils.
g. Attractive sights.
h. A population of about 500,000 inhabitants.

FIFTH CLAUSE—The referred studies of the aforesaid clause will involve the compilation of basic maps, mosaics and overlays in which will be represented, for each area, essential information concerning geology, showing type and occurrences of rock formation and unconsolidated deposits, as well as depth of soil over underlying rock, beside the elements concerning drainage conditions, land use and land classification, ground and surface water resources, location of sources of construction materials, potential sites for hydro-electric developments, airports and highways.

SIXTH CLAUSE—Based on these studies, which will be accompanied by special report, Donald J. Belcher & Associates, Incorporated will select the 5 best sites inside the total area for the location of the New Federal Capital, and in these five selected sites, each one with an approximate area of 1,000 sq.km., will proceed detailed studies on the conditions already described in the preceding clause, beside the topographic surveys to be performed by stereophotogrammetric plotting based on a triangulation network, radial plot and careful barometric leveling, having mosaic maps compiled in proper scale as a result.

SEVENTH CLAUSE—At the conclusion of the work within the contractual time Donald J. Belcher & Associates, Inc. will furnish, beside the base maps, mosaics, overlays and special reports on each one of the selected sites, a general report with all basic data concerning the various sites and accompanied of relief models and oblique photographs so that the comparison of the different features of each site can be possible and a precise judgment can be exercised in the final selection of that one which shows the best conditions for the location of the New Federal Capital.

EIGHTH CLAUSE—The ordinary inspection of the work will be charged to persons for that purpose named by the Superintendent Director of the Comissão do Vale do São Francisco, beside the members who may be named by the President of the Comissão de Localização da Nova Capital Federal for the same purpose.

NINTH CLAUSE—The total price for the performance of all the studies specified in this contract includes two amounts one in U.S. currency at a value of US$350000.00 and the other in Brazilian currency at a value of Cr$6,400,000.00, both of which will be paid by fractions to Donald J. Belcher & Associates, Incorporated in the following way:

a. 15% of the total American dollars will be deposited at the Chase National Bank of New York in favor of Donald J. Belcher & Associates, Inc. within 30 days after the validity of this contract;
b. 15% of the total in cruzeiros will be deposited at the Banco do Brasil in favor of Donald J. Belcher & Associates, Inc. Within 30 days after the validity of this contract;
c. The balance in U.S. currency as well as in Brazilian currency will be paid to Donald J. Belcher & Associates, Inc. in ten equal monthly installments on, or before 10th working day of each month, starting from the second month after the validity of this contact, the deposit of these amounts being made in the same way stated in the immediate above letters of this clause.

TENTH CLAUSE—Donald J. Belcher & Associates, Inc. as a guaranty of this contract made an initial deposit of Cr$100,000.00 at the Tesouro Nacional (National Treasury), as proved by receipt filed at the Comissão do Vale do São Francisco and from each payment must be deducted an amount of 5% of the total of each invoice as a reinforcement of the initial guaranty so that the deductions plus the initial guaranty sum up to a total of 5% of the value of this contract which amounts to Cr$12,987,000.00, the American amount being computed at the official rate of Cr$18.82 per dollar.

ELEVENTH CLAUSE—The total value of the guaranty referred in the aforesaid clause will be refunded to Donald J. Belcher & Associates, Inc. in 90 days after the final delivery of the whole material, for that purpose being given, within 60 days after the delivery, the necessary opinion of the Comissão de Localização da Nova Capital Federal on the acceptance and approval of the services so that the performance bond may be released.

TWELFTH CLAUSE—The present contract is signed in accordance with the proposal presented by Donald J. Belcher & Associates, Inc. duly signed by the President of the Comissão de Localização da Nova Capital Federal and by the Superintendent Director of the Comissão do Vale do São Francisco, with the same effect as though it has been incorporated to this contract even though not transcribed herein and free of public competition, in accordance with the authorization of the President of Brasil as per his decision written in the Exposicão de Motivos (Public Explanation) no. PR 3068/54 – 120 of January 20, 1954 of the Ministerio da Fazenda (published at the Diario Oficial (Official Journal) of the same month and year all the expenses of the current fiscal year being initially made on account of the resources established in the §1/item 1/sub-consignation 03/consignation 9/account 3 anex no. 9 of law no. 2135 of December 14, 1953, being immediately deposited for this purpose, the amount of Cr$3,396,100.00 (three million eight hundred ninety six thousand one hundred cruzeiros), as per disbursement document no._____.

THIRTEENTH CLAUSE—The payment connected to the balance will be made on account of the special credit of Cr20,000,000.00 authorized by item eight of Law 1,803 of January 5, 1953; for this, the Comissão de Localização da Nova Capital Federal at the proper time must put at the disposal of Comissão do Vale do São Francisco the amounts corresponding to the referred payment, by requisition.

FOURTEENTH CLAUSE—The time for the completion and delivery of all the services which constitute the subject of this contract, including the aerial photogrammetric survey, is ten months from the day of its registry at the Federal Accounting Tribunal (Tribunal de Contas da União) except "force majeur" as foreseen by law.

FIFTEENTH CLAUSE—Any delay by the Comissão do Vale do São Francisco in delivering to Donald J. Belcher & Associates, Inc. the items mentioned in the third clause of this contract, within the time established in the said clause will automatically postpone the time for the final delivery of the work by a number of days corresponding to the total time of delay in delivering the above mentioned auxiliary items.

SIXTEENTH CLAUSE—The above said in the preceding clause is applicable in the event of the payment referred in the ninth clause of this contract suffer, by any reason, a delay of more than 30 days.

SEVENTEENTH CLAUSE—The Comissão do Vale do São Francisco is not responsible for any indemnity to Donald J. Belcher & Associates, Incorporated if, by any reason, the Accounting Tribunal refuses to register this contract.

EIGHTEENTH CLAUSE—In the event of Donald J. Belcher & Associates, Incorporated do not finish the contracted work within the time established in this contract, will be submitted to a fine of 1,000 cruzeiros per each exceeding day.

NINETEENTH CLAUSE—The value of the fines eventually due will be deducted from the deposited guaranty, according to the tenth clause of this contract, if Donald J. Belcher & Associates, Incorporated do not make their payment to the National Treasury within the non extensible time of 5 days, having the duty, in this case, of completing, within 10 days, also non extensible, the value of the deposited guaranty.

TWENTIETH CLAUSE—The Comissão do Vale do São Francisco will pay all the cooperation according to its possibilities, to Donald J. Belcher & Associates, Inc. in the way of:

a. To import the working items which are necessary to the performance of this contract.
b. To bring into the country specialized technicians.
c. To get the necessary permission for the Donald J. Belcher & Associates, Inc.'s technicians and agents to enter the public and private properties located inside the area subjected to study, while they act according to their functions and in the exclusive interest of the job conferred to them by this contract.

TWENTY FIRST CLAUSE—The fluctuations of salaries, taxes, including income tax, fees and other tributes which occur during the validity of this contract being a result of acts of the Brazilian Government, will be subject of a readjustment between the contracting parties since those fluctuation surpass 5% of the actual rates which were the basis to the proposal mentioned in the 12th clause of this contract.

TWENTY SECOND CLAUSE—Comissão do Vale do São Francisco will not give to Donald J. Belcher & Associates, Inc. any readjustment based on an eventual depreciation of currency.

TWENTY THIRD CLAUSE—Taking into consideration the considerable interest in preparing and employing a maximum of Brazilian engineers as far as the modern technique of airphoto-analysis and foto-interpretation is concerned, are Donald J. Belcher & Associates, Inc. by this contract, authorized to sub-contract with engineers, technicians and Brazilian enterprises, partial services which may be performed by the same, under Donald J. Belcher & Associates, Inc.'s judgment and supervision, with their thorough technical and financial responsibility for the partial services in that way performed.

TWENTY FOURTH CLAUSE—Donald J. Belcher & Associates, Inc. bind themselves to allow, without expenses for their organization, civil and military technicians, for that purpose named by the President of the Comissão de Localização da Nova Capital Federal, to follow, in every detail the field parties and indoor services subject of this contract, without any interference in the said services.

TWENTY FIFTH CLAUSE—Comissão do Vale do São Francisco and Comissão de Localização da Nova Capital Federal will not enter into other contracts with third parties for the same area and with the same scope of this contract during its duration.

TWENTY SIXTH CLAUSE—This contract will exclusively cover those items of geology referred in the fifth clause of this contract, that are precisely necessary and pertinent to the selection of the most suitable sites for the location of the new Federal Capital, but in no way are to be interpreted to include oil or minerals and other subsurface conditions that are not specifically involved in or in any way connected with the principal purpose of this contract, which is that of determining the most satisfactory location of the new Federal Capital.

TWENTY SEVENTH CLAUSE—If Donald J. Belcher & Associates, Inc. during the studies and research by air-photo analysis, according to the principal purpose of this contract, find out accidentally any significant information on the surface and subsurface of the area concerning unexplored and unknown mineral resources, will be obliged to inform the Brazilian Government about those discoveries and they are forbidden under the punishments of Brazilian laws, to give the same information to third parties, no matter whom they are, the respective studies being the subject of a new contract if they are of some interest to the Brazilian Government.

TWENTY EIGHTH CLAUSE—Any question or disagreements arising out of the meaning of the clauses of this contract shall be decided by a Board of Arbitration consisting of three members: one member selected by the President of the Comissão de Localização da Nova Capital Federal, one representing Donald J. Belcher & Associates, Inc. and a third member mutually appointed by the duly authorized representatives of the contracting parties, no appeal being possible on the decisions of the Board of Arbitration, except for the cases of ordinary justice.

TWENTY NINTH CLAUSE—It is thoroughly understood that all taxes, fees, indemnities, including income tax set up by Brazilian Legislation as well as due by Donald J. Belcher & Associates, Inc. to their country, are the full and exclusive responsibility of the latter.

THIRTIETH CLAUSE—In case of withdrawal or unjustified abrogation, and also by unaccomplishment of mutual obligations foreseen in this contract, the damaged party shall have the right of an indemnity to be paid by the other party and in accordance with the following way:

a. If the damaged parties are the Comissão de Localização da Nova Capital Federal and the Comissão do Vale do São Francisco, Donald J. Belcher & Associates, Inc. shall pay them a compensatory fine composed of two amounts, one of Cr$500,00.00. . . . (quinhentos mil cruzeiros) and another in the amount of US$25,000.00 (twenty-five thousand dollars)—beside the obligation of refunding all sums already received in payment of services accomplished, if those services have no actual use for the intent foreseen in this contract.

b. If the damaged parties are Donald J. Belcher & Associates, Inc., the Comissão de Localização da Nova Capital Federal and Comissão do Vale do São Francisco shall pay them the same compensatory fine referred in the preceding letter of this clause, plus all disbursements already made with material purchasing, salaries, travel and general expenses, up to the date of the event, increased of a 10% administration tax, from that total being deducted the contractual amounts already paid.

THIRTY FIRST CLAUSE—For the effect of that established in the aforesaid clause, the contracting party that, at any time, consider itself damaged, shall make a written communication to the other party explaining its reasons, the other party having 15 days to accept or refuse them, otherwise being applied the punishment foreseen in the preceding clause of this instrument.

THIRTY SECOND CLAUSE—This contract shall be abrogated with the consequent loss of the guaranty and, by judgment of the Comissão do Vale do São Francisco, the loss of capacity to go on dealing with the Brazilian Government, independently of any judicial action or appeal, in case of:

a. Donald J. Belcher & Associates, Inc. become bankrupt, enter into a concordat, dissolve or transfer at the whole or part of this contract without a previous permission of the Comissão do Vale do São Francisco.
b. Donald J. Belcher & Associates, Inc. do not observe the specifications which constitute this contract and further details mentioned in their proposal, after a written warning of the Comissão do Vale do São Francisco.
c. Fines due by Donald J. Belcher & Associates, Inc. are not paid within the time foreseen in the 19th clause of this contract, or if there is non observance of any the clauses of this contract.

THIRTY THIRD CLAUSE—The contracting parties will submit this to the laws of Rio de Janeiro.

THIRTY FOURTH CLAUSE—This contract is exempt of payment of proportional stamps according to the concession established by the Circular no. 23 of August 6, 1948, of the Finance Minister, published in the Diario Oficial (Official Journal) of August 12, 1948 and according to the Resolução do Tribunal de Contas (Decision of the Accounting Tribunal) of September 10, 1948.

So being fully in agreement the contracting parties, I Analia Luz, Chief of the Section of Comissão do Vale do São Francisco, by myself, have written this contract on the proper book, pages 160 verso to 161 verso that is at the Administration Bureau of the Comissão do Vale do São Francisco, which, after being read and considered consistent is signed by General Aguinaldo Caiado de Castio, President of the Comissão de Localização da Nova Capital Federal, by Civil Engineer Paulo Peltier de Quiros, Superintendent Director of the Comissão do Vale do São Francisco and by the duly authorized representative in Brasil, of Donald J. Belcher & Associates, Incorporated, Civil Engineer Edson de Alencar Cabral, in the presence of the following witness.

Rio de Janeiro, 25 de Fevereiro de 1954.

Aguinaldo Caiado de Castro— President (Comissão de Localização da Nova Capital Federal)
Paulo Peltier de Queiroz
Edson de Alancar Cabral
Irineu Bornhausen
Antonio Gallotti
Jeronimo Coimbra Bueno

From Hollister Kent. "Vera Cruz: Brazil's New Federal Capital." PhD diss., Cornell University, 1956. Appendix, 8–16.

International Space Station. Courtesy of NASA.

31
NASA and
the Apollo Program

United States

DID YOU KNOW . . . ?

➤ The idea for NASA was launched at a cocktail party.
➤ What started as a space race between the Soviet Union and the United States evolved into a joint International Space Station.
➤ NASA was officially established in 1958.
➤ By 1965 NASA had a workforce of 36,000 civil-service employees and 376,000 contract workers, with an annual operating budget of over $5 billion.

> That's one small step for a man, one giant leap for mankind.
> —*Neil Armstrong, 1969, taking his first step onto the moon.*

The NASA program evolved as an American response to the Soviet Union's successful launch of Sputnik, the world's first artificial satellite. It was the size of a basketball, weighed only 183 pounds, and took about 98 minutes to orbit the earth on an elliptical path. That event ushered in new political, military, technological, and scientific developments that marked the start of the space age. In time, it led to the Apollo moon landing, an event of extraordinary worldwide interest. The landing of the first human being on the moon was far more than a U.S. achievement; it was a species achievement, a moment in evolution.

HISTORY

NASA was born during the cold war, a time of icy tensions, inflamed rhetoric, and fragile peace between the United States and the Soviet Union after World War II. As the world's two superpowers competed for political and military pre-eminence, a space race unfolded with potential rewards of military advantage and heightened national prestige. Humankind's entry into space promised a bonanza for virtually all fields of scientific research.

As early as 1954 a scientific conference in Rome, preparing for the 1957–58 International Geophysical Year (IGY), a worldwide endeavor to explore space noteworthy as a cooperative pooling effort to develop and expand science, recommended serious consideration of small satellite vehicles. The United States mounted such a program, called Project Vanguard, but the Soviets launched their artificial satellite first—Sputnik in October 1957, followed by Sputnik 2, carrying a dog, the next month. The first Vanguard rocket, launched in December 1957, exploded soon after takeoff.

The evidence of Soviet superiority in rocketry stunned the American public. The United States finally managed to put a satellite into orbit at the end of January 1958, but it used a launch vehicle set on a military missile; the first Vanguard did not get into orbit until March.

Amid widespread calls for action lest the United States lose the space race, in mid-1958 Congress passed the National Aeronautics and Space Act (reproduced here as Document I). Among other things, it created the National Aeronautics and Space Administration (NASA), charged with "preservation of the role of the United States as a leader in aeronautical and space science and technology."

In its initial years, NASA was not given sufficient funding, and the Soviets continued to lead in the space race. In 1959, for example, they achieved the first impact on the moon by an artificial object, and the first pictures of the moon's far side. In April 1961, Soviet cosmonaut Yuri Gagarin became the first human to orbit the earth; the first U.S. piloted space flight three weeks later was only suborbital.

President John F. Kennedy took office in January 1961, and in late May he appealed to Congress to approve a major escalation in the U.S. space program, the centerpiece of which would be a manned voyage to the moon and back before the end of the decade. It was time, he said, for the United States "to take a clearly leading role in space achievement, which in many ways may hold the key to our future on earth." The relevant section of Kennedy's speech to a joint session of Congress is reproduced here as Document II. In his speech, the president warned that his proposal would "carry very heavy costs" and would likely divert resources from other important national priorities. He stressed that it would require "a degree of dedication, organization and discipline which have not always characterized our research and development efforts."

To sustain the Apollo program, by 1965 NASA had ramped up to a workforce of 36,000 civil-service employees plus more than 376,000 contract workers, with an annual operating budget of over $5 billion. By its conclusion in 1972 the program had, during several launches, landed 12 astronauts on the moon, led by Neil Armstrong on July 20, 1969. The number of Soviet cosmonauts to land on the moon remained at zero. Kennedy's goal of the United States taking a leading role in space exploration had been achieved.

NASA's space-exploration projects in subsequent years were lower in profile with skimpier budgets. Highlights included the launch of the Spacelab space station, the landing of two unmanned Viking spacecraft on Mars in the mid-1970s, and the launch of the first space shuttle in 1981. During the 1980s the Soviets again occupied the limelight by sending successful unmanned missions to Venus and, in 1986, launching the core section of what became the first permanently manned space station, Mir. At the end of the decade, NASA again achieved success in interplanetary space by initiating a series of unmanned missions.

The *Apollo 11* flight team, (left to right) Neil A. Armstrong, Michael Collins, and Edwin E. Aldrin, Jr. Courtesy of NASA.

CULTURAL CONTEXT

Although the U.S. space program is usually associated with Kennedy, it was actually Dwight D. Eisenhower who in the 1940s initiated research in the areas of rocket science and space that would prove to be strategically important in the postwar era. That technical background was needed to launch and orbit a satellite—America's "science project"—for the 1957–58 IGY.

On a Friday night in October 1957, the embassy of the Soviet Union in Washington, DC, was hosting a reception. Dr. John Hagan, head of Project Vanguard, came to the party early, hoping that perhaps under the influence of their national schnapps some of the Soviet scientists would tell him if their IGY project was as far behind in time and cost overruns as his own beleaguered satellite. Just before 6:00 P.M., Walter Sullivan from the *New York Times* took a telephone call, then came back to whisper into the ear of Richard Porter of the American IGY committee, that Sputnik 1, the first satellite to orbit the earth, had been launched by the Soviet Union from their desert test site in Tyuratam in Kazakhstan. Thus began a competition that changed the world. That launch of a satellite by the Soviets in celebration of the IGY introduced a new catchphrase into the world lexicon: *space race*. The Americans were not nearly as ready as they had been when the Manhattan Project under President Roosevelt had prepared its countermoves. Senate Majority Leader Lyndon B. Johnson was also hosting a party that same night Sputnik launched. Johnson's aide George Reedy commented when everyone rushed from Johnson's ranch house to the front lawn to peer into the sky, "The simple fact is that we can no longer consider the Russians to be behind us in technology. It took them four years to catch up to our atomic bomb and nine months to catch up to our hydrogen bomb. Now we are trying to catch up to their satellite" (NASA Web site).

It was evidence of the world's mood and the intensity of the ongoing cold war at the time that some Americans suspected the radio signals Sputnik was sending to track its orbital location were instead spying on the United States and other countries. U.S. pop culture of the time was producing James Bond movies that dramatized the cold war with hot cars, airplanes, and secret weapons.

American scientists became even more concerned when the first two Vanguard launches were spectacular public failures. On December 6, 1957, John Hagan invited the entire press corps and the general media to view the launch of the Vanguard booster. It rose three feet off the ground, vibrated ominously, then crashed, taking the country's hopes with it. A subsequent launch try on February 5, 1958, went four miles up before bursting into flames.

America's fortunes began to change with the appearance of Werner von Braun. Just as Albert Einstein, a German refugee, had contributed scientific insight to developing the atomic bomb, so too did von Braun, a German engineer who headed a team of scientists who had immigrated to the United States after World War II. Von Braun sorted through a pile of untried plans that had been proposed for the IGY and pulled out one plan, Project Explorer, that he felt sure would work. On January 31, 1958, at 10:55 P.M. Pacific daylight time, a perfect launch by the Juno I booster rocket sent Explorer I aloft. There was the moment of collective agony when von Braun stopped a Pentagon meeting at exactly 12:41 A.M. Wrinkling his brow in a puzzled expression, he asked where the signal was that Explorer 1 was to send when passing California. A collective sigh of relief came at 12:49 when the signal came through, eight minutes late, because the Juno booster had fired so strongly that Explorer 1 was somewhat higher in orbit than expected. This success went a long way toward renewing U.S. confidence in its space programs.

PLANNING

NASA was not a new idea. Its forerunner, the National Advisory Committee for Aeronautics (NACA) had been around since 1915, its mission to improve and develop aviation. It was a group filled with rocket scientists and no politicians, it was civilian, and it was low profile, so it could stay away from the watchful eyes of cold-war spies.

In February 1958, there was sufficient public confidence and political willingness to create a permanent agency for space exploration. President Eisenhower reviewed a number of proposals, then sought the advice of his science advisor, James R. Killian, who convened the President's Science Advisory Committee. The committee recommended that all nonmilitary space efforts be grouped under a single entity, once again known as the National Advisory Committee for Aeronautics (NACA). However, this new NACA was charged with (1) planning, directing, and conducting aeronautical and space activities, (2) working with the American scientific establishment in such activities, and (3) releasing and spreading information about space science. One sign of the expansion of the old NACA into the new agency was Eisenhower's clever renaming of the entity, which changed its mission but not its pronunciation: NASA. In July 1958, Congress passed into law the National Aeronautics and Space Act, which became fully operational on October 1, 1958. It was just one year since Sputnik 1 had appeared, and in a spirit of competition, NASA announced its first initiative: manned space exploration.

BUILDING

Once announced, the manned space-exploration program went through a series of incremental steps and several leaps. If humans were to go into space, then everything about science, technology, and physiology had to be questioned, even the basics such as gravity and walking.

In 1961, Project Mercury proved that humans could withstand the rigors of space flight, as Alan B. Shepard Jr. became the first American to enter suborbital space, for 15 minutes. In 1962, John H. Glenn Jr. became the first American to orbit the earth. Project Gemini accommodated teams of two astronauts each on 10 separate missions. The fourth mission, on June 3, 1965, hosted the extravehicular activity (EVA) that would later become critical to building and repairs in space.

All these efforts and successes added to Kennedy's confidence, as the new president captured the country's imagination with his daring challenge to place a person on the moon within the decade. It took $25.4 billion—a cost comparable to the Panama Canal and the Manhattan Project. Apollo 11 left the Cape Canaveral space center in Florida at 9:32 A.M. on July 16, 1969. A strong Saturn rocket was the ignition launcher, which catapulted the Apollo 11 spacecraft into orbit around the earth one and a half times before Saturn's third stage ignited to boost the rocket onward toward the moon. The command/service module, waiting in the third stage, disconnected from Saturn, turned, and connected to the Eagle lunar module. The world waited.

Three days later, on July 19, Apollo 11 neared the moon and fired its engines to bring it into lunar orbit. After a full 24-hour day in lunar orbit, it was time for the greatest human journey in history. Astronauts Neil Armstrong and Buzz Aldrin crawled into the relatively tiny Eagle module and separated from Apollo at 1:28 P.M. Eastern standard time, edging ever closer to the lunar surface. At 4:18 P.M., the Eagle landed, but it was not until several hours later, at 8:56 P.M., that Neil Armstrong stepped down the Eagle's ladder and, as the world held its collective breath, stepped onto a new planet, and spoke the words heard around the world: "That's one small step for a man, one giant leap for mankind." It was all the more a world event as the connective experience of television beamed live shots from the moon via a camera set up by Armstrong and Aldrin, who had followed Armstrong out of the Eagle and onto the moon's surface.

All too soon it was over, and the astronauts returned to earth. But Apollo returned to the moon five more times. *Apollo 17* astronaut Harrison H. "Jack" Schmitt was the first geologist to visit the moon; he later shared his findings with the Massachusetts Institute of Technology in Cambridge, Massachusetts, where he was a professor, and later became a U.S. senator.

After the Apollo program, the United States continued to develop and launch space missions. However, three disasters clouded a program of considerable overall achievement. The first manned Apollo mission ended in disaster in January 1967 when a fire during a prelaunch test killed all three crew members. The *Challenger*, the second orbiter to become operational at Kennedy Space Center, had flown nine successful space-shuttle missions. On January 28, 1986, the *Challenger* and its seven-member crew were lost 73 seconds after launch when a booster failure resulted in the breakup of the vehicle. The space shuttle

Planting the U.S. flag on the surface of the moon during the first *Apollo 11* lunar landing mission. Here, astronaut Neil A. Armstrong, commander, stands on the left at the flag's staff. Courtesy of NASA.

Columbia, unlike the other shuttles in its lineage, did not visit the International Space Station or conduct construction in space, but instead was devoted entirely to scientific experiments in orbit. On February 1, 2003, the *Columbia* broke up during reentry into the earth's atmosphere, and all seven astronauts aboard perished.

The first cooperative effort between NASA and the Soviet Union began in 1975 when the two countries sent spaceships that met in space and docked together to conduct joint experiments. From 1984 to 1993 many ideas were put forward for a space station, and the quest was resolved by the Russians, who partnered with the United States to create the International Space Station. In a truly cooperative effort, Russia volunteered its own Mir, and NASA built the shuttles that rotated from earth to the space station. The International Space Station is expected to be the largest international scientific project in history.

IMPORTANCE IN HISTORY

The successes of the NASA space-exploration program lead inexorably to the question, why can't we generate the same kind of excitement for terrestrial projects that will bring extraordinary benefit? Could the world do both land and space energy exploration and development? For instance, the Tar Sands in Canada could go far toward solving the world's demand for oil: there are greater petro-leum reserves in Canada than in Saudi Arabia. In the past, it was not known how to retrieve oil from the Tar Sands and refine it, but progress is now being made.

Among politicians, who are in office for relatively short periods, the unfor-tunate response to such long-term issues is "Let somebody else deal with it." Most large-scale government initiatives are undertaken only for reasons of survival or war; the Roman aqueducts were built when the Tiber River was threatened with poisonous pollution.

The world needs an institution to train people in all aspects of the technol-ogy of space flight, including management, economics, and social engineering. Since there has been no national decision about the path to be pursued by the United States in the future, it is entirely possible we will suddenly encounter an opportunity and there will not be sufficient people trained to take advantage of it. One option is the use of the moon as a base for transmitting solar energy to the earth and for other science-based programs. Another is to use the moon as a base for exploring Mars. A third is using the moon as a potential source of raw materials. A fourth piques the imagination: establish a University of the Moon, a research and educational institute physically located on the lunar surface, where generations of young people can experience life in space. With par-ticipation from universities on every continent, study abroad would take on a whole new meaning. These programs can only be launched through coop-eration, as was made clear by the transformation of the space race into the International Space Station.

FOR FURTHER REFERENCE

Books and Articles

Bijker, Wiebe E., Thomas P. Hughes, and Trevor J. Pinch, eds. *The Social Construction of Technological Systems: New Directions in the Sociology and History of Technology.* Cambridge, MA: MIT Press, 1987.

Davidson, Frank P. *Macro: A Clear Vision of How Science and Technology Will Shape Our Future.* New York: William Morrow, 1983. Includes the first suggestion for a University of the Moon.

Glaser, Peter E., Frank P. Davidson, and Katinka I. Csigi. *Solar Power Satellites: A Space Energy System for Earth.* Chichester, England: John Wiley and Sons/Praxis Publishing, 1998.

Hughes, Thomas P. *American Genesis: A Century of Invention and Technological Enthusiasm.* New York: Penguin Books, 1989.

Kuhn, Thomas W. *The Structure of Scientific Revolutions.* Chicago: University of Chicago Press, 1970.

Lambright, W. Henry. *Powering Apollo: James E. Webb of NASA.* Baltimore: Johns Hopkins University Press, 1995.

Lusk Brooke, Kathleen, and George H. Litwin. "Organizing and Managing Satellite Solar Power." *Space Policy: an International Journal* 16, no. 3 (July, 2000): 145–56.

Rosenstock-Huessy, Eugen. *Out of Revolution.* Norwich, VT: Argo Books, 1993.

Internet

For the history of NASA and the Apollo mission, see http://history/nasa.gov. For historical records of the International Geophysical Year, send an e-mail to the Archives of the National Academy of Sciences—National Research Council at archives@nas.edu.

For histories of Sputnik, NASA, and other space initiatives, and articles and monographs by Roger D. Launius, NASA's chief historian, see http://www.ksc.nasa.gov/nasadirect/archives/KSCDirect/launius.htm.

To learn more about the International Space Station, see http://www.shuttlepresskit.com/ISS_OVR/.

For information about the concept of a "space elevator," see http://www.spaceelevator.com.

To explore the work of Eugen Rosenstock-Huessy, see http://www.valley.net/~transnat/erh.html.

For space travel in popular culture, see the Web sites for the television show *Star Trek* and the comic strip *Buck Rogers in the 25th Century* (published in more than 400 world newspapers in 18 languages from 1929 to 1967): http://www.startrek.com/ and http://www.buck-rogers.com/.

Film and Television

2001: A Space Odyssey. Directed by Stanley Kubrick. Burbank: Warner Studios, 1968.

Music

Beethoven, Ludwig van. Piano Sonata No. 14 in C Sharp Minor, Opus 27, No. 2, "Moonlight." Rudolf Serkin, pianist. Sony 37219, 1990.

Debussy, Claude. "Clair de lune." From *Suite Bergamasque*. To hear this piano composition, go to http://www.supcropera.com/mp3/debussy/debussy.htm.

Documents of Authorization—I

THE NATIONAL AERONAUTICS AND SPACE ACT

Pub. L. No. 85–568

July 29, 1958

[H.R.12575]

AN ACT

To provide for research into problems of flight within and outside the earth's atmosphere, and for other purposes.

Be it enacted by the Senate and House of Representative of the United States of America in Congress assembled.

TITLE I—SHORT TITLE, DECLARATION OF POLICY, AND DEFINITIONS

SHORT TITLE

SEC. 101. This Act may be cited as the "National Aeronautics and Space Act of 1958".

DECLARATION OF POLICY AND PURPOSE

SEC. 102. (a) The Congress hereby declares that it is the policy of the United States that activities in space should be devoted to peaceful purposes for the benefit of all mankind.

(b) The Congress declares that the general welfare and security of the United States require that adequate provision be made for aeronautical and space activities. The Congress further declares that such activities shall be the responsibility of, and shall be directed by, a civilian agency exercising control over aeronautical and space activities sponsored by the United States, except that activities peculiar to or primarily associated with the development of weapons systems, military operations, or the defense of the United States (including the research and development necessary to make effective provision for the defense of the United States) shall be the responsibility of, and shall be directed by, the Department of Defense; and that determination as to which such agency has responsibility for and direction of any such activity shall be made by the President in conformity with section 201 (e).

(c) The aeronautical and space activities of he United States shall be conducted so as to contribute materially to one or more of the following objectives:(1) The expansion of human knowledge of phenomena in the atmosphere and space;

(2) The improvement of the usefulness, performance, speed, safety, and efficiency of aeronautical and space vehicles;

(3) The development and operation of vehicles capable of carrying instruments, equipment, supplies, and living organisms through space;

(4) The establishment of long-range studies of the potential benefits to be gained from, the opportunities for, and the problems involved in the utilization of aeronautical and space activities for peaceful and scientific purposes;

(5) The preservation of the role of the United States as a leader in aeronautical and space science and technology and n the application thereof to the conduct of peaceful activities within and outside the atmosphere;

(6) The making available to agencies directly concerned with national defense of discoveries that have military value or significance, and the furnishing by such agencies, to the civilian agency established to direct and control nonmilitary aeronautical and space activities, of information as to discoveries which have value or significance to that agency;

(7) Cooperation by the United States with other nations and groups of nations in work done pursuant to this Act and in the peaceful application of the results thereof; and

(8) The most effective utilization of the scientific and engineering resources of the United States, with close cooperation among all interested agencies of the United States in order to avoid unnecessary duplication of effort, facilities, and equipment.

(d) It is the purpose of this Act to carry out and effectuate the policies declared in subsections (a), (b), and (c).

DEFINITIONS

Sec. 103. As used in this Act—

(1) the term "aeronautical and space activities" means (A) research into, and the solution of, problems of flight within and outside the Earth's atmosphere, (B) the development, construction, testing, and operation for research purposes of aeronautical and space vehicles, and (C) such other activities as may be required for the exploration of space; and

(2) the term "aeronautical and space vehicles" means aircraft, missiles, satellites, and other space vehicles, manned and unmanned, together with related equipment, devices, components, and parts.

TITLE II—COORDINATION OF AERONAUTICAL AND SPACE ACTIVITIES

NATIONAL AERONAUTICS AND SPACE COUNCIL

Sec. 201. (a) There is hereby established the National Aeronautics and Space Council (hereinafter called the "Council") which shall be composed of—

(1)   the President (who shall preside over meetings of the Council);
(2)   the Secretary of State;
(3)   the Secretary of Defense;
(4)   the Administrator of the National Aeronautics and Space Administration;
(5)   the Chairman of the Atomic Energy Commission;
(6)   not more than one additional member appointed by the President from the departments and agencies of the Federal Government; and
(7)   not more than three other members appointed by the President, solely on the basis of established records of distinguished achievement, from among individuals in private life who are eminent in science, engineering, technology, education, administration, or public affairs.

(b) Each member of the Council from a department or agency of the Federal Government may designate another officer of his department or agency to serve on the Council as his alternate in his unavoidable absence.

(c) Each member of the Council appointed or designated under paragraphs (6) and (7) of subsection (1), and each alternate member designated under subsection (b), shall be appointed or designated to serve as such by and with the advice and consent of the Senate, unless at the time of such appointment or designation he holds an office in the Federal Government to which he was appointed by and with the advice and consent of the Senate.

(d) It shall be the function of the Council to advise the President with respect to the performance of the duties prescribed in subsection (e) of this section.

(e) In conformity with the provisions of section 102 of this Act, it shall be the duty of the President to—

(1)   survey all significant aeronautical and space activities, including the policies, plans, programs and accomplishments of all agencies of the United States engaged in such activities;

(2)   develop a comprehensive program of aeronautical and space activities to be conducted by agencies of the Unites States;

(3)   designate and fix responsibility for the direction of major aeronautical and space activities;

(4)   provide for effective cooperation between the National Aeronautics and Space Administration and the Department of Defense in all such activities, and specify which of such activities may be carried on concurrently by both such agencies notwithstanding the assignment of primary responsibility therefore to one or the other of such agencies; and

(5)   resolve differences arising among departments and agencies of the United States with respect to aeronautical and space activities under this Act, including differences as to whether a particular project is an aeronautical and space activity.

(f) The Council may employ a staff to be headed by a civilian executive secretary who shall be appointed by the President by and with the advice and consent of the Senate and shall receive compensation at the rate of $20,000 a year. The executive secretary, subject to the direction of the Council, is authorized to appoint and fix the compensation of such personnel, including not more than three persons who may be appointed without regard to the civil service laws or the Classification Act of 1949 and compensated at the rate of not more than $19,000 a year, as may be necessary to perform such duties as may be prescribed by the Council in connection with the performance of its functions. Each appointment under this subsection shall be subject to the same security requirements as those established for personnel of the National Aeronautics and Space administration appointed under section 203 (b) (2) of this Act.

(g) Members of the Council appointed from private life under subsection (a) (7) may be compensated at a rate not to exceed $100 per diem, and may be paid travel expenses and per diem in lieu of subsistence in accordance with the provisions of section 5 of the Administrative Expenses Act of 1946 (5 U. S. C. 73b-2) relating to persons serving without compensation.

NATIONAL AERONAUTICS AND SPACE ADMINISTRATION

SEC. 202. (a) There is hereby established the National Aeronautics and Space Administration (hereinafter called the "Administration"). The Administration shall be headed by an Administrator, who shall be appointed from civilian life by the President by and with the advice and consent of the Senate, and shall receive compensation at the rate of $22,500 per annum. Under the supervision and direction of the President, the Administrator shall be responsible for the exercise of all powers and the discharge of all duties of the Administration, and shall have authority and control over all personnel and activities thereof.

(b) There shall be in the Administration a Deputy Administrator, who shall be appointed from civilian life by the President by and with the advice and consent of the Senate, who shall receive compensation at the rate of $21,500 per annum, and shall perform such duties and exercise such powers as the Administrator may prescribe. The Deputy Administrator shall act for, and exercise the powers of, the Administrator during his absence or disability.

(c) The Administrator and the Deputy Administrator shall not engage in any other business, vocation, or employment while serving as such.

FUNCTIONS OF THE ADMINISTRATION

SEC. 203. (a) The Administration, in order to carry out the purpose of this Act, shall—

(1)   plan, direct, and conduct aeronautical and space activities;
(2)   arrange for participation by the scientific community in planning scientific measurements and observations to be made through use of aeronautical and space vehicles, and conduct or arrange for the conduct of such measurements and observations; and
(3)   provide for the widest practicable and appropriate dissemination of information concerning its activities and the results thereof.

 (b) In the performance of its functions the Administration is authorized—

(1) to make, promulgate, issue, rescind, and amend rules and regulations governing the manner of its operations and the exercise of the powers vested in it by law;

(2) to appoint and fix the compensation of such officers and employees as may be necessary to carry out such functions. Such officers and employees shall be appointed in accordance with the civil-service laws and their compensation fixed in accordance with the Classification Act of 1949, except that (A) to the extent the Administrator deems such action necessary to the discharge of his responsibilities, he may appoint and fix the compensation (up to a limit of $19,000 a year, or up to a limit of $21,000 a year for a maximum of ten positions) of not more than two hundred and sixty of the scientific, engineering,

and administrative personnel of the Administration without regard to such laws, and (B) to the extent the Administrator deems such action necessary to recruit specially qualified scientific and engineering talent, he may establish the entrance grade for scientific and engineering personnel without previous service in the Federal Government at a level up to two grades higher than the grade provided for such personnel under the General Schedule established by the Classification Act of 1949, and fix their compensation accordingly;

(3) to acquire (by purchase, lease, condemnation, or otherwise), construct, improve, repair, operate, and maintain laboratories, research and testing sites and facilities, aeronautical and space vehicles, quarters and related accommodations for employees and dependents of employees of the Administration, and such other real and personal property (including patents), or any interest therein, as the Administration deems necessary within and outside the continental United States; to acquire by lease or otherwise, through the Administrator of General Services, buildings or parts of buildings in the District of Columbia for the use of the Administration for a period not to exceed ten years without regard to the Act of March 3, 1877 (40 U.S.C. 34); to lease to others such real and personal property; to sell and otherwise dispose of real and personal property (including patents and rights thereunder) in accordance with the provisions of the Federal Property and Administrative Services Act of 1949, as amended (40 U.S.C. 471 et seq.); and to provide by contract or otherwise for cafeterias and other necessary facilities for the welfare of employees of the Administration at its installations and purchase and maintain equipment therefor;

(4) to accept unconditional gifts or donations of services, money, or property, real, personal, or mixed, tangible or intangible;

(5) without regard to section 3648 of the Revised Statutes, as amended (31 U.S.C. 529), to enter into and perform such contracts, leases, cooperative agreements, or other transactions as may be necessary in the conduct of its work and on such terms as it may deem appropriate, with any agency or instrumentality of the United States, or with any State, Territory, or possession, or with any political subdivision thereof, or with any person, firm, association, corporation, or educational institution. To the maximum extent practicable and consistent with the accomplishment of the purposes of this Act, such contracts, leases, agreements, and other transactions shall be allocated by the administrator in a manner which will enable small-business concerns to participate equitably and proportionately in the conduct of the work of the Administration;

(6) to use, with their consent, the services, equipment, personnel, and facilities of Federal and other agencies with or without reimbursement, and on a similar basis to cooperate with other public and private agencies and instrumentalities in the use of services, equipment, and facilities. Each department and agency of the Federal Government shall cooperate fully with the Administration in making its services, equipment, personnel, and facilities available to the Administration, and any such department or agency is authorized, notwithstanding any other provision of law, to transfer to or to receive from the Administration, without

reimbursement, aeronautical and space vehicles, and supplies and equipment other than administrative supplies or equipment;

(7) to appoint such advisory committees as may be appropriate for purposes of consultation and advice to the Administration in the performance of its functions;

(8) to establish within the Administration such offices and procedures as may be appropriate to provide for the greatest possible coordination of its activities under this Act with related scientific and other activities being carried on by other public and private agencies and organizations;

(9) to obtain services as authorized by section 15 of the Act of August 2, 1946 (5 U. S. C. 55a), at rates not to exceed $100 per diem for individuals;

(10) when determined by the Administrator to be necessary, and subject to such security investigations as he may determine to be appropriate, to employ aliens without regard to statutory provisions prohibiting payment of compensation to aliens;

(11) to employ retired commissioned officers of the armed forces of the United States and compensate them at the rate established for the positions occupied by them within the administration, subject only to the limitations in pay set forth in section 212 of the Act of June 30, 1932, as amended (5 U. S. C. 59a);

(12) with the approval of the President, to enter into cooperative agreements under which members of the Army, Navy, Air Force, and Marine Corps may be detailed by the appropriate Secretary for services in the performance of functions under this Act to the same extent as that to which they might be lawfully assigned in the Department of Defense; and

(13) (A) to consider, ascertain, adjust, determine, settle, and pay, on behalf of the United States, in full satisfaction thereof, any claim for $5,000 or less against the United States for bodily injury, death, or damage to or loss of real or personal property resulting from the conduct of the Administration's functions as specified in subsection (a) of this section, where such claim is presented to the Administration in writing within two years after the accident or incident out of which the claim arises; and

(B) if the Administration considers that a claim in excess of $5,000 is meritorious and would otherwise be covered by this paragraph, to report the facts and circumstances thereof to the Congress for its consideration.

CIVILIAN-MILITARY LIAISON COMMITTEE

SEC. 204. (a) There shall be a Civilian-Military Liaison Committee consisting of—

(1) a Chairman, who shall be the head thereof and who shall be appointed by the President, shall serve at the pleasure of the President, and shall receive compensation (in the manner provided in subsection (d)) at the rate of $20,000 per annum;

(2) one or more representatives from the Department of Defense, and one or more representatives from each of the Departments of the Army, Navy, and Air Force, to be assigned by the Secretary of Defense to serve on the Committee without additional compensation; and

(3) representatives from the Administration, to be assigned by the Administrator to serve on the Committee without additional compensation, equal in number to the number of representatives assigned to serve on the Committee under paragraph (2).

(b) The Administration and the Department of Defense, through the Liaison Committee, shall advise and consult with each other on all matters within their respective jurisdictions relating to aeronautical and space activities and shall keep each other fully and currently informed with respect to such activities.

(c) If the Secretary of Defense concludes that any request, action, proposed action, or failure to act on the part of the Administrator is adverse to the responsibilities of the Department of Defense, or the Administrator concludes that any request, action, proposed action, or failure to act on the part of the Department of Defense is adverse to the responsibilities of the Administration, and the Administrator and the Secretary of Defense are unable to reach an agreement with respect thereto, either the Administrator or the Secretary of Defense may refer the matter to the President for his decision (which shall be final) as provided in section 201 (e).

(d) Notwithstanding the provisions of any other law, any active or retired officer of the Army, Navy, or Air Force may serve as Chairman of the Liaison Committee without prejudice to his active or retired status as such officer. The compensation received by any such officer for his service as Chairman of the Liaison Committee shall be equal to the amount (if any) by which the compensation fixed by subsection (a) (1) for such Chairman exceeds his pay and allowances (including special and incentive pays) as an active officer, or his retired pay.

INTERNATIONAL COOPERATION

Sec. 205. The Administration, under the foreign policy guidance of the President, may engage in a program of international cooperation in work done pursuant to this Act, and in the peaceful application of the results thereof, pursuant to agreements made by the President with the advice and consent of the Senate.

REPORTS TO CONGRESS

Sec. 206. (a) The Administration shall submit to the President for transmittal to the Congress, semiannually and at such other times as it deems desirable, a report of its activities and accomplishments.

(b) The President shall transmit to the Congress in January of each year a report, which shall include (1) a comprehensive description of the programmed activities and the accomplishments of all agencies of the United States in the field of aeronautics and space activities during the preceding calendar year, and (2) an evaluation of such activities and accomplishments in terms of the attainment of, or the failure to attain, the objectives described in section 102(c) of this Act.

(c) Any report made under this section shall contain such recommendations for additional legislation as the Administrator or the President may consider necessary or desirable for the attainment of the objectives described in section 102(c) of this Act.

(d) No information which has been classified for reasons of national security shall be included in any report made under this section, unless such information has been declassified by, or pursuant to authorization given by, the President.

TITLE III—MISCELLANEOUS

NATIONAL ADVISORY COMMITTEE FOR AERONAUTICS

Sec. 301. (a) The National Advisory Committee for Aeronautics, on the effective date of this section, shall cease to exist. On such date all functions, powers, duties, and obligations, and all real and personal property, personnel (other than members of the Committee), funds, and records of that organization, shall be transferred to the Administration.

(b) Section 2302 of title 10 of the United States Code is amended by striking out "or the Executive Secretary of the National Advisory Committee for Aeronautics." and inserting in lieu thereof "or the Administrator of the National Aeronautics and Space Administration."; and section 2303 of such title 10 is amended by striking out "The National Advisory Committee for Aeronautics" and inserting in lieu thereof "The National Aeronautics and Space Administration."

(c) The first section of the Act of August 26, 1950 (5 U.S.C. 22–1), is amended by striking out "the Director, National Advisory Committee for Aeronautics" and inserting in lieu thereof "the Administrator of the National Aeronautics and Space Administration", and by striking out "or National Advisory Committee for Aeronautics" and inserting in lieu thereof "or National Aeronautics and Space Administration."

(d) The Unitary Wind Tunnel Plan Act of 1949 (50 U.S.C. 511–515) is amended (1) by striking out "The National Advisory Committee for Aeronautics (hereinafter referred to as the 'Committee')" and inserting in lieu thereof "The Administrator of the National Aeronautics and Space Administration (hereinafter referred to as the 'Administrator')", (2) by striking out "Committee" or "Committee's" wherever they appear and inserting in lieu thereof "Administrator" and "Administrator's", respectively, and (3) by striking out "its" wherever it appears and inserting in lieu thereof "his".

(e) This section shall take effect ninety days after the date of the enactment of this Act, or on any earlier date on which the Administrator shall determine, and announce by proclamation published in the Federal Register, that the Administration has been organized and is prepared to discharge the duties and exercise the powers conferred upon it by this Act.

TRANSFER OF RELATED FUNCTIONS

SEC. 302. (a) Subject to the provisions of this section, the President, for a period of four years after the date of enactment of this Act, may transfer to the Administration any functions (including powers, duties, activities, facilities, and parts of functions) of any other department or agency of the United States or of any officer or organizational entity thereof, which relate primarily to the functions, powers, and duties of the Administration as prescribed by section 203 of this act. In connection with any such transfer, the President may, under this section or other applicable authority, provide for appropriate transfers of records, property, civilian personnel, and funds.

(b) Whenever any such transfer is made before January 1, 1959, the President shall transmit to the Speaker of the House of Representatives and the President pro tempore of the Senate a full and complete report concerning the nature and effect of such transfer.

(c) After December 31, 1958, no transfer shall be made under this section until (1) a full and complete report concerning the nature and effect of such proposed transfer has been transmitted by the President to the Congress, and (2) the first period of sixty calendar days of regular session of the Congress following the date of receipt of such report by the Congress has expired without the adoption by the Congress of a concurrent resolution stating that the Congress does not favor such transfer.

ACCESS TO INFORMATION

SEC. 303. Information obtained or developed by the Administrator in the performance of his functions under this Act shall be made available for public inspection; except (A) information authorized or required by Federal statute to be withheld, and (B) information classified to protect the national security: *Provided*, That nothing in this Act shall authorize the withholding of information by the Administrator from the duly authorized committees of the Congress.

SECURITY

SEC. 304. (a) The Administrator shall establish such security requirements, restrictions, and safeguards as he deems necessary in the interest of the national security. The Administrator may arrange with the Civil Service Commission for the conduct of such security or other personnel investigations of the Administration's officers, employees, and consultants, and its contractors and subcontractors and their officers and employees, actual or prospective, as he deems appropriate; and if any such investigation develops any data reflecting that the individual who is the subject thereof is of questionable loyalty the matter shall be referred to the Federal Bureau of Investigation for the conduct of a full field investigation, the results of which shall be furnished to the Administrator.

(b) The Atomic Energy Commission may authorize any of its employees, or employees of any contractor, prospective contractor, licensee, or prospective licensee of the Atomic Energy Commission or any other person authorized to have access to Restricted Data by the Atomic Energy Commission under subsection 145b. of the Atomic Energy Act of 1954 (42 U.S.C. 2165 (b)), to permit any member, officer, or employee of the Council, or the Administrator, or any officer, employee, member of an advisory committee, contractor, subcontractor, or officer or employee of a contractor or subcontractor of the Administration, to have access to Restricted Data relating to aeronautical and space activities which is required in the performance of his duties and so certified by the Council or the Administrator, as the case may be, but only if (1) the Council or Administrator or designee thereof has determined, in accordance with the established personnel security procedures and standards of the Council or Administration, that permitting such individual to have access to such Restricted Data will not endanger the common defense and security, and (2) the Council or Administrator or designee thereof finds that the established personnel and other security procedures and standards of the Council or Administration are adequate and in reasonable conformity to the standards established by the Atomic Energy Commission under section 145 of the Atomic Energy Act of 1954 (42 U.S.C. 2165). Any individual granted access to such Restricted Data pursuant to this subsection may exchange such data with any individual who (A) is an officer or employee of the Department of Defense, or any department or agency thereof, or a member of the armed forces, or a contractor or subcontractor of any such department, agency, or armed force, or an officer or employee of any such contractor or subcontractor, and (B) has been authorized to have access to Restricted Data under the provisions of section 143 of the Atomic Energy Act of 1954 (42 U.S.C. 2163).

(c) Chapter 37 of title 18 of the United States Code (entitled Espionage and Censorship) is amended by—

(1) adding at the end thereof the following new section:

"§799. Violation of regulations of National Aeronautics and Space Administration

"Whoever willfully shall violate, attempt to violate, or conspire to violate any regulation or order promulgated by the Administrator of the National Aeronautics and Space Administration for the protection or security of any laboratory, station, base or other facility, or part thereof, or any aircraft, missile, spacecraft, or similar vehicle, or part hereof, or other property or equipment in the custody of the Administration, or any real or personal property or equipment in the custody of any contractor under any contract with the administration, or any real or personal property or equipment in the custody of any contractor under any contract with the administration or any subcontractor of any such contractor, shall be fined not more than $5,000, or imprisoned not more than one year, or both."

(2) adding at the end of the sectional analysis thereof the following new item:

"799. Violation of regulations of National Aeronautics and Space Administration."

(d) Section 1114 of title 18 of the United States Code is amended by inserting immediately before "while engaged in the performance of his official duties" the following: "or any officer or employee of the National Aeronautics and Space Administration directed to guard and protect property of the United States under the administration and control of the National Aeronautics and Space Administration."

(e) The Administrator may direct such of the officers and employees of the Administration as he deems necessary in the public interest to carry firearms while in the conduct of their official duties. The Administrator may also authorize such of those employees of the contractors and subcontractors of the Administration engaged in the protection of property owned by the United States and located at facilities owned by or contracted to the United States as he deems necessary in the public interest, to carry firearms while in the conduct of their official duties.

PROPERTY RIGHTS IN INVENTIONS

Sec. 305. (a) Whenever any invention is made in the performance of any work under any contract of the Administration, and the Administrator determines that—

(1) the person who made the invention was employed or assigned to perform research, development, or exploration work and the invention is related to the work he was employed or assigned to perform, or that it was within the scope of his employment duties, whether or not it was made during working hours, or with a contribution by the Government of the use of Government facilities, equipment, materials, allocated funds, information proprietary to the Government, or services of Government employees during working hours; or

(2) the person who made the invention was not employed or assigned to perform research, development, or exploration work, but the invention is nevertheless related to the contract, or to the work or duties he was employed or assigned to perform, and was made during working hours, or with a contribution from the Government of the sort referred to in clause (1),

such invention shall be the exclusive property of the United States, and if such invention is patentable a patent therefore shall be issued to the United States upon application made by the Administrator, unless the Administrator waives all or any part of the rights of the United States to such invention in conformity with the provisions of subsection (f) of this section.

(b) Each contract entered into by the Administrator with any party for the performance of any work shall contain effective provisions under which such party shall furnish promptly to the Administrator a written report containing full and complete technical information concerning any invention, discovery, improvement, or innovation which may be made in the performance of any such work.

(c) No patent may be issued to any applicant other than the Administrator for any invention which appears to the Commissioner of Patents to have significant utility in the conduct of aeronautical and space activities unless the applicant files with the Commissioner, with the application or within thirty days after request therefor by the Commissioner, a written statement executed under oath setting forth the full facts concerning the circumstances under which such invention was made and stating the relationship (if any) of such invention to the performance of any work under any contract of the Administration. Copies of each such statement and the application to which it relates shall be transmitted forthwith by the Commissioner to the Administrator.

(d) Upon any application as to which any such statement has been transmitted to the Administrator, the Commissioner may, if the invention is patentable, issue a patent to the applicant unless the Administrator, within ninety days after receipt of such application and statement, requests that such patent be issued to him on behalf of the United States. If, within such time, the Administrator files such a request with the Commissioner, the Commissioner shall transmit notice thereof to the applicant, and shall issue such patent to the Administrator unless the applicant within thirty days after receipt of such notice requests a hearing before a Board of Patent Appeals and Interferences on the question whether the Administrator is entitled under this section to receive such patent. The Board may hear and determine, in accordance with rules and procedures established for interference cases, the question so presented, and its determination shall be subject to appeal by the applicant or by the Administrator to the United States Court of Appeals for the Federal Circuit in accordance with procedures governing appeals from decisions of the Board of Patent Appeals and Interferences in other proceedings.

(e) Whenever any patent has been issued to any applicant in conformity with subsection (d), and the Administrator thereafter has reason to believe that the statement filed by the applicant in connection therewith contained any false representation of any material fact, the Administrator within five years after the date of issuance of such patent may file with the Commissioner a request for the transfer to the Administrator of title to such patent on the records of the Commissioner. Notice of any such request shall be transmitted by the Commissioner to the owner of record of such patent, and title to such patent shall be so transferred to the Administrator unless within thirty days after receipt of such notice such owner of record requests a hearing before a Board of Patent Appeals and Interferences on the question whether any such false representation was contained in such statement. Such question shall be heard and determined, and determination thereof shall be subject to review, in the manner prescribed by subsection (d) for questions arising thereunder. No request made by the Administrator under this subsection for the transfer of title to any patent, and to prosecution for the violation of any criminal statute, shall be barred by any failure of the Administrator to make a request under subsection (d) for the issuance of such patent to him, or by any notice previously given by the Administrator stating that he had no objection to the issuance of such patent to the applicant therefor.

(f) Under such regulations in conformity with this subsection as the Administrator shall prescribe, he may waive all or any part of the rights of the United States under this section with respect to any invention or class of inventions made or which may be made by any person or class of persons in the performance of any work required by any contract of the Administration if the Administrator determines that the interests of the United States will be served thereby. Any such waiver may be made upon such terms and under such conditions as the Administrator shall determine to be required for the protection of the interests of the United States. Each such waiver made with respect to any invention shall be subject to the reservation by the Administrator of an irrevocable, nonexclusive, nontransferable, royalty-free license for the practice of such invention throughout the world by or on behalf of the United States or any foreign government pursuant to any treaty or agreement with the United States. Each proposal for any waiver under this subsection shall be referred to an Inventions and Contribution Board which shall be established by the Administrator within the Administration. Such Board shall accord to each interested party an opportunity for hearing, and shall transmit to the Administrator its findings of fact with respect to such proposal and its recommendations for action to be taken with respect thereto.

(g)   The Administrator shall determine, and promulgate regulations specifying, the terms and conditions upon which licenses will be granted by the Administration for the practice by any person (other than agency of the United States) of any invention for which the Administrator holds a patent on behalf of the United States. (h) The Administrator is authorized to take all suitable and necessary steps to protect any invention or discovery to which he has title, and to require that contractors or persons who retain title to inventions or discoveries under this section protect the inventions or discoveries to which the Administration has or may acquire a license of use.

(i) The Administration shall be considered a defense agency of the United States for the purpose of chapter 17 of title 35 of the United States Code.

(j) As used in this section—

(1) the term "person" means any individual, partnership, corporation, association, institution, or other entity;

(2) the term "contract" means any actual or proposed contract, agreement, understanding, or other arrangement, and includes any assignment, substitution of parties, or subcontract executed or entered into thereunder; and

(3) the term "made", when used in relation to any invention, means the conception or first actual reduction to practice of such invention.

CONTRIBUTIONS AWARDS

SEC. 306. (a) Subject to the provisions of this section, the Administrator is authorized, upon his own initiative or upon application of any person, to make a monetary award, in such amount and upon such terms as he shall determine to be warranted, to any person (as defined by section 305) for any scientific or technical contribution to the Administration which is determined by the Administrator to have significant value in the conduct of aeronautical and space activities. Each application made for any such award shall be referred to the Inventions and Contributions Board established under section 305 of this Act. Such Board shall accord to each such applicant an opportunity for hearing upon such application, and shall transmit to the Administrator its recommendation as to the terms of the award, if any, to be made to such applicant for such contribution. In determining the terms and conditions of any award the Administrator shall take into account—

(1) the value of the contribution to the United States;

(2) the aggregate amount of any sums which have been expended by the applicant for the development of such contribution;

(3) the amount of any compensation (other than salary received for services rendered as an officer or employee of the Government) previously received by the applicant for or on account of the use of such contribution by the United States; and

(4) such other factors as the Administrator shall determine to be material.

(b) If more than one applicant under subsection (a) claims an interest in the same contribution, the Administrator shall ascertain and determine the respective interests of such applicants, and shall apportion any award to be made with respect to such contribution among such applicants in such proportions as he shall determine to be equitable. No award may be made under subsection (a) with respect to any contribution—

(1) unless the applicant surrenders, by such means as the Administrator shall determine to be effective, all claims which such applicant may have to receive any compensation (other than the award made under this section) for the use of such contribution or any element thereof at any time by or on behalf of the United States, or by or on behalf of any foreign government pursuant to any treaty or agreement with the United States, within the United States or at any other place;

(2) in any amount exceeding $100,000, unless the Administrator has transmitted to the appropriate committees of the Congress a full and complete report concerning the amount and terms of, and the basis for, such proposed award, and thirty calendar days of regular session of the Congress have expired after receipt of such report by such committees.

APPROPRIATIONS

Sec. 307. (a) There are hereby authorized to be appropriated such sums as may be necessary to carry out this Act, except that nothing in this Act shall authorize the appropriation of any amount for (1) the acquisition or condemnation of any real property, or (2) any other item of a capital nature (such as plant or facility acquisition, construction, or expansion) which exceeds $250,000. Sums appropriated pursuant to this subsection for the construction of facilities, or for research and development activities, shall remain available until expended.

(b) Any funds appropriated for the construction of facilities may be used for emergency repairs of existing facilities when such existing facilities are made inoperative by major breakdown, accident, or other circumstances and such repairs are deemed by the Administrator to be of greater urgency than the construction of new facilities.

Approved July 29, 1958.

Documents of Authorization—II

Excerpt from Speech by President John F. Kennedy, May 25, 1961

Finally, if we are to win the battle that is now going on around the world between freedom and tyranny, the dramatic achievements in space which occurred in recent weeks should have made clear to us all, as did the Sputnik in 1957, the impact of this adventure on the minds of men everywhere, who are attempting to make a determination of which road they should take. Since early in my term, our efforts in space have been under review. With the advice of the Vice President, who is Chairman of the National Space Council, we have examined where we are strong and where we are not, where we may succeed and where we may not. Now it is time to take longer strides, time for a great new American enterprise, time for this nation to take a clearly leading role in space achievement, which in many ways may hold the key to our future on earth.

I believe we possess all the resources and talents necessary. But the facts of the matter are that we have never made the national decisions or marshalled the national resources required for such leadership. We have never specified long-range goals on an urgent time schedule, or managed our resources and our time so as to insure their fulfillment.

Recognizing the head start obtained by the Soviets with their large rocket engines, which gives them many months of lead time, and recognizing the likelihood that they will exploit this lead for some time to come in still more impressive successes, we nevertheless are required to make new efforts on our own. For while we cannot guarantee that we shall one day be first, we can guarantee that any failure to make this effort will make us last. We take an additional risk by making it in full view of the world, but as shown by the feat of astronaut Shepard, this very risk enhances our stature when we are successful. But this is not merely a race. Space is open to us now, and our eagerness to share its meaning is not governed by the efforts of others. We go into space because whatever mankind must undertake, free men must fully share.

I therefore ask the Congress, above and beyond the increases I have earlier requested for space activities, to provide the funds which are needed to meet the following national goals:

First, I believe that this nation should commit itself to achieving the goals, before this decade is out, of landing a man on the moon and returning him safely to the earth. No single space project in this period will be more impressive to mankind, or more important for the long-range exploration of space; and none will be so difficult or expensive to accomplish. We propose to accelerate the development of the appropriate lunar spacecraft. We propose to develop alternate liquid and solid fuel boosters, much larger than any now being developed, until certain, which is superior. We propose additional funds for other engine development and for unmanned explorations—explorations which are particularly important for one purpose which this nation will never overlook: the survival of the man who first makes this daring flight. But in a very real sense, it will not be one man going to the moon—if we make this judgment affirmatively, it will be an entire nation. For all of us must work to put him there.

Secondly, an additional 23 million dollars, together with 7 million dollars already available, will accelerate development of the Rover nuclear rocket. This gives promise of some day providing a means for even more exciting and ambitious exploration of space, perhaps beyond the moon, perhaps to the very end of the solar system itself.

Third, an additional 50 million dollars will make the most of our present leadership, by accelerating the use of space satellites for world-wide communications.

Fourth, an additional 75 million dollars—of which 53 million dollars is for the Weather Bureau—will help give us at the earliest possible time a satellite system for world-wide weather observation.

Let it be clear—and this is a judgment which the Members of the Congress must finally make—let it be clear that I am asking the Congress and the country to accept a firm commitment to a new course of action—a course which will last for many years and carry very heavy costs: 531 million dollars in fiscal 1962—an estimated seven to nine billion dollars additional over the next five years. If we are to go only half way, or reduce our sights in the face of difficulty, in my judgment it would be better not to go at all.

Now this is a choice which this country must make, and I am confident that under the leadership of the Space Committees of the Congress, and the Appropriating Committees, that you will consider the matter carefully.

It is a most important decision that we make as a nation. But all of you have lived through the last four years and have seen the significance of space and the adventures in space, and no one can predict with certainty what the ultimate meaning will be of mastery of space.

I believe we should go to the moon. But I think every citizen of this country as well as the Members of the Congress should consider the matter carefully in making their judgment, to which we have given attention over many weeks and months, because it is a heavy burden, and there is no sense in agreeing or desiring that the United States take an affirmative position in outer space, unless we are prepared to do the work and bear the burdens to make it successful. If we are not, we should decide today and this year.

This decision demands a major national commitment of scientific and technical manpower, material and facilities, and the possibility of their diversion from other important activities where they are already thinly spread. It means a degree of dedication, organization and discipline which have not always characterized our research and development efforts. It means we cannot afford undue work stoppages, inflated costs of material or talent, wasteful interagency rivalries, or a high turnover of key personnel.

New objectives and new money cannot solve these problems. They could in fact, aggravate them further—unless every scientist, every engineer, every serviceman, every technician, contractor, and civil servant gives his personal pledge that this nation will move forward, with the full speed of freedom, in the exciting adventure of space.

From *Public Papers of the Presidents of the United States: John F. Kennedy, Containing the Public Messages, Speeches, and Statements of the President, January 20 to December 31, 1961* (Washington, DC: United States Government Printing Office, 1962), 403–4.

Documents of Authorization—III

1962 NASA AUTHORIZATION

Tuesday, May 9, 1961
House of Representatives
Committee on Science and Astronautics,
Washington, D.C.

[H.R. 6874, 87th Congress, 1st sess.]

A BILL To authorize appropriations to the National Aeronautics and Space Administration for salaries and expenses, research and development, construction of facilities, and for other purposes.

Be it enacted by the Senate and House of Representatives of the United States of America in Congress assembled, That there is hereby authorized to be appropriated to the National Aeronautics and Space Administration for the fiscal year 1962 the sum of $1,376,900,000, as follows:

(a) For "Salaries and expenses", $199,286,000.
(b) For "Research and development", $1,023,539,000.
(c) For "Construction of facilities", $139,075,000, as follows:
(1) Langley Research Center, Hampton, Virginia, $3,980,000.
(2) Ames Research Center, Moffett Field, California, $5,680,000.
(3) Lewis Research Center, Cleveland, Ohio, $3,590,000.
(4) Goddard Space Flight Center, Greenbelt, Maryland, $9,212,000.
(5) Wallops Station, Wallops Island, Virginia, $6,313,000.

(6) Jet Propulsion Laboratory, Pasadena, California, $3,642,000.

(7) Marshall Space Flight Center, Huntsville, Alabama, $12,891,000.

(8) Atlantic Missile Range, Cape Canaveral, Florida, $49,583,000.

(9) Pacific Missile Range, Point Agruello, California, $998,000.

(10) At locations to be determined, for nuclear rocket test and maintenance facilities, $15,000,000.

(11) Various locations: Tracking facilities, including land acquisition; propulsion development facilities; sounding rocket facilities; and damage repair construction, $23,186,000.

(12) Facility planning and design not otherwise provided for, $5,000,000.

(d) For emergency "Construction of facilities", in accordance with the provisions of section 3, $15,000,000.

(e) Appropriations for "Research and development" may be used for any items of a capital nature (other than acquisition of land) which may be required for the performance of research and development contracts: *Provided,* That none of the funds appropriated for "Research and development" pursuant to this Act may be used for construction of any major facility, the estimated cost of which, including collateral equipment, exceeds, $250,000, unless the Administrator or his designee notified the Committee on Science and Astronautics of the House of Representatives and the Committee on Aeronautical and Space Sciences of the Senate of the nature, location, and estimated cost of such facility.

(f) When so specified in an appropriation Act any amount appropriated for "Research and development' and for "Construction of facilities" may remain available without fiscal year limitation.

(g) Appropriations other than "Construction of facilities" may be used, but not to exceed $20,000, for scientific consultations or extraordinary expenses upon the approval or authority of the Administrator and his determination shall be final and conclusive upon the accounting officers of the Government.

SEC. 2. Authorization is hereby granted whereby any of the amounts prescribed in subparagraphs (1), (2), (3), (4), (5), (6), (7), (8), (9), (10), (11), or (12) of subsection 1 (c) may, in the discretion of the Administrator of the National Aeronautics and Space Administration, be varied upward 5 per centum to meet unusual cost variations, but the total cost of all work authorized under such subparagraphs shall not exceed a total of $139,075,000.

SEC. 3. The sum authorized by subsection 1 (d) for emergency "Construction of facilities", and any amount not to exceed $5,000,000 of the funds appropriated pursuant to subsection 1 (c) hereof, shall be available for expenditure to construct, expand, or modify laboratories and other installations if (1) the Administrator determines such action to be necessary because of changes in the national program of aeronautical and space activities or new scientific or engineering developments and (2) he determines that deferral of such action until the enactment of the next authorization Act would be inconsistent with the interest of the Nation in aeronautical and space activities. The funds so made available may be expended to acquire, construct, convert, rehabilitate, or install permanent or temporary public works, including land acquisition, site preparation, appurtenances, utilities, and equipment. No portion of such sums may be obligated for expenditure and expended to construct, expand, or modify laboratories and other installations until the Administrator or his designee has transmitted to the Committee on Science and Astronautics of the House of Representatives and to the Committee on Aeronautical and Space Sciences of the Senate a written report containing a full and complete statement concerning (1) the nature of such construction, expansion, or modification, (2) the cost thereof, including the cost of any real estate action pertaining thereto, and (3) the reason why such construction, expansion, or modification is necessary in the national interest. No such funds may be used for any construction, expansion, or modification if authorization for such construction, expansion, or modification previously has been denied by the Congress.

(Whereupon the committee proceeded to further business.)

From Hearings before the Committee on Science and Astronautics, U.S. House of Representatives, 87th Cong., 1st sess. on HR 6874), P. L. 87-98, May 9, 1961,1033–34.

32
The High Dam at Aswan

Egypt

DID YOU KNOW . . . ?

➤ There are two dams: the Low Dam and the High Dam.

➤ The Low Dam was built by Mohammed Ali, the founder of modern Egypt, in 1843.

➤ The High Dam was financed by Soviet Russia and built under the supervision of its engineers.

➤ The High Dam was built between 1960 and 1970 at an estimated cost of $450 million.

➤ The High Dam is 12,562 feet (3,829 meters) long and 364 feet (111 meters) high.

➤ Thousands of people were moved from a huge area that was flooded to create the water-storage lake.

➤ Only four percent of Egypt's soil was naturally irrigated by the Nile prior to the High Dam.

➤ Egypt is the most populous country in the Arab world, and the second-most populous in Africa, after Nigeria.

➤ Agriculture accounts for 17 percent of Egypt's gross domestic product.

The Aswan High Dam illustrates the role of hydropolitics in today's world. *Hydropolitics* was added to the international lexicon after the dam was proposed and bid on—first by the Egyptians, then jointly by the British and Americans, then by the World Bank, and finally by the Soviet Union. The story of the Aswan Dam may well prefigure the pattern of hydropolitics for the ensuing century. Therefore, it is important to understand the political aspects of the High Dam at Aswan.

The Aswan High Dam. © The Art Archive / Dagli Orti.

HISTORY

The Nile River in Africa is one of the world's longest rivers, and is revered as a fountain of life that has sustained Egypt for thousands of years. It has always been a source of transport, commerce, irrigation, and inspiration, nourishing the land through which it passes and bringing silt to riverbank lands during summer floods and renewed nutrients as it arrives downstream. But floods were just as often disastrous, as were the alternating times of drought. Even if the flow of water was regular and steady, it was helpful to only four percent of Egypt's soil, nourishing thin ribbons of earth on each side of the river but leaving the rest of the country a desert.

Interventions in the Nile can be found as far back as 1843 when the Albanian founder of modern-day Egypt, Mohammed Ali, tried to improve the river with a *barrage* (an artificial obstruction in a waterway, especially for the purpose of irrigation). As the population of Egypt grew, so did demands upon the river, and there were attempts to construct dams. From 1898 to 1902, the Egyptians and British worked together to build a sizable dam on the Nile River at Aswan, the country's southernmost city, to collect water for irrigation and power generation. Referred to as the Aswan Low Dam, it measures about 7,000 feet (2,100 meters) long and, after additional heightening in 1907–12 and 1929–34, it is now 125 feet (38 meters) high. The Low Dam has 180 sluices through which silt-laden water can pass, which proved to be a huge boon to the land near the riverbanks.

But the river proved stronger than man, and the need for another, larger dam was already apparent when the first one was completed in 1933.

The Low Dam is dwarfed by a larger dam, known as the Aswan High Dam, built between 1960 and 1970, about four miles (six kilometers) upstream at an estimated cost of more than $1 billion. The newer structure gave a substantial boost to the Egyptian economy, played a major role in the cold war, and has had complex effects—both negative and positive—on the surrounding environment.

Even before it was built, the politics of building a dam were affecting the geopolitical scene during the cold war. Britain and the United States had pledged to finance its construction but changed their minds, perhaps because of a growing closeness between Egypt and the Soviet Union. In retaliation, Egyptian president Gamal Abdel Nasser nationalized the Suez Canal in July 1956, declaring that canal tolls would be used to fund the cost of building a dam.

Later that year Israel, and then Britain and France, invaded Egypt. International reaction included Soviet threats of intervention, but all forces were eventually withdrawn. Nasser found that his political stature had improved dramatically as a result of the incident. Ties between the Soviets and Egypt were strengthened, leading to an agreement by the Soviet Union to help finance the dam and to provide technical specialists to help with its construction (reproduced here as Document I). (Note that the documents refer to the "United Arab Republic," the name of a union formed by Egypt and Syria from 1958 to 1961, and then, for another 10 years, to Egypt alone.) Nikita Khrushchev requested an invitation from President Nasser to the opening ceremony for the High Dam, becoming the first Soviet general secretary to visit an Arab or African country. When Nasser died in 1970, the Egyptian portion of the dam's reservoir was named after him.

CULTURAL CONTEXT

The High Dam's impact on the Egyptian economy has been immense, permitting the reclamation of huge amounts of land for agriculture, raising the country's electricity-production capacity by more than 2,100 megawatts, controlling the Nile's flooding, and improving navigation on the river. The storage lake behind the dam was stocked with fish, fostering development of the local fishing industry.

Some of the dam's effects on the environment are harmful. The fertility of the lands along the river diminished as a result of flood-control measures, as the area received far less of the rich silt formerly deposited by annual floods. Also diminished were the nutrients feeding into the Mediterranean Sea from the Nile, which led to a sharp falloff in coastal fishing, although the catch seemed to have rebounded by the late 1980s. The decline in the flow of freshwater contributed to increased salinity in the soil of the Nile delta, which also fell prey to erosion since the river no longer deposited as much sediment at its mouth. Finally, conditions favoring the spread of the potentially fatal disease schistosomiasis, caused by blood flukes carried by snails that previously were washed out to sea, now flourished in the reservoir and irrigation channels.

The dam had a major social cost: tens of thousands of people were moved from the huge area that was flooded to create the water-storage lake. There was also a related archaeological impact: the area of inundation contained temples and monuments dating from ancient times, only some of which could be relocated in the 1960s before the waters covered the remainder.

PLANNING

Planning for the High Dam arose in response to demographics: there were more mouths to feed every year than the arid land could support. Agricultural production for the region depended entirely on the Nile River and its contiguous land. To meet the basic needs of the growing population, it was decided to build a higher dam that would store more water for reliable irrigation, which would create additional arable land.

First proposed in 1948, with a price tag of approximately $1 billion, the project was presented to the World Bank, which declined. Why did the World Bank reject the offer? Water projects of this magnitude are very costly. Egypt was able to get started on its own because it managed to collect some taxes and reduce government expenses. Thus, it had the money to begin some aspects of the project, but needed foreign assistance to complete the program. The World Bank had helped Egypt get through the trauma of nationalizing the Suez Canal. During that event, Nasser in effect took the canal away from its principal investors. To calm the situation, the World Bank compensated shareholders of the Suez Canal Company, so the canal became Egyptian but the Suez Canal Company was transformed into the Suez Financial Company. The shareholders were well cared for thanks to arrangements negotiated by Jean-Paul Calon, general counsel of the Suez Canal Company. There is speculation that the World Bank, having already been involved in an Egyptian program that had ramifications for the organization, this time decided to decline the offer of participation.

The United States offered to help with a loan of $270 million, in part to mend strained relations with Nasser. However, in July 1956 the United States rejected the deal via a letter delivered to the Egyptian ambassador in Washington, DC. Nasser knew the possibility of U.S. participation was fragile, so he had other alternatives under consideration, such as a plan to use revenues from the Suez Canal to pay for the Aswan High Dam project.

Why did the Egyptians turn to the Soviet Union for help? It was an obvious move because the Egyptians were well aware that the Soviets were eager to develop influence in the African continent. The High Dam would be a very visible project, the first Soviet-assisted project in Africa. As early as June 1956, Soviet Foreign Minister Dmitri Shepilov contacted Nasser to float an offer of 400 million rubles; Nasser laughed at the figure because he knew it was inadequate. One month later, another offer was discussed, but it took quite a bit longer to reach an agreement.

The technology was easily available, with the science of the Russian Zuk Hydroproject Institute. A decision was made at the highest levels of Soviet political circles to construct the High Dam at Aswan as a "gift" in 1958, and the contract as here presented, dated December 27, 1958, is an agreement between the USSR and the UAR mentioning "friendly relations . . . and the full respect for the dignity and national sovereignty of each of the two countries."

BUILDING

Construction began in 1960; by 1964 the reservoir had begun to fill, and it reached its full capacity in 1976. The High Dam is 12,562 feet (3,829 meters) long at its crest and 364 feet (111 meters) high. Made of earth and rock fill on a core of cement and clay, the dam is over 3,000 feet (almost 1,000 meters) thick at its base and 130 feet (40 meters) at the top. The reservoir created by the stored water is one of the largest in the world, with a capacity of approximately 131 million acre-feet (162 billion cubic meters) of water. The lake, measuring about 300 feet (90 meters) deep and 300 miles (480 kilometers) long, extends into Sudan, which by treaty with Egypt is allocated a portion of the Nile's annual discharge. The overall cost for the High Dam was $450 million, which was paid back in the astonishingly short time of two years using annual revenues generated.

The High Dam brought with it many benefits, including electricity for every city, town, and village in Egypt, as well as agricultural improvement through irrigation, fewer disastrous floods, and better navigation on the powerful Nile. But there were negatives as well. The portion of the Nile located downstream of the dam lost much of its power. Prior to the dam's construction, the river brought along 12 million tons of rich silt full of nutrients; following completion of the dam, silt now collects behind the dam. The silt used to bring fertilizing nitrogen to the land; now the nutrients must be added via lime-nitrate fertilizers. As is the case with so many of the world's dams, salinity is also a major problem.

Flooding to create a reservoir entails difficult decisions regarding relocation of villages; such was the case when developing the Tennessee Valley Authority in the United States and the Three Gorges Dam in China. But Egypt's situation was especially difficult since the areas designated for flooding contained treasures from the days of the pharaohs as well as ancient Nubian settlements. At enormous cost, a massive relocation program began, moving Egyptian relics and monuments, temples, and tombs. The cost for transferring objects from just one site, Abu Simbel, was $40 million; in the end, many relics were lost in the waters that quickly filled the new Lake Nasser. Also lost were islands in the Nile River itself; prior to the dam there were 150 islands, but today only 36 remain.

The Aswan Low Dam. Courtesy of the Library of Congress.

IMPORTANCE IN HISTORY

Hydropolitics has always been a key factor in the history of Egypt. Napoleon observed, at a time when the French were occupying Egypt, "If I were to rule a country like Egypt, not a single drop of water would be allowed to flow into the Mediterranean!" (Biswas, 25).

The High Dam suffered from bad press and challenging public relations. When the United States pulled out of the construction project in 1956, American reporters and others wrote environmental reports that questioned many aspects of the dam. Certainly concern about the environment was becoming recognized as a global issue, and Aswan was an easy target. However, in the 1980s, the Canadian International Development Agency (CIDA) reported that the Aswan High Dam actually received high marks for performance and benefits.

Throughout the centuries, Egyptians have looked to the Nile for their needs: agriculture, transport, communications, and, more recently, electric power. In the future, as the population continues to grow, can the Egyptians continue to depend on the Nile, or will they have to look elsewhere? In California, the Imperial Valley now suffers from increased salinity in the soil. In the Netherlands and in Australia's Snowy Mountains project, the same problem emerges. Is this an opportunity for a special international research program to design alternative methods of reducing salinity before water reaches the land to be irrigated?

Because 40 percent of all irrigated land is damaged by salinity, the outcomes from such a program would add immeasurably to world sustainable agriculture.

While the World Bank opted not to get involved in building the High Dam, it has built more dams than any other organization in history: 527 dam-building loans totaling $58 billion (1993 dollars). Altogether, more than 604 dams have been built in 93 countries with World Bank funding. In the late 1990s the Manibeli Declaration, endorsed by groups from 44 countries and many nongovernmental organizations, suggested a halt to World Bank funding of more dams until a worldwide body was established to review environmental implications and articulate policies that will heighten the benefits and diminish unwanted side effects to the regions in which dams are built.

The time is approaching to develop a comprehensive study of macroengineering projects that would be most effective for supplying water to the Middle East and North Africa. Is an "ever-normal water supply" a goal that the world can now work toward?

FOR FURTHER REFERENCE

Books and Articles

Biswas, Asit K. "Aswan Dam Revisited: The Benefits of a Much Maligned Dam." *D+C (Development and Cooperation)*, Monograph No. 6 (November/December 2002): 25–27. http://www.inwent.org/E+Z/1997-2002/de602-11.htm.

Fahim, Hussein M. *Egyptian Nubians: Resettlement and Years of Coping.* Salt Lake City: University of Utah Press, 1983.

Failure, Jacob M. *The Everything Middle East Book: The Nations, Their Histories, and Their Conflicts.* Avon, MA: Adams Media, 2004.

Gorlov, Alexander. "Tidal Power." *Encyclopedia of Ocean Sciences.* London: Academic Press, 2001.

Hourani, Albert. *A History of the Arab Peoples.* New York: Warner Books, 1991.

Macaulay, David. *Building Big.* Boston: Houghton Mifflin, 2000.

Steele, James. "The Effect of the Aswan High Dam upon Village Life in Upper Egypt." IASTE 2nd International Conference. "First World-Third World: Duality and Coincidence in Traditional Dwellings and Settlements." University of California at Berkeley, CA, USA, October 1990.

Whittington, Dale, and Giorgio Guariso. *Water Management Models in Practice: A Case Study at the Aswan High Dam.* Amsterdam and New York: Elsevier Scientific Publishing, 1983.

Internet

For summary information about the Aswan Dam, see http://www.thebestlinks.com/Aswan_High_Dam.html. (Note the link to an article by Sayed El-Sayed and Gert L. van Dijken relating to the effects of the dam on the Mediterranean.)

For a drawing showing both the Low Dam and the High Dam, as well as the cooperative efforts between Egypt and the Soviet Union, see http://carbon.cudenver.edu/stc-link/aswan/organi.htm.

To read a report from the International Rivers Network on the World Bank's role in funding dams, see http://www.irn.org/pubs/wp/damming.html.

For statistics on increased acreage and a helpful chart illustrating land reclamation as a result of the Aswan Dam, see http://www.fao.org/ag/agl/swlwpnr/reports/y_nf/egypt/e_lcover.htm.

Documents of Authorization—I

Agreement with the U.S.S.R. Concerning the Project for the Implementation of the High Dam, Approved by Decree No. 8 of January 9, 1959 (Official Gazette No. 2n.)

December 27, 1958

The governments of the U.A.R. and the U.S.S.R., impelled by the friendly relations which exist between them, and in their desire to strengthen economic and technical cooperation between them on a basis of equality and non-intervention in internal affairs and full respect for the dignity and national sovereignty of each of the two countries, and in view of the great importance of the High Dam project at Aswan to the national economy of the U.A.R., have agreed upon the following:

1. In answer to the desire of the Government of the U.A.R. to develop its national economy, the Government of the U.S.S.R. expresses its readiness to cooperate with the Government of the U.A.R. in constructing the first stage of the High Dam at Aswan.

The first stage comprises the construction of the front part of the main dam, with a height of 50 meters and a length of 600 meters, and the downstream coffer dam with a height of 27 meters and a length of 600 meters; together with work on the diversion of the waters and the sluices, as well as the supply of equipment, and instruments necessary for this work. The two parties will agree on measures in the course of study of the details, or whenever the need arises in the course of implementation.

The first stage also includes projects for converting the basins and the projects of irrigation and land reclamation for the purpose of utilizing the surplus waters resulting from this stage. The volume of assistance offered by the Soviet side will be determined by the agreement of the two parties as regards the implementation of these projects, after the U.A.R. has completed the studies necessary for the execution of these projects.

It is agreed that all the expenses which will be assumed by the Soviet side, whether for the construction of the Dam itself or for implementation of the works of irrigation and the conversion of the basins, imputed to the loan, will be covered within the limit of the loan offered, according to Article 5 of this agreement.

2. In implementation of the cooperation stipulated in Article 1:

a) The Government of the U.S.S.R. undertakes, through the medium of Soviet organisms, to prepare for the execution of the work as well as the necessary research which will be agreed upon by the two parties, with the aim of introducing amendment or modification to the details of the drawings whenever it is deemed necessary, conforming to the agreement with the competent U.A.R. authorities, on condition that this should take place in the shortest time possible and according to the conditions and hydraulic specifications and the basic information drawn up by the U.A.R., on condition that these modifications agree with the final plans for the Dam.

b) The Government of the U.S.S.R. will supply the sluices, machinery and equipment—with a supply of necessary spare parts—as well as the material required for the construction and functioning of the first stage, and the projects related to it, in a perfect manner—material which is not available in the U.A.R.

c) The Government of the U.S.S.R. will offer the technical aid necessary for construction. To accomplish this it will send the required number of Soviet experts to the U.A.R., according to the agreement concluded between the two parties.

3. The Government of the U.A.R. will create a special organism for the management of the project. It will be entrusted with questions of an administrative, technical and financial nature. The implementation of the work which the Government of the U.A.R. will demand within the framework of the first stage will be entrusted to contractors agreed upon by the two parties, and this on the basis of the employment of Soviet equipment and the cooperation of Soviet specialists and experts.

The contract to be concluded between the Government of the U.A.R. and the contractors will include, apart from the plans and specifications of the work, all the obligations of the contractors and the services and facilities which the Government of the U.A.R. will extend.

The above-mentioned organism will supervise the contractors to make sure of the implementation of the obligations required of them according to the contract, and this organism will be charged with the services and facilities stipulated in the article.

4. The Soviet organisms will be responsible for the technical management relating to the construction work of the first stage of the High Dam at Aswan and the execution of all work in a perfect manner, and will also be responsible for the fitting of the machines and their operation within the time limit agreed upon by both parties, and on condition that the special organism and the contractors mentioned in Article 3 fulfill their obligations as to the operations of research, fitting and construction, according to the implementation programme with regard to similar operations following the agreement between the two countries.

To this effect the Soviet organisms have delegated to the U.A.R. a highly efficient Soviet expert as well as the necessary number of engineers, technicians and skilled Soviet workers, according to the agreement concluded between the two parties.

The Soviet expert, with the collaboration of the organism mentioned in Article 3, will attend to the organisation of work between the Soviet and U.A.R. specialists for the technical supervision of the said work.

5. The Government of the U.S.S.R. offers the U.A.R. Government a loan of 400 million roubles (the rouble is equal to 0.222168 grammes of pure gold) to cover cost of the operations to be carried out by the Soviet organisms for all matters relating to the execution of the projects as well as the studies and researches, the delivery of machinery, equipment and material on the basis of Soviet port prices free of charge (FOB), and the travel fares of Soviet experts from the U.S.S.R. to the U.A.R. and back, in accordance with Article 2.

In the event the total value of the above-mentioned machinery, sluices, equipment and material estimated on the basis of Soviet port prices free of charges, plus the transportation expenses of Soviet specialists and the expenses of Soviet organisms which comprise the technical assistance included in the framework of this agreement, exceeds the amount of the loan, that is, 400 million roubles, the Government of the U.A.R. will pay the excess amount to the U.S.S.R. by supplying it with U.A.R. merchandise, according to the trade and payments agreement in force between the U.A.R. (Egyptian Region) and the U.S.S.R.

6. The Government of the U.A.R. will reimburse the utilized sums of the loan, granted to it in accordance with Article 5 of this agreement, in twelve equal annuities commencing one year after the complete execution of the work on the first stage of the High Dam at Aswan and the filling up of the basin, on condition this is not later than January 1, 1964. The date of the utilization of the loan with regard to machinery, equipment and material will be that of the acknowledgment of the receipt of consignment. For expenses relating to the plans, studies and research, and the expenses of the delegation of Soviet experts and specialists sent to the U.A.R., the date of the utilization of the loan will be that of the vouchers.

The rate of interest on the loan is 2.5 per cent per annum, starting from the date of the utilization of each part of the loan, and will be settled in the course of the first three months of the year following that of their falling due.

7. The Government of the U.A.R. will reimburse the loan and its interest by depositing in Egyptian pounds (the rate of the Egyptian pound being 2.55187 grammes of pure gold) the sums due in a special account opened at the U.A.R. Central Bank (Egyptian Region) on behalf of the U.S.S.R. State Bank.

The price of the rouble with regard to the Egyptian pound is estimated on the basis of the gold balance between the two foreign currencies on the day of payment.

The Soviet organisms will use all the sums deposited to their account to buy articles from the U.A.R. (Egyptian Region) in accordance with the trade and payments agreement in force between the U.A.R. (Egyptian Region) and the U.S.S.R.

It is also possible to transfer all sums deposited in this account to sterling pounds or to whatever other transferable foreign exchange following agreement between both parties. If the rate of the Egyptian pound is changed, the evaluation of the balance opened on behalf of the U.S.S.R. State Bank will be referred to the date of this change of the Central Bank of the Egyptian Region of the U.A.R. in accordance with the modifications with the gold contained in the Egyptian pound.

8. The Central Bank of the Egyptian Region of the U.A.R. and the U.S.S.R. State Bank will open special accounts to register the operations relating to the loan offered in accordance with this agreement and its execution, as well as the interest due by virtue of this agreement; and the two banks will agree upon the financial and technical measures necessary for its implementation.

9. The Government of the U.A.R. will pay for the Soviet party all the expenses incurred by the Soviet organisms relative to the expenses of food and accommodation, as well as the travel expenses of Soviet experts—inside the U.A.R.—delegated to offer their technical services according to this agreement and by virtue of the conditions stipulated in the special contracts. The settlement of these expenses will be made by adding these sums in Egyptian pounds to the "collect" account opened at the Central Bank of the U.A.R. (Egyptian Region) in favor of the U.S.S.R. State Bank, by virtue of the payments agreement in force between the U.A.R. (Egyptian Region) and the U.S.S.R.

10. The supply of machinery, equipment and material, as well as the preparations for the project, the studies and research and the dispatch of Soviet specialists to the U.A.R., will take place in accordance with the agreements to be concluded between the U.A.R. and the competent Soviet organisms, in conformity with Article 2.

The contracts will determine in particular the sums, dates, prices and guarantees concerning each kind of material and machinery, and the responsibility of each party with regard to the circumstances independent of the will of each of them, as well as the violation of the provisions of invention patents, and the provisions and conditions relating to the implementation of obligations of the Soviet party according to this agreement.

The cost of equipment, machinery and material delivered to the U.A.R. by the U.S.S.R. by virtue of this agreement will be determined on the basis of prices in effect on world markets.

11. Without prejudice to the provisions of Article 5 concerning the utilization of the loan for covering the cost of material, machinery and equipment, Soviet ports delivery free of charge on the basis of FOB prices, the furnishing of equipment, material and machinery presented by the U.S.S.R. will be covered by an insurance policy (CIF) in the ports of the U.A.R. (Egyptian Region).

The expenses for shipment and insurance will be paid separately on the basis of the actual value, in accordance with the trade and payments agreement in force between the U.A.R. (Egyptian Region) and the U.S.S.R.

The maritime transportation of the above-mentioned equipment, machinery and material will be effected in accordance with the agreement of maritime transportation concluded between the two countries on September 18, 1958.

12. In case of litigation or contention between the competent authorities in the U.A.R. and the Soviet organisms concerning any question relating to this agreement or its execution, representatives of the governments of the U.A.R. and the U.S.S.R. will consult together so as to come to an understanding on the subject of the above-mentioned contention or litigation.

This agreement will be in force after its ratification, on condition that this is done at the soonest possible time. It will become effective with the exchange of documents of ratification in Moscow.

Documents of Authorization—II

Agreement Concerning the Economic and Technical Assistance Accorded by the U.S.S.R. to the U.A.R., to Complete the High Dam at Aswan in its Final Form

August 27, 1960

The Government of the United Arab Republic and the Government of the Union of Soviet Socialist Republics in their desire to extend the friendly relations between the two countries.

And the creation of economic and technical cooperation between them on a basis of equality, non-interference in internal affairs and complete respect for the national dignity and sovereignty of each of the two countries.

And owing to the great economic, national importance of the creation of the Aswan High Dam with regard to the Government of the United Arab Republic.

And in execution of the agreements concluded in the letters exchanged on January 15 and 17, 1960, between the President of the United Arab Republic and the Premier of the Union of Soviet Socialist Republics concerning the participation of the Soviet Union in the completion of the High Dam project at Aswan Have decided on the following:

Article 1. The Government of the Union of Soviet Socialist Republics, inspired by the desire to assist in the economic development of the United Arab Republic and in answer to the will of the Government of the United Arab Republic, has agreed to collaborate with the Government of the United Arab Republic in completing the final stage of the High Dam project at Aswan.

This stage includes the following work:

a) Completing the construction of the Dam in its final form with an overall height of 111 meters from the river bed.

b) Installing a hydro-electric power station in the course of the diversion canal on the eastern bank of the river, with a capacity of 2.1 million kw.

c) Constructing an overflow channel allowing a discharge of 200 million cubic meters daily, so that the maximum level of storage waters does not exceed 182 meters.

d) Establishing two transmission lines transmitting electric power from the High Dam in Aswan to Cairo, each with a tension of 400–500 kilovolts and a length of 900 kilometers, including three or four transformer stations. Other transmission lines will also be established with a tension of 132–220 kilovolts and approximately 1,000 kilometers long, including from 10 to 20 transformer stations.

e) Projects of irrigation and reclamation in the lands depending on the High Dam waters, whose area is approximately two million feddans, including the lands depending on the waters resulting from the first stage of the High Dam.

It should be noted that this information is preliminary and will be agreed upon by the two parties in the course of the discussion concerning the details of the design or when the need arises during the process of execution.

Article 2. In realization of the collaboration mentioned in Article 1 of this agreement, the Government of the U.S.S.R. undertakes the following:

a) It undertakes, by means of the Soviet organisations, to draw up the complete designs, blueprints, specifications and the list of quantities, in conformity with the hydraulic circumstances and information supplied by the U.A.R., and when the need arises it undertakes to carry out the necessary researches and studies. Furthermore it undertakes to draw up the plans for implementing the work necessary for the completion of the High Dam in its final form, in accordance with the agreement reached.

All these aforementioned jobs must be completed in the shortest possible time, thereby making it possible to complete the construction of the Dam at a level of 155 meters by 1967 and completing it in its final form in 1968.

b) Designing, manufacturing, supplying and installing all the sluice gates with the mechanical and electrical annexes necessary for their operation; and also supplying all the necessary spare parts.

c) Designing, manufacturing, supplying, installing, testing and operating all the equipment necessary for the hydro-electric power station and the sluice-gates necessary for it, so that the station units and sluice-gates will be completed, installed and ready for operation according to the following schedule.

First—the first three units	1967
Second—the second three units	1968
Third—the third three units	1969
Fourth—the fourth three units	1970

Also designing, manufacturing, supplying, installing and testing all the equipment necessary for the two transmission lines each with a tension of 400/500 kilovolts and 900 kilometers long, going from Aswan to Cairo (with the exception of the construction and installation of pylons for supporting the electric line), including three or four transformer stations equipped with computers for regulating the tension. Furthermore it will install transmission lines with a tension of 132/220 kilovolts and approximately 1,000 kilometers long, including from 10 to 12 transformer stations comprising communications and precautionary instruments and centres for distributing pressure which operate according to the system of sonic waves. All this will be undertaken according to the agreement reached by the two parties in a manner that will ensure the operation of a line with a tension of 400/500 kilovolts which is approximately 1,000 kilometers long, in the course of 1967. As for the second line with a tension of 400/500 kilovolts, it will begin operating in 1968.

Furthermore it undertakes to provide sufficient quantities of spare parts for all the above-mentioned equipment.

d) Supplying and installing the additional construction instruments necessary to complete the final stage of the High Dam project, in addition to the materials necessary for the completion of the High Dam project and which are unavailable in the United Arab Republic, according to the dates agreed upon.

e) Extending the technical assistance necessary for construction. For this purpose the required number of Soviet experts will be sent, according to the agreement reached by the two parties.

f) Extending the technical assistance necessary for training Arab technicians, in the Soviet Union or the United Arab Republic, with regard to the work connected with the High Dam project, if the Government of the United Arab Republic so desires.

g) The Soviet organisations will undertake the work requiring special experience, and whose nature and the basis on which they will be carried out will be determined in the letters exchanged between the two parties at the signing of this agreement.

h) Carrying out the necessary tests to insure the soundness of the Dam and also carrying out final tests on the sluice gates and the hydro-electric power station when the storage waters reach their maximum level, which is 182 meters. It is understood that this will be realized not later than 1975.

i) Supplying, installing and operating the mechanical and electrical equipment necessary for the irrigation and land reclamation projects mentioned in Article 1 of this agreement.

It is understood that the dates mentioned above are founded on the basis that the U.A.R. side will furnish the required information and will undertake the obligations stipulated in this agreement and all that will be agreed upon by the two parties.

Article 3. The U.S.S.R. Government will grant the Government of the U.A.R. a loan of approximately 900 million roubles (the rouble contains 0.222168 grammes of pure gold) to cover the expenses of the Soviet organisations with regard to the designs of the project, research work, studies, and supplying and installing the sluice-gates and hydro-electric power generating units, equipment and materials according to Article 2 of this agreement on the basis of Soviet ports delivery prices (FOB), and the transportation expenses of Soviet experts supplied to offer technical assistance according to this agreement, to and from the United Arab Republic.

If the total expenditure mentioned above exceeds the loan, fixed in this Article at 900 million roubles, the Government of the U.A.R. will reimburse the balance to the Government of the U.S.S.R. in the form of merchandise from the United Arab Republic, in conformity with the trade and payments agreements in force between the United Arab Republic (Egyptian Region) and the Government of the U.S.S.R.

Article 4. The U.A.R. Government will reimburse the sums used from the loan granted to it in accordance with Article 3 of this agreement in twelve equal annuities, starting one year after the completion of the High Dam in its final form, and the implementation of the hydro-electric power station which will be ready to generate not less than 10 million kilowatts of electricity, provided that this is not done later than January 1st, 1970. As for the part of the loan which will be utilized from January 1st, 1969, to execute the remaining parts of the project, it will be reimbursed according to the same conditions of payment, one year after the execution of all these jobs, provided this is not done later than January 1st, 1972.

The rate of interest on the loan is 2.5 per cent annually. Interest is payable from the date of the utilization of every part of the loan, and is to be paid in the course of the first three months of the year following upon the year in which they fall due. The date of the utilization of the loan with regard to equipment, instruments, and materials is considered as the date of the shipping voucher. As regards the expenditures connected with the work involved in designing, research and studies, and also the expenses incurred by sending Soviet experts to the United Arab Republic, the date of the utilization of the loan is considered as the date of the bills concerning these matters.

Article 5. Apart from the stipulations present in this agreement, the stipulations of Articles (3), (4), (7), (8), (9), (10), (11), (12) of the agreement concluded between the Union of Soviet Socialist Republics and the U.A.R. in which the U.S.S.R. undertakes to extend technical and economic assistance with regard to the construction of the first stage of the High Dam at Aswan and which was signed on December 27, 1958, are in force. Also in force are the stipulations and conditions mentioned in the letters exchanged between the two parties on the same date as the signing of the said agreement, and will be effective as of the date of this agreement.

Article 6. This agreement shall be ratified within the shortest time possible and is considered as an integral part of the exchange of the ratification documents in Cairo.

This agreement was written in Moscow on August 27, 1960, in two copies, one in Arabic and one in Russian, each of which has the same legal power.

MOUSSA ARAFA,
(representing the Government of the United Arab Republic)
Y.F. ARKHIBOFF,
(representing the Government of the Union of Soviet Socialist Republics)

Related Cultural Documents

"The Mountain Inside the River"

High Dam at Aswan
Interview with Dr. Alexander Gorlov
January 6, 2005

I was involved with the project for two or three years when it just started, 1961–1963. President Nasser from Egypt sent to Russia the first documents of the project. These documents were from 1925 to 1927 and had been done by an English company, Alexander Gibb & Company, a design company in London.

They originated the design and conveyed all their drawings to Egypt, but nothing happened after that.

It was a conceptual design, not detailed design. They provided about 25 or 30 original drawings and conceptual designs.

Then in 1958–59, the project was conveyed to Russia. The Moscow Institute was assigned to make a detailed design of the entire project. This included the construction of the High Dam and hydropower plant underground. I was involved in the design of some of the elements of the hydropower plant.

We worked very hard at that time, I remember we worked through the weekends all the time. From those additional 25 or 30 drawings, the Moscow Institute developed several thousand; it was a very detailed project. All the drawings are now in Egypt.

The Dam itself and the power plant are very complicated designs. The power plant had—big tunnels, sixteen of them. These tunnels were 5.5 meters in diameter, just to build the cofferdam that empties the water of the Nile River out of the construction site. The site was protected from water, dried out, and the Dam and power plant were built.

I did not participate in the construction itself. Just the design. I was in Russia all the time; I was not allowed to visit the construction site. We were supposed to have a team assembled to visit the site, but there was nothing there at that time. But for some reason we were not allowed to go.

The first time I saw the Dam was when I was traveling there from the United States, years later when I became a professor in Boston. I was on a trip with my wife, and we went to the site. I was excited to see what it finally looked like.

The Egyptian government and experts demanded that the Dam itself be very strong and able to withstand force because it is located in an area that has experienced many wars. They said that the Dam should be strong enough to withstand severe bombardment. So for that reason the Dam is huge. It has trapezoidal sections—"V"s that are 120 meters high. So it is a mountain inside the river.

After I left the project, there were a number of changes and some remodeling of the project itself, including a redesign of the power plant to accommodate much bigger turbines than were originally installed.

That is how it happened, as I remember.

Later I traveled to Egypt where I saw the environmental impact of the Dam, which is so immense that it had flattened huge areas, replaced large villages, and caused problems with fish migration.

Dr. Alexander Gorlov
Professor Emeritus, Mechanical Engineering
Northeastern University
Boston, MA

Dr. Gorlov has been a teacher for 30 years concentrating on research in the area of renewable energy. Dr. Gorlov has developed turbines that extract power from moving water without the need to construct dams.

33
The Communications Satellite (COMSAT)

United States

DID YOU KNOW . . . ?

➤ In 1945, Arthur C. Clarke forecast the use of television signals beamed around the world by orbiting satellites.

➤ In 1955, Bell Labs published a plan by David J. Whalen of NASA for a communications-satellite system.

➤ The COMSAT Corporation was established as a publicly traded company in 1963.

➤ COMSAT joined with more than 12 agencies to form INTELSAT.

➤ INTELSAT satellites launched in 1997 could carry 22,500 two-way telephone calls and three color television broadcasts simultaneously.

➤ COMSAT may foreshadow a mixed-economy corporation to promote worldwide distribution of solar power.

In 1865, Jules Verne predicted the launch of a spaceship supported by an international cooperative league of many nations. At the time, it was science fiction, from his book *From the Earth to the Moon*; more than 100 years later it has become fact.

In 1945, a young Royal Air Force electronics officer and proud member of the British Interplanetary Society wrote an article for the journal *Wireless World*. His vision? Television signals beamed around the world. How? By orbiting satellites. The article caused little stir, but its author, Arthur C. Clarke, later became famous as a renowned writer of science fiction.

Newspaper headlines surrounding *Early Bird,* the world's first commercial communications satellite. Courtesy of COMSAT.

HISTORY

Although Verne and Clarke planted the concept of space communications in readers' minds, it was John R. Pierce, a scientist at AT&T's Bell Labs, who became known as the father of satellite communications. Pierce published an article in 1955 that discussed a plan by David J. Whalen of NASA for a "communications 'mirror' in space, a medium-orbit repeater and a 14-hour-orbit repeater" (http://www.hq.nasa.gov/office/pao/History/satcomhistory.html). Pierce compared the capability of a mirror/repeater system that could carry 1,000 telephone calls simultaneously with the first transatlantic telephone cable, which carried only 36.

Plans for the practical use of earth satellites took shape quickly after Sputnik was launched by the Soviet Union in 1957. Two years later Congress authorized (in legislation reproduced here) the Communications Satellite Act of 1962, which mandated the formation of a private corporation to establish and manage a U.S. commercial communications-satellite system as part of a global system including other countries. The form of a government-regulated private corporation was chosen as a compromise between advocates of privatization, who argued that the public good would more greatly benefit from a private corporation's ability to capitalize on entrepreneurial opportunities offered by satellite technology, and supporters of government ownership, who feared that privatization would give too much control over the airwaves to private interests.

The resulting entity was the COMSAT Corporation, incorporated as a publicly traded company in 1963, with shares of stock sold in equal proportion to public and private communications companies. COMSAT's purview extended to the development of satellite technology and the operation of ground stations; satellite channels were to be marketed to U.S. communications carriers and "other authorized entities," including governments.

In August 1964, COMSAT, along with agencies from more than a dozen other nations, formed a worldwide commercial network known as the International Telecommunications Satellite Consortium, or INTELSAT, which today has 143 member countries and signatories. COMSAT also represented the United States in the International Maritime Satellite Organization, or INMARSAT, established in 1979. INMARSAT's coverage later expanded to land, mobile, and aeronautical communications, and it was renamed the International Mobile Satellite Organization.

COMSAT launched satellites for both domestic and international use. INTELSAT launched the world's first commercial communications satellite, *Early Bird*, or *INTELSAT 1*, in 1965. *Early Bird* could transmit one color TV channel or 240 simultaneous telephone calls between North America and Europe. More INTELSAT satellites followed. By 1969, INTELSAT had created a communications system that covered the entire globe except for the polar regions, positioning satellites in high synchronous, or stationary, orbit over the Indian Ocean to work together with similar INTELSAT satellites over the Atlantic and Pacific. (Writer Arthur C. Clarke pointed out as early as 1945 that three satellites, spaced equal distances apart in high, synchronous orbits, could provide continuous line-of-sight transmission of radio and television signals to most of the earth's surface.) Satellites in the INTELSAT 8 series, first launched in 1997, could carry 22,500 two-way telephone calls and three color television broadcasts at the same time; with advanced equipment, the number of telephone calls handled was raised to 112,500.

In 2000, Lockheed Martin purchased COMSAT and absorbed it into Lockheed's Global Telecommunications unit. When Congress agreed to the purchase, it rescinded COMSAT's exclusive right to access the INTELSAT network and called for privatization of INTELSAT. The following year, with a score of satellites in orbit and a membership of 145 countries, INTELSAT was privatized.

Here is a selective chronology of the evolution of the communications satellite, from David J. Whalen's "Communications Satellites: Making the Global Village Possible":

- 1945 Arthur C. Clarke article: "Extra-Terrestrial Relays"
- 1955 John R. Pierce article: "Orbital Radio Relays"
- 1956 First Trans-Atlantic Telephone Cable: TAT-1
- 1957 Sputnik: Russia launches the first earth satellite.
- 1960 1st Successful DELTA Launch Vehicle

- 1960 AT&T applied to FCC for experimental satellite communications license
- 1961 Formal start of TELSTAR, RELAY, and SYNCOM Programs
- 1962 TELSTAR and RELAY launched
- 1962 Communications Satellite Act (U.S.)
- 1963 SYNCOM launched
- 1964 INTELSAT formed
- 1965 COMSAT's EARLY BIRD: 1st commercial communications satellite
- 1969 INTELSAT-III series provided global coverage
- 1972 ANIK: 1st Domestic Communications Satellite (Canada)
- 1974 WESTAR: 1st U.S. Domestic Communications Satellite
- 1975 INTELSAT-IVA: 1st use of dual-polarization
- 1975 RCA SATCOM: 1st operational body-stabilized comm. satellite
- 1976 MARISAT: 1st mobile communications satellite
- 1976 PALAPA: 3rd country (Indonesia) to launch domestic comm. satellite
- 1979 INMARSAT formed.
- 1988 TAT-8: 1st Fiber-Optic Trans-Atlantic telephone cable

CULTURAL CONTEXT

During the cold war, the space race was foremost on the minds of scientists and presidents—Soviet and American. But U.S. efforts to develop an

John F. Kennedy signs the Communications Satellite Act of 1962 into law. Courtesy of COMSAT.

orbiting satellite lagged behind those of the Soviet scientists who launched Sputnik. Soon thereafter, rocket expert Werner von Braun was recruited to the American scientific team, and the United States moved rapidly into the Explorer program.

President John F. Kennedy put forward a new vision for space that spoke of the cooperative nature of large-scale initiatives: "I now invite all nations—including the Soviet Union—to join with us in developing a weather prediction program, in a new communications satellite program and in preparation for probing the distant planets of Mars and Venus, probes which may someday unlock the deepest secrets of the universe" (Jan. 30, 1961 State of the Union address).

PLANNING

On August 31, 1962, the Communications Satellite Act of 1962 became law and set a new tone of inclusiveness that would transform the space race with greater multinational, public/private cooperation. While it was a government program authorized by the president, the Communications Satellite Act of 1962 launched "a private corporation, subject to appropriate governmental regulation." The initial capitalization was $200 million.

Expanded communication was an obvious application since satellites held great promise of easily relaying telephone, television, and other types of electronic data streams between distant points on the earth's surface. The first voice retransmission from space was achieved by the satellite *Score*, placed in orbit by the United States in 1958. In 1960 the United States launched the first communications satellites—*Echo 1*, a "passive" satellite that merely reflected received signals back to earth, and *Courier 1B*, an "active-repeater" satellite that amplified and retransmitted incoming signals.

In the document presented here, the purpose is well defined: "It is the policy of the United States to establish, in conjunction and in cooperation with other countries, as expeditiously as practicable, a commercial communications satellite system, as part of an improved global communications network, which will be responsive to public needs and objectives, which will serve the communication needs of the United States and other countries and which will contribute to world peace and understanding."

The vision went beyond the countries funding the effort: "Care and attention will be directed toward providing such services to economically less developed countries."

BUILDING

COMSAT took six months to incorporate, becoming official on February 1, 1963. The former chairman of the board of the Standard Oil Company, Leo D. Welch, was its first chairman, and Joseph V. Charyl, former undersecretary of the U.S. Air Force, was its president.

When INTELSAT was established in 1964, one country was noticeably missing from the list of 11 founding partners: the Soviet Union. The reason? A Soviet scientist named Sergei Korolov, who had contributed to the initial triumph of Sputnik, was working on an important innovation: an elliptical 12-hour orbit with a perigee of 300 kilometers and an apogee of 40,000 kilometers, or what came to be known as *the highly elliptical orbit*. This innovation meant that signals could be received in the remote northern regions of the Soviet Union that were missed by geostationary orbits. Korolov's new satellite series, *Molniya* (lightning) circled the globe every 12 hours; therefore three satellites would provide continuous coverage. By 1967, festivities marking the 50th anniversary of the Bolshevik revolution were broadcast on a network of 16 *Molniya* satellites that carried TV signals to the entire Soviet Union. With that triumph, combined with a growing dissatisfaction with INTELSAT's activities, the Soviet Union authorized its own parallel organization called INTERSPUTNIK, whose members included the USSR, Bulgaria, Cuba, Czechoslovakia, Hungary, Mongolia, Poland, and Romania. Experts agree that the satellite achievements of both INTERSPUTNIK and INTELSAT were milestones in technology (Labrador and Galace).

Despite the absence of the Soviets, INTELSAT continued to grow, finally achieving full global coverage on July 1, 1969, when *INTELSAT III* lit up a large swath over the Indian Ocean. It was an era of tremendous strides, for without INTELSAT, the 500 million people who watched worldwide on television would never have been able to witness in real time the historic moment when Neil Armstrong set foot on the moon—July 20, 1969.

INTELSAT now has more members than the United Nations, and provides the world with such efficient telephone circuits that the cost of a domestic, long-distance telephone call has been greatly reduced. Following John Kennedy's vision, the service is provided not only to the large industrialized nations but to all countries around the globe.

Another cooperative venture involves the United States and Canada working together to create the North American Mobile Satellite system (MSAT). An extension of COMSAT resulted in INMARSAT (the International Maritime Satellite Organization), which offers mobile communications to the U.S. Navy and others worldwide. While initially conceived as a maritime service that would provide telephone communications at sea, the system also gave birth to the satellite phone in a suitcase used by journalists on assignment in far-flung areas of the world.

IMPORTANCE IN HISTORY

COMSAT serves as a precedent for the possibility of globally transmitting electrical power by microwave. With the cooperation of private and public interests, and the cooperation and partnership of many countries, this initiative is the kind of economic model the world needs for conceiving wide-ranging agreements.

With the proliferation of cell phones worldwide, satellites are now the main-stay of world communications, and low-earth-orbit (LEO) systems are the foundation of personal-communications-system (PCS) devices. Early efforts at the low orbit went through the Van Allen radiation belt around the earth, but advances have produced systems that can fly under the Van Allen limit.

Solar-power-satellite (SPS) technology may be many years away. However, it takes 75 years to transition to a new energy technology, so it is never too early to begin, especially when fossil fuels now used for electricity are increasingly elusive and expensive.

FOR FURTHER REFERENCE

Books and Articles

Clarke, Arthur C. *The Exploration of Space*. New York: Harper, 1951.
———. "Extra-Terrestrial Relays." *Wireless World*, October 1945, 305–8. For a facsimile of the article pages, see http://www.lsi.usp.br/~rbianchi/clarke/ACC.ETR2.gif

Garrels, Anne. *Naked in Baghdad: The Iraq War as Seen by NPR's Correspondent*. New York: Farrar, Straus and Giroux, 2003. This book was written for National Public Radio using a satellite phone in a suitcase.

Glaser, Peter E., Frank P. Davidson, and Katinka I. Csigi. *Solar Power Satellites: A Space Energy System for Earth*. Chichester, England: John Wiley and Sons/Praxis Publishing, 1998.

Lusk Brooke, Kathleen, and George H. Litwin. "Organizing and Managing Satellite Solar Power." *Space Policy* 16 (2000): 145–56.

Mueller, Milton. *Universal Service: Competition, Interconnection, and Monopoly in the Making of the American Telephone System*. Cambridge, MA: MIT Press, 1997.

Pierce, John Robinson. *The Beginnings of Satellite Communications*. History of Technology Monograph. Berkeley, CA: San Francisco Press, 1968.

Sellers, Wallace O. "Financing 'Orbital Power & Light, Inc.'" In *Solar Power Satellites: A Space Energy System for Earth*, by Peter E. Glaser, Frank P. Davidson, and Katinka Csigi. Chichester, England: John Wiley and Sons/Praxis Publishing, 1998.

Verne, Jules. *De la terre à la lune* [From the Earth to the Moon]. 1865. New York: Bantam Classic, 1993. Lowell Bair, trans. There have also been various editions in many languages.

Whalen, David J. *The Origins of Satellite Communications, 1945–1965*. Washington, DC: Smithsonian Institute Press, 2002.

Internet

For information on satellite, broadband, cable, broadcasting, and multimedia industries, see http://www.satnews.com.

For David J. Whalen's "Communications Satellites: Making the Global Village Possible," which includes a selective satellite chronology from 1945 to 1988, see http://www.hq.nasa.gov/office/pao/History/satcomhistory.html.

For information about the transatlantic cable and Morse Code, see http://www.oldcablehouse.com/cablestations/history.html.

For the history of wireless, see http://www.pbs.org/tesla/res/res_radtime.html.

For U.S. Supreme Court opinions regarding the Marconi-Tesla decision to resolve the proper patentholder for wireless transmission (radio), see http://www.justia.us/us/320/1/case.html.

Documents of Authorization

AN ACT

To provide for the establishment, ownership, operation, and regulation of a commercial communications satellite system, and for other purposes.

Be it enacted by the Senate and House of Representatives of the United States of America in Congress assembled,

TITLE I—SHORT TITLE, DECLARATION OF POLICY AND DEFINITIONS

SHORT TITLE

Sec. 101. This Act may be cited as the "Communications Satellite Act of 1962".

DECLARATION OF POLICY AND PURPOSE

Sec. 102. (a) The Congress hereby declares that it is the policy of the United States to establish, in conjunction and in cooperation with other countries, as expeditiously as practicable a commercial communications satellite system, as part of an improved global communications network, which will be responsive to public needs and national objectives, which will serve the communication needs of the United States and other countries, and which will contribute to world peace and understanding.

(b) The new and expanded telecommunication services are to be made available as promptly as possible and are to be extended to provide global coverage at the earliest practicable date. In effectuating this program, care and attention will be directed toward providing such services to economically less developed countries and areas as well as those more highly developed, toward efficient and economical use of the electromagnetic frequency spectrum, and toward the reflection of the benefits of this new technology in both quality of services and charges for such services.

(c) In order to facilitate this development and to provide for the widest possible participation by private enterprise, United States participation in the global system shall be in the form of a private corporation, subject to appropriate governmental regulation. It is the intent of Congress that all authorized users shall have nondiscriminatory access to the system; that maximum competition be maintained in the provision of equipment and services utilized by the system; that the corporation created under this Act be so organized and operated as to maintain and strengthen competition in the provision of communications services to the public; and that the activities of the corporation created under this Act and of the persons or companies participating in the ownership of the corporation shall be consistent with the Federal antitrust laws.

(d) It is not the intent of Congress by this Act to preclude the use of the communications satellite system for domestic communication services where consistent with the provisions of this Act nor to preclude the creation of additional communications satellite systems, if required to meet unique governmental needs or if otherwise required in the national interest.

DEFINITIONS

Sec. 103. As used in this Act, and unless the context otherwise requires—

(1) the term "communications satellite system" refers to a system of communications satellites in space whose purpose is to relay telecommunication information between satellite terminal stations, together with such associated equipment and facilities for tracking, guidance, control, and command functions as are not part of the generalized launching, tracking, control, and command facilities for all space purposes;

(2) the term "satellite terminal station" refers to a complex of communications equipment located on the earth's surface, operationally connected with one or more terrestrial communication systems, and capable of transmitting telecommunications to or receiving telecommunications from a communications satellite system;

(3) the term "communications satellite" means an earth satellite which is intentionally used to relay telecommunication information;

(4) the term "associated equipment and facilities" refers to facilities other than satellite terminal stations and communications satellites, to be constructed and operated for the primary purpose of a communications satellite system, whether for administration and management, for research and development, or for direct support of space operations;

(5) the term "research and development" refers to the conception, design, and first creation of experimental or prototype operational devices for the operation of a communications satellite system, including the assembly of separate components into a working whole, as distinguished from the term "production," which relates to the construction of such devices to fixed specifications compatible with repetitive duplication for operational applications; and

(6) the term "telecommunication" means any transmission, emission or reception of signs, signals, writings, images, and sounds or intelligence of any nature by wire, radio, optical, or other electromagnetic systems;

(7) the term "communications common carrier" has the same meaning as the term "common carrier" as when used in the Communications Act of 1934, as amended, and in addition includes, but only for purposes of sections 303 and 304, any individual, partnership, association, joint-stock company, trust,

corporation, or other entity which owns or controls, directly or indirectly, or is under direct or indirect common control with, any such carrier; and the term "authorized carrier", except as otherwise provided for purposes of section 304 by section 304(b) (1), means a communications common carrier which has been authorized by the Federal Communications Commission under the Communications Act of 1934, as amended, to provide services by means of communications satellites;

(8) the term "corporation" means the corporation authorized by title III of this Act;

(9) the term "Administration" means the National Aeronautics and Space Administration; and

(10) the term "Commission" means the Federal Communications Commission.

TITLE II—FEDERAL COORDINATION, PLANNING, AND REGULATION

Implementation of Policy

Sec. 201. In order to achieve the objectives and to carry out the purposes of this Act—

(a) the President shall—

(1) aid in the planning and development and foster the execution of a national program for the establishment and operation, as expeditiously as possible, of a commercial communications satellite system;

(2) provide for continuous review of all phases of the development and operation of such a system, including the activities of a communications satellite corporation authorized under title III of this Act;

(3) coordinate the activities of governmental agencies with responsibilities in the field of telecommunication, so as to insure that there is full and effective compliance at all times with the policies set forth in this Act;

(4) exercise such supervision over relationships of the corporation with foreign governments or entities or with international bodies as may be appropriate to assure that such relationships shall be consistent with the national interest and foreign policy of the United States;

(5) insure that timely arrangements are made under which there can be foreign participation in the establishment and use of a communications satellite system;

(6) take all necessary steps to insure the availability and appropriate utilization of the communications satellite system for general governmental purposes except where a separate communications satellite system is required to meet unique governmental needs, or is otherwise required in the national interest; and

(7) so exercise his authority as to help attain coordinated and efficient use of the electromagnetic spectrum and the technical compatibility of the system with existing communications facilities both in the United States and abroad.

(b) the National Aeronautics and Space Administration shall—

(1) advise the Commission on technical characteristics of the communications satellite system;

(2) cooperate with the corporation in research and development to the extent deemed appropriate by the Administration in the public interest;

(3) assist the corporation in the conduct of its research and development program by furnishing to the corporation, when requested, on a reimbursable basis, such satellite launching and associated services as the Administration deems necessary for the most expeditious and economical development of the communications satellite system;

(4) consult with the corporation with respect to the technical characteristics of the communications satellite system;

(5) furnish to the corporation, on request and on a reimbursable basis, satellite launching and associated services required for the establishment, operation, and maintenance of the communications satellite system approved by the Commission; and

(6) to the extent feasible, furnish other services, on a reimbursable basis, to the corporation in connection with the establishment and operation of the system.

(c) the Federal Communications Commission, in its administration of the provisions of the Communications Act of 1934, as amended, and as supplemented by this Act, shall—

(1) insure effective competition, including the use of competitive bidding where appropriate, in the procurement by the corporation and communications common carriers of apparatus, equipment, and service required for the establishment and operation of the communications satellite system and satellite terminal stations; and the Commission shall consult with the Small Business Administration and solicit its recommendations on measures and procedures which will insure that small business concerns are given an equitable opportunity to share in the procurement program of the corporation for property and services, including but not limited to research, development, construction, maintenance, and repair;

(2) insure that all present and future authorized carriers shall have nondiscriminatory use of, and equitable access to, the communications satellite system and satellite terminal stations under just and reasonable charges, classifications, practices, regulations, and other terms and conditions and regulate the manner in which available facilities of the system and stations are allocated among such users thereof;

(3) in any case where the Secretary of State, after obtaining the advice of the Administration as to technical feasibility, has advised that commercial communication to a particular foreign point by means of the communications satellite system and satellite terminal stations should be established in the national interest, institute forthwith appropriate proceedings under section 214 (d) of the Communications Act of 1934, as amended, to require the establishment of such communication by the corporation and the appropriate common carrier or carriers;

(4) insure that facilities of the communications satellite system and satellite terminal stations are technically compatible and interconnected operationally with each other and with existing communications facilities;

(5) prescribe such accounting regulations and systems and engage in such ratemaking procedures as will insure that any economies made possible by a communications satellite system are appropriately reflected in rates for public communication services;

(6) approve technical characteristics of the operational communications satellite system to be employed by the corporation and of the satellite terminal stations; and

(7) grant appropriate authorizations for the construction and operation of each satellite terminal station, either to the corporation or to one or more authorized carriers or to the corporation and one or more such carriers jointly, as will best serve the public interest, convenience, and necessity. In determining the public interest, convenience, and necessity the Commission shall authorize the construction and operation of such stations by communications common carriers or the corporation, without preference to either;

(8) authorize the corporation to issue any shares of capital stock, except the initial issue of capital stock referred to in section 304 (a), or to borrow any moneys, or to assume any obligation in respect of the securities of any other person, upon a finding that such issuance, borrowing, or assumption is compatible with the public interest, convenience, and necessity and is necessary or appropriate for or consistent with carrying out the purposes and objectives of this Act by the corporation;

(9) insure that no substantial additions are made by the corporation or carriers with respect to facilities of the system or satellite terminal stations unless such additions are required by the public interest, convenience, and necessity;

(10) require, in accordance with the procedural requirements of section 214 of the Communications Act of 1934, as amended, that additions be made by the corporation or carriers with respect to facilities of the system or satellite terminal stations where such additions would serve the public interest, convenience, and necessity; and

(11) make rules and regulations to carry out the provisions of this Act.

TITLE III—CREATION OF A COMMUNICATIONS SATELLITE CORPORATION

Creation of Corporation

Sec. 301. There is hereby authorized to be created a communications satellite corporation for profit which will not be an agency or establishment of the United States Government. The corporation shall be subject to the provisions of this Act and, to the extent consistent with this Act, to the District of Columbia Business Corporation Act. The right to repeal, alter, or amend this Act at any time is expressly reserved.

Process of Organization

Sec. 302. The President of the United States shall appoint incorporators, by and with the advice and consent of the Senate, who shall serve as the initial board of directors until the first annual meeting of stockholders or until their successors are elected and qualified. Such incorporators shall arrange for an initial stock offering and take whatever other actions are necessary to establish the corporation, including the filing of articles of incorporation, as approved by the President.

Directors and Officers

Sec. 303. (a) The corporation shall have a board of directors consisting of individuals who are citizens of the United States, of whom one shall be elected annually by the board to serve as chairman. Three members of the board shall be appointed by the President of the United States, by and with the advice and consent of the Senate, effective the date on which the other members are elected, and for terms of three years or until their successors have been appointed and qualified, except that the first three members of the board so appointed shall continue in office for terms of one, two, and three years, respectively, and any member so appointed to fill a vacancy shall be appointed only for the unexpired term of the director whom he succeeds. Six members of the board shall be elected annually by those stockholders who are communications common carriers and six shall be elected annually by the other stockholders of the corporation. No stockholder who is a communications common carrier and no trustee for such a stockholder shall vote, either directly or indirectly, through the votes of subsidiaries or affiliated companies, nominees, or any persons subject to his direction or control, for more than three candidates for membership on the board. Subject to such limitation, the articles of incorporation to be filed by the incorporators designated under section 302 shall provide for cumulative voting under section 27(d) of the District of Columbia Business Corporation Act (D.C. Code, sec. 29-911 (d)).

(b) The corporation shall have a president, and such other officers as may be named and appointed by the board, at rates of compensation fixed by the board, and serving at the pleasure of the board. No individual other than a citizen of the United States may be an officer of the corporation. No officer of the corporation shall receive any salary from any source other than the corporation during the period of his employment by the corporation.

FINANCING THE CORPORATION

Sec. 304. (a) The corporation is authorized to issue and have outstanding, in such amounts as it shall determine, shares of capital stock, without par value, which shall carry voting rights and be eligible for dividends. The shares of such stock initially offered shall be sold at a price not in excess of $100 for each share and in a manner to encourage the widest distribution to the American public.

Subject to the provisions of subsections (b) and (d) of this section, shares of stock offered under this subsection may be issued to and held by any person.

(b) (1) For the purposes of this section the term "authorized carrier" shall mean a communications common carrier which is specifically authorized or which is a member of a class of carriers authorized by the Commission to own shares of stock in the corporation upon a finding that such ownership will be consistent with the public interest, convenience, and necessity.

(2) Only those communications common carriers which are authorized carriers shall own shares of stock in the corporation at any time, and no other communications common carrier shall own shares either directly or indirectly through subsidiaries or affiliated companies, nominees, or any persons subject to its direction or control. Fifty per centum of the shares of stock authorized for issuance at any time by the corporation shall be reserved for purchase by authorized carriers and such carriers shall in the aggregate be entitled to make purchases of the served shares in a total number not exceeding the total number of nonreserved shares of any issue purchased by other persons. At no time after the initial issue is completed shall the aggregate of the shares of voting stock of the corporation owned by authorized carriers directly or indirectly through subsidiaries or affiliated companies, nominees, or any persons subject to their direction or control exceed 50 per centum of such shares issued and outstanding.

(3) At no time shall any stockholder who is not an authorized carrier, or any syndicate or affiliated group of such stockholders, own more than 10 per centum of the shares of voting stock of the corporation issued and outstanding.

(c) The corporation is authorized to issue, in addition to the stock authorized by subsection (a) of this section, nonvoting securities, bonds, debentures, and other certificates of indebtedness, as it may determine. Such nonvoting securities, bonds, debentures, or other certificates of indebtedness of the corporation as a communications common carrier may own shall be eligible for inclusion in the rate base of the carrier to the extent allowed by the Commission. The voting stock of the corporation shall not be eligible for inclusion in the rate base of the carrier.

(d) Not more than an aggregate of 20 per centum of the shares of stock of the corporation authorized by subsection (a) of this section which are held by holders other than authorized carriers may be held by persons of the classes described in paragraphs (1), (2), (3), (4), and (5) of section 310 (a) of the Communications Act of 1934, as amended (47 U.S.C. 310).

(e) The requirement of section 45 (b) of the District of Columbia Business Corporation Act (D.C. Code, sec. 29-920 (b)) as to the percentage of stock which a stockholder must hold in order to have the rights of inspection and copying set forth in that subsection shall not be applicable in the case of holders of the stock of the corporation, and they may exercise such rights without regard to the percentage of stock they hold.

(f) Upon application to the Commission by any authorized carrier and after notice and hearing, the Commission may compel any other authorized carrier which owns shares of stock in the corporation to transfer to the applicant, for a fair and reasonable consideration, a number of such shares as the Commission determines will advance the public interest and the purposes of this Act. In its determination with respect to ownership of shares of stock in the corporation, the Commission, whenever possible will oversee distribution of stock among the authorized carriers.

PURPOSES AND POWERS OF THE CORPORATION

Sec. 305. (a) In order to achieve the objectives and to carry out the purposes of this Act, the corporation is authorized to—

(1) plan, initiate, construct, own, manage, and operate itself or in conjunction with foreign governments or business entities a commercial, communications satellite system;

(2) furnish, for hire, channels of communication to United States communications common carriers and to other authorized entities, foreign and domestic; and

(3) own and operate satellite terminal stations when licensed by the Commission under section 201 (c) (7).

(b) Included in the activities authorized to the corporation for accomplishment of the purposes indicated in subsection (a) of this section, are, among others not specifically named—

(1) to conduct or contract for research and development related to its mission;

(2) to acquire the physical facilities, equipment and devices necessary to its operations, including communications satellites and associated equipment and facilities, whether by construction, purchase, or gift;

(3) to purchase satellite launching and related services from the United States Government;

(4) to contract with authorized users, including the United States Government, for the services of the communications satellite system; and

(5) to develop plans for the technical specifications of all elements of the communications satellite system.

(c) To carry out the foregoing purposes, the corporation shall have the usual powers conferred upon a stock corporation by the District of Columbia Business Corporation Act.

TITLE IV—MISCELLANEOUS

APPLICABILITY OF COMMUNICATIONS ACT OF 1934

Sec. 401. The corporation shall be deemed to be a common carrier within the meaning of section 3 (h) of the Communications Act of 1934, as amended, and as such shall be fully subject to the provisions of title II and title III of that Act. The provision of satellite terminal station facilities by one communication common carrier to one or more other communications common carriers shall be deemed to be a common carrier activity fully subject to the Communications

Act. Whenever the application of the provisions of this Act shall be inconsistent with the application of the provisions of the Communications Act, the provisions of this Act shall govern.

NOTICE OF FOREIGN BUSINESS NEGOTIATIONS

Sec. 402. Whenever the corporation shall enter into business negotiations with respect to facilities, operations, or services authorized by this Act with any international or foreign entity, it shall notify the Department of State of the negotiations, and the Department of State shall advise the corporation of relevant foreign policy considerations. Throughout such negotiations the corporation shall keep the Department of State informed with respect to such considerations. The corporation may request the Department of State to assist in the negotiations, and that Department shall render such assistance as may be appropriate.

SANCTIONS

Sec. 403. (a) If the corporation created pursuant to this Act shall engage in or adhere to any action, practices, or policies inconsistent with the policy and purposes declared in section 102 of this Act, or if the corporation or any other person shall violate any provision of this Act, or shall obstruct or interfere with any activities authorized by this Act, or shall refuse, fail, or neglect to discharge his duties and responsibilities under this Act, or shall threaten any such violation, obstruction, interference, refusal, failure, or neglect, the district court of the United States for any district in which such corporation or other person resides or may be found shall have jurisdiction, except as otherwise prohibited by law, upon petition of the Attorney General of the United States, to grant such equitable relief as may be necessary or appropriate to prevent or terminate such conduct or threat.

(b) Nothing contained in this section shall be construed as relieving any person of any punishment, liability, or sanction which may be imposed otherwise than under this Act.

(c) It shall be the duty of the corporation and all communications common carriers to comply insofar as applicable, with all provisions of this Act and all rules and regulations promulgated thereunder.

REPORTS TO THE CONGRESS

Sec. 404. (a) The President shall transmit to the Congress in January of each year a report which shall include a comprehensive description of the activities and accomplishments during the preceding calendar year under the national program referred to in section 201 (a) (1), together with an evaluation of such activities and accomplishments in terms of the attainment of the objectives of this Act and any recommendations for additional legislative or other action which the President may consider necessary or desirable for the attainment of such objectives.

(b) The corporation shall transmit to the President and the Congress, annually and at such other times as it deems desirable, a comprehensive and detailed report of its operations, activities, and accomplishments under this Act.

(c) The Commission shall transmit to the Congress, annually and at such other times as it deems desirable, (i) a report of its activities and actions on anti-competitive practices as they apply to the communications satellite programs; (ii) an evaluation of such activities and actions taken by it within the scope of its authority with a view to recommending such additional legislation which the Commission may consider necessary in the public interest; and (iii) an evaluation of the capital structure of the corporation so as to assure the Congress that such structure is consistent with the most efficient and economical operation of such corporation.

Approved August 31, 1962, 9:51 a.m.

From Public Law 87-624(August 31, 1962).

A hiking trail through lichen-covered trees to Takelma Gorge on the Rogue River in Oregon. Courtesy of Shutterstock.

34
The National Trails System

United States

DID YOU KNOW . . . ?

➤ The Appalachian Trail owes it existence to a 1921 article by forester and environmentalist Benton MacKaye.

➤ The Appalachian Trail was created in the 1920s, and was walkable from end to end by 1938.

➤ It stretches more than 2,000 miles (3,200 kilometers) from central Maine to northern Georgia.

➤ By 2000, there were almost 50,000 miles (80,470 kilometers) of biking, hiking, and horseback-riding trails in the United States.

➤ The Silk Road has been an intercontinental trail for 2,000 years.

➤ The French government has approved Philippe Bernard's proposal of a Paris-to-Moscow bikeway.

➤ The Rails-to-Trails Conservancy has built more than 11,000 miles (18,000 kilometers) of bikeways utilizing abandoned railway routes.

What actually started the United States down the path of developing a network of national trails was a 1921 article written by an avid forester and environmentalist, Benton MacKaye. While not official legislation, it was MacKaye's broad outline of a plan of active trails that generated enthusiasm for the project and led to the formal legislation.

HISTORY

The origin of the National Trails System can be traced back to the early twentieth century when outdoors enthusiasts and government agencies undertook the creation of the Appalachian Trail in the eastern part of the United States and the Pacific Crest National Scenic Trail in the western part. The Appalachian Trail was created in the 1920s, and was walkable from end to end by 1938, stretching more than 2,000 miles (3,200 kilometers) from central Maine to northern Georgia. Segments of the western trail were built in the 1930s; today the Pacific Crest Trail essentially links the Mexican and Canadian borders.

During the 1940s Congress discussed the possibility of creating a national system of hiking trails. Action was slow in coming, however. Finally, recognition of the need for a national trails system became driven by threats to the Appalachian Trail. By the 1960s vacation homes, pipelines, highways, and other land use changes threatened the trail's integrity as a wilderness-like experience. The Appalachian Trail Conference appealed to Congress, on both recreational and environmental grounds, to help protect this valuable resource by asking the

A signpost on the Appalachian Trail. Courtesy of Shutterstock.

federal government to acquire threatened segments. In 1965 President Johnson launched the idea of spreading volunteer-maintained trails nationwide as a replication of the Appalachian Trail. In 1966 the Department of Interior issued its seminal "Trails for America," which provided a vision of a national system of trails.

In 1968, Congress passed the National Trails System Act, "in order to provide for the ever-increasing outdoor recreation needs of an expanding population and in order to promote public access to, travel within, and enjoyment and appreciation of the open-air, outdoor areas of the Nation." The act, reproduced here, mandates three types of trails: recreation trails that need to be "reasonably accessible to urban areas," scenic trails, and "connecting or side trails" (section 3). Amendments to the 1968 law subsequently added national historic trails to the mix, which, like scenic trails, can be established only by Congressional action. By the end of the twentieth century, the United States had 20 national trails—12 historic and 8 scenic—stretching nearly 40,000 miles (64,000 kilometers). There also were about 800 national recreation trails, covering more than 9,000 miles (14,000 kilometers).

CULTURAL CONTEXT

The Silk Road could be thought of as one of the world's earliest trails still in use. It was a connected route used by merchants and travelers to move silk and other precious commodities between China and western Europe.

Many famous hikers used ancient trails. Did Petrarch walk through Italy and France to spread the word of the Renaissance? Certainly Jesus and some of his disciples followed numerous trails throughout the Middle East as they fanned outward from Jerusalem toward imperial Rome. In Europe, the tradition of the "grand tour" was an invitation to the young to hike and ride while learning about other cultures as a means of broadening their education.

The Trans-Siberian Railway was built along sleigh and dogsled routes, as was the Alaska Highway. Swiss mule paths through the Alps laid down the first transitways used by the Celts and were later studied by builders of tunnels, including the Mont Blanc. The Greek colony of Cyrene was founded following an overnight hike hosted by the Libyans to help the pioneering Therans find a spot where water was plentiful.

The U.S. National Trails System is an outgrowth of the push to explore ever further westward as early Americans sought to learn more about the land they knew lay beyond the cultivated eastern seaboard. Frederick J. Turner's books explore the American instinct to go west. Chief among such explorers were Meriwether Lewis and William Clark, who were commissioned by President Thomas Jefferson to explore the western portion of the United States and report back to him. On May 14, 1804, the two set out on an amazing expedition that, among other things, raised American awareness of the great outdoors and the magnificent land available to all Americans. Somewhat later, when the West was being opened to pioneers and settlements, the Pony Express began operations in April 1860 to carry letters and packages. It operated until November 1861, and today that route is preserved as a national historic trail.

PLANNING

The National Trails System was envisioned as a network of legally protected pathways running through government-owned and private lands in urban, rural, and wilderness settings. Portions of the system are directly administered by federal agencies such as the National Park Service; some trails are overseen by state or local agencies or private organizations with technical or perhaps financial assistance from the federal government.

Planning for trails often begins as a local initiative. People in a specific area would like to define a trail, or convert a railroad bed to use as a trail; such plans are decided at the regional level. There has always been a combination of completely private people with local governments, states, and federal agencies. It is an interdisciplinary or cross-departmental enthusiasm. One of the challenges in planning is how to connect up all these small sections.

There are four categories of trails: National Scenic (created in 1968), National Recreation (1968), Connecting or Side trails (1968), and National Historic (1978). National Scenic Trails (NSTs) are long-distance trails (over 100 miles/161 kilometers) that link superb resources in a continuous corridor for nonmotorized recreation (usually hiking and sometimes horseback). They must be created by Congress. To date, eight of this type of trail have been created:

Appalachian NST	1968
Pacific Crest NST	1968
Continental Divide NST	1978
North Country NST	1980
Ice Age NST	1980
Florida NST	1980
Potomac Heritage NST	1983
Natchez Trace NST	1983

National Recreation Trails (NRT) are more local, diverse types of trails ranging from touring motor routes to backcountry water trails. They do not require Congressional action, but do need formal recognition by either the secretary of agriculture or the secretary of the interior. There are more than 800 of these trails (over 400 in national forests).

Connecting or Side trails, can link to any of the other types of trail created in the National Trails System. Two were created in 1990: the 86-mile Anvik Connector in Alaska, which links to the Iditarod National Historic Trail, and the 14-mile Timm's Hill Trail in Wisconsin, which connects to the Ice Age NST.

In 1978, Congress added a fourth category—National Historic Trails (NHTs)—in order to commemorate major routes of historic (and prehistoric) travel throughout the United States. These trails do not have to be continuous but have to satisfy three criteria: be nationally significant, have a documented route (usually through maps or journals), and provide for significant outdoor recreation. To date, 12 have been created (although many more have been studied).

Trail Name	Type	Year created
Oregon Trail	NHT	1978
Mormon Pioneer	NHT	1978
Lewis & Clark	NHT	1978
Iditarod	NHT	1978
Overmountain Victory	NHT	1980
Nez Perce (Nee-Me-Poo)	NHT	1986
Santa Fe	NHT	1987
Trail of Tears	NHT	1987
Juan Bautista de Anza	NHT	1990
California	NHT	1992
Pony Express	NHT	1992
Selma to Montgomery	NHT	1996

Most of the trails in the national system are historic in the sense that, to a greater or lesser degree, they embody the spirit of an earlier era before mechanical transport became the norm. A number of trails directly follow, at least in part, paths traversed on foot or horseback in bygone days. But those classified as historic are specifically associated with exploration, migration, or military activities. Historic trails include both water and land segments and need not be continuous. Examples include the Iditarod National Historic Trail in Alaska and the Trail of Tears National Historic Trail. The Iditarod route incorporates trails used by prospectors and their dog teams during the gold rush around the turn of the twentieth century. The Trail of Tears memorializes the U.S. Army's relocation in the 1830s of more than 15,000 Cherokee Native Americans from their ancestral lands in the southeastern United States to what is now Oklahoma—a forced migration in which more than 4,000 died from exposure, disease, and starvation.

On June 9, 1998, Public Law 105-178, the Transportation Equity Act for the 21st Century (TEA-21), was enacted, which reauthorizes the Intermodal Surface Transportation Efficiency program and helps U.S. states fund multi-use trails.

BUILDING

Building a trail is not like a traditional construction project. Each type of trail—hiking, biking, horseback riding, canoeing, etc.—requires a different approach. In the case of a multi-use trail, many engineering aspects must be coordinated. Abandoned railroad lines are attractive candidates because they

Lyndon B. Johnson called for the replication of "the great Appalachian Trail in all parts of our country." Courtesy of the Library of Congress.

have already been granted right-of-way permissions. Grading may not be needed because the railroad beds were originally constructed to remain within specified incline limits. Using a former railway line also involves deconstruction of cross-ties and iron tracks before a usable trail can emerge. The program Rails-to-Trails has done an excellent job of pursuing this approach to trail building.

The Appalachian National Scenic Trail, beginning at Mount Katahdin in Maine and stretching 2,160 miles to Springer Mountain, Georgia, was the first trail in the national system. A longer trail, the Pacific Crest National Scenic Trail, stretched 2,665 miles from Canada to Mexico. By 1978, there were four historic trails with a combined length exceeding 9,000 miles (14,000 kilometers).

After each trail is built, it must be managed and administered. The National Trails System Act authorizes and provides funding to be administered by the parties responsible—the secretary of the interior or the secretary of agriculture. The National Park Service administers 15 of the 20 trails in the National Trails System, the Forest Service tends to 4, and the Bureau of Land Management is responsible for 1. One could question whether it is time for a coordinated effort that brings the management of all trails into the purview of one management entity.

But there is more to trail creation than clearing brush or developing transitways. Before the first trail was constructed, Benton MacKaye proffered the idea, in a 1921 article in the *Journal of the American Institute of Architects* (included here as a Related Cultural Document), that trails be linked to destinations or what he called "recreational camps." It was MacKaye's notion that "the ability to cope with nature directly—unshielded by the weakening wall of civilization—is one of the admitted needs of modern times." MacKaye proposed farm hostels to be located at periodic intervals. It was his seminal article that piqued interest in the possibilities of the Appalachian Trail and led to the legislation that launched it.

For that trail, he suggested that "each stage of its construction be made of immediate strategic value in preventing and fighting forest fires," coordinating

A Pony Express rider passes by men stringing telegraph wires, 1876. The route is preserved as a national historic trail. Courtesy of the Library of Congress.

with federal and state authorities. Second, MacKaye advocated shelter camps, equipped for sleeping and cooking, set one day's walk from each other along the trail. Blazing the trail and construction of shelter camps was to be undertaken by volunteers, thus stimulating cooperation and community spirit. From the idea of shelter camps, MacKaye then envisioned 100-acre communities focused on nonindustrial activities such as scientific or travel study for "every line of outdoor non-industrial endeavor." Finally, MacKaye urged the establishment of food and farming camps, as well as those supplying logs, fuelwood, and lumber. He cited Camp Tamiment in Pennsylvania, which had 2,000 acres and a sawmill, and community housing built with timber produced by their own mill. The development of such enterprises would allow "permanent, steady, healthy employment in the open."

IMPORTANCE IN HISTORY

Among the offshoots of the National Trails System Act was a "rails-to-trails" program intended to create greenways from abandoned railroad corridors, a goal facilitated by a 1983 amendment to the 1968 law. According to the Rails-to-Trails Conservancy, a nonprofit organization promoting the transformation of old rail rights-of-way, by early 2001 the United States had more than 1,100 rail trails, stretching over 11,000 miles (18,000 kilometers). The perception that railway lines, once so important to American transport and commerce, were now being abandoned was foremost in many environmentalists' minds. The Rails-to-Trails Conservancy realized the opportunity and seized it, signaling a transformation in American travel habits from one century to another.

Around 1996, George Krassovsky retired from the tourism industry and in his late 70s decided to mount his bicycle in Bordeaux and ride to Paris, then from Paris to Moscow, and finally from Moscow to Ekaterinburg. His objective was to call attention to the need for amity and cooperation in Europe. He publicized his trip in advance so that whenever he stopped for the night, the local village or town would offer him some kind of accommodation or place to spread his sleeping pack. On his way back to Paris, 100 Russian war veterans accompanied him on their bicycles. Krassovsky's efforts helped inspire France to approve a proposal by Phillipe Bernard, formerly of the World Bank and retired head of the Humanities Division of the Ecole Polytechnique, for a Paris-to-Moscow Bikeway. France has officially sanctioned the bikeway, and the minister of the environment has agreed that the bikeway can be developed to the Belgian border.

Trails are not just for hikers. The Trans Canada Trail sponsors a national Horse Week when the Provincial Horse Association and Equine Canada promote the event celebrated across the nation. In the United States, the Southern New England Trunkline Trail and the American Endurance Ride Conference sponsor similar rides that make use of local trails. Canoe and kayak trails are also well used: the Everglades Canoe Trails in Florida traverse 10,000 islands and mangrove coves.

Trails have had an important use in medical and disaster relief. Roads are not always available; even if they are they may not be usable, or fuel supplies may be cut off. During World War II, the Swiss Army maintained a bicycle troop, although recently the Swiss government decided this mode of military transport was no longer needed. Nevertheless, the world might want to reconsider those times of disaster, natural or human-made, when electricity or fuel is unavailable and roads may be difficult to travel; bicycle trails could be a feasible way to deliver rescue and medical services.

The sportsway concept is one that bears future exploration and development. Why make new trails solely for partisans of a single sport when multi-use trails can serve a wider purpose? And, if the trails above ground are paralleled by below-surface transport and communications, the use expands many fold. This was the principle of the Champlain Corridor, proposed by Dr. Hollister Kent to Governor Philip H. Hoff of Vermont in the 1960s. Ideally, the sportsway should be combined with an updated version of the Civilian Conservation Corps (CCC), which would help build and maintain sportsways and trails. The concept of sportsways could be of use to other countries, especially China, which has more bicycles than any other country in the world.

The building of recreational trails has been one of the activities of organizations like the CCC, devoted in part to the sustenance and training of unemployed youth. The CCC was itself an application by President Franklin D. Roosevelt of the proposal of "an army against nature" in William James's "The Moral Equivalent of War." James asserted the need for protection against eruptions of nature that threatened human life; the CCC was often called upon to combat forest fires, floods, and other natural disasters.

A developmental group met at the Rensselaerville Institute in 1986 to consider an action plan for creating an interconnected network of trails across America. Among those present were Darrell Lewis, chief of the Natural Resource Management Branch of the U.S. Army Corps of Engineers; John Landis, president of the American Society for Macro-Engineering and senior vice president of Stone and Webster Engineering Corporation; and David Burwell, founder and president of the Rails-to-Trails Conservancy in Washington, DC. The conference took place just 50 years after the Appalachian Trail was inaugurated in 1936. The Rensselaerville conference brought together people from the public and private sectors who were interested in trails. Already the Rails-to-Trails Conservancy had built over 11,000 miles (18,000 kilometers) of bikeways and trails.

Millions of dollars of private funding, both from individual donors and from foundations, have been crucial to the creation of national trails. Some of the trails that extend across the Midwestern states were privately funded. In the future, fund-raising programs may offer opportunities for individuals to have a section of a trail named after themselves or their family in recognition of a donation. Similar programs have been used to raise private donor funds to replace storm-damaged trees on the grounds of the Chateau de Versailles.

The National Park Service published its *National Trails Assessment* in 1986. Over 125 pages in length, the study was commissioned by a 1983 amendment to the National Trails Systems Act to verify the extent of a total national system and network of connected trails. The study looked into the habits of those who used the existing trails, considering which aspects were most enjoyed, how frequently individuals and families vacationed or hiked on the trails, and whether there were models that might suggest what could be done in the future to develop and maintain wilderness access via trails. Authored by William Penn Mott Jr., director of the National Park Service, the *National Trails Assessment* contains 19 tables conveying status and use data, as well as an extensive map of the trail system.

On the first Saturday of June each year, more than 3,000 trail organizations, agencies, and businesses across the country host a variety of events as part of National Trails Day. The annual celebration features new trail dedications, workshops, educational exhibits, equestrian and mountain-bike rides, trail-maintenance projects, and hikes on backcountry trails. National systems of trails, wherever they are located, are unique and valuable to each country's citizens. As the developing areas of the world lay down more concrete infrastructure, it is hoped that the story of the U.S. National Trails System will help governments and citizen groups make decisions about how best to preserve their wilderness areas and provide access to those areas for national and international enjoyment.

FOR FURTHER REFERENCE

Books and Articles

Davidson, Frank P. "An Action Plan for Creating an Interconnected Network of Trails across America." Report of a conference held at the Rensselaerville Institute, Rensselaerville, NY, October 1986.

James, William. "The Moral Equivalent of War." In *American Youth: An Enforced Reconnaissance*, edited by Thacher Winslow and Frank P. Davidson. Cambridge, MA: Harvard University Press, 1940.

MacKaye, Benton. "An Appalachian Trail: A Project in Regional Planning." *Journal of the American Institute of Architects* 9 (October 1921): 325–30.

Pearse, Innes H., and Lucy H. Crocker. *The Peckham Experiment: A Study in the Living Structure of Society.* London: Allen and Unwin, 1943. Reprinted, Edinburgh: Scottish Academic Press, 1985.

Taylor, Eugene, and Robert H. Wosniak, eds. *Pure Experience: The Response to William James.* St. Augustine, FL: St. Augustine Press, 1996.

Turner, Frederick J. *The Frontier in American History.* New York: Henry Holt, 1921.

U.S. Department of the Interior. National Park Service. *National Trails Assessment.* 1986. Includes maps.

Internet

For an overview of the National Trails System that includes a table with the authorization dates and public laws of trails, see http://www.americantrails.org/resources/feds/NatTrSysOverview.html.

For information on the background and history of the National Trails System, see http://usparks.about.com/library/weekly/aa060599.htm.

For the Rails-to-Trails Conservancy, see http://www.railtrails.org/.

For the National Council for Science and the Environment, see http://www.cnie.org/.

For information about the Iditarod National Historic Trail, see http://www.iditarod.com.

For information about equestrian trails in Canada, see http://www.equinecanada.ca/.

For information about the Trans Canada Trail, see http://www.tctrail.ca/.

Film and Television

The California Conservation Corps. Videocassette. Sacramento: California Conservation Corps. Contains an interview with Page Smith, founder of the California Conservation Corps and the first manager of Camp William James.

Music

Yo Yo Ma and the Silk Road Ensemble (various artists). Two CDs: *Silk Road Journeys: When Strangers Meet.* New York: Sony Classical #089782. Release date 4/16/2002, and *The Silk Road: A Musical Caravan.* Washington, DC: Smithsonian Folkways Recording. SFW40438. Release date 2002. The Silk Road was an overland route from China across Asia to the Middle East and Europe.

Documents of Authorization

Public Law 90–543
90th Congress, S. 827
October 2, 1968
An Act
[82 Stat. 919]
To establish a national trails system, and for other purposes.

Be it enacted by the Senate and House of Representatives of the United States of America in Congress assembled,

SHORT TITLE

Section 1. This Act may be cited as the "National Trails System Act."

STATEMENT OF POLICY

Sec. 2. (a) In order to provide for the ever-increasing outdoor recreation needs of an expanding population and in order to promote public access to, travel within, and enjoyment and appreciation of the open-air, outdoor areas of the Nation, trails should be established (i) primarily, near the urban areas of the Nation, and (ii) secondarily, within established scenic areas more remotely located.

(b) the purpose of this Act is to provide the means for attaining these objectives by instituting a national system of recreation and scenic trails, by designating the Appalachian Trail and the Pacific Crest Trail as the initial components of that system, and by prescribing the methods by which, and standards according to which, additional components may be added to the system.

NATIONAL TRAILS SYSTEM

Sec. 3. The national system of trails shall be composed of—

(a) National recreation trails, established as provided in section 4 of this Act, which will provide a variety of outdoor recreation uses in or reasonably accessible to urban areas.

(b) National scenic trails, established as provided in section 5 of this Act, which will be extended trails so located as to provide for maximum outdoor recreation potential and for the conservation and enjoyment of the national significant scenic, historic, natural, or cultural qualities of the areas through which such trails may pass.

(c) Connecting or side trails, established as provided in section 6 of this Act, which will provide additional points of public access to national recreation or national scenic trails or which will provide connections between such trails.

The Secretary of the Interior and the Secretary of Agriculture, in consultation with appropriate governmental agencies and public and private organizations, shall establish a uniform marker for the national trails system.

NATIONAL RECREATION TRAILS

Sec. 4. (a) The Secretary of the Interior, or the Secretary of Agriculture where lands administered by him are involved, may establish and designate national recreation trails, with the consent of the Federal agency, State, or political subdivision having jurisdiction over the lands involved, upon finding that—

(i) such trails are reasonably accessible to urban areas, and, or

(ii) such trails meet the criteria established in this Act and such supplementary criteria as he may prescribe.

(b) As provided in this section, trails within park, forest, and other recreation areas administered by the Secretary of the Interior or the Secretary of Agriculture or in other federally administered areas may be established and designated as "National Recreation Trails" by the appropriate Secretary and, when no Federal land acquisition is involved—

(i) trails in or reasonably accessible to urban areas may be designated as "National Recreation Trails" by the Secretary of the Interior with the consent of the States, their political subdivisions, or other appropriate administering agencies, and

(ii) trails within park, forest, and other recreation areas owned or administered by States may be designated as "National Recreation Trails" by the Secretary of the Interior with the consent of the State.

NATIONAL SCENIC TRAILS

Sec. 5. (a) National scenic trails shall be authorized and designated only by act of Congress. There are hereby established as the initial National Scenic Trails:

(1) The Appalachian Trail, a trail of approximately two thousand miles extending generally along the Appalachian Mountains from Mount Katahdin, Maine, to Springer Mountain, Georgia. Insofar as practicable, the right-of-way for such trail shall comprise the trail depicted on the maps identified as "Nationwide System of Trails, Proposed Appalachian Trail, NST-AT-101-May 1967", which shall be on file and available for public inspection in the office of the Director of the National Park service. Where practicable, such rights-of-way shall include lands protected for it under agreements in effect as of the date of enactment of this Act, to which Federal agencies and States were parties. The Appalachian Trail shall be administered primarily as a footpath by the Secretary of the Interior, in consultation with the Secretary of Agriculture.

(2) The Pacific Crest Trail, a trail of approximately two thousand three hundred fifty miles, extending from the Mexican-California border northward generally along the mountain ranges of the west coast States to the Canadian-Washington border near Lake Ross, following the route as generally depicted on the map, identified as "Nationwide System of Trails, Proposed Pacific Crest Trail, NST-PC-103-May 1967" which shall be on file and available for public inspection in the office of the Chief of the Forest Service. The Pacific Crest Trail

shall be administered by the Secretary of Agriculture, in consultation with the Secretary of the Interior.

(3) The Secretary of the Interior shall establish an advisory council for the Appalachian National Scenic Trail, and the Secretary of Agriculture shall establish an advisory council for the Pacific Crest National Scenic Trail. The appropriate Secretary shall consult with such council from time to time with respect to matters relating to the trail, including the selection of rights-of-way, standards of the erection and maintenance of markers along the trail, and the administration of the trail. The members of each advisory council, which shall not exceed thirty-five in number, shall serve without compensation or expense to the Federal Government for a term of five years and shall be appointed by the appropriate Secretary as follows:

(i) A member appointed to represent each Federal department or independent agency administering lands through which the trail route passes and each appointee shall be the person designated by the head of such department or agency;

(ii) A member appointed to represent each State through which the trail passes and such appointments shall be made from recommendations of the Governors of such States;

(iii) One or more members appointed to represent private organizations, including landowners and land users, that, in the opinion of the Secretary, have an established and recognized interest in the trail and such appointment shall be made from recommendations of the heads of such organizations: Provided, That the Appalachian Trail Conference shall be represented by a sufficient number of persons to represent the various sections of the country through which the Appalachian Trail passes and

(iv) The Secretary shall designate one member to be chairman and shall fill vacancies in the same manner as the original appointment.

(b) The Secretary of the Interior, and the Secretary of Agriculture where lands administered by him are involved, shall make such additional studies as are herein or may hereafter be authorized by the Congress for the purpose of determining the feasibility and desirability of designating other trails as national scenic trails. Such studies shall be made in consultation with the heads of other Federal agencies administering lands through which such additional proposed trails would pass and in cooperation with interested interstate, State, and local governmental agencies, public and private organizations, and landowners and land users concerned. When completed, such studies shall be the basis of appropriate proposals for additional national scenic trails which shall be submitted from time to time to the President and to the Congress. Such proposals shall be accompanied by a report, which shall be printed as a House or Senate document, showing among other things—

(1) the proposed route of such trail (including maps and illustrations);

(2) the areas adjacent to such trails, to be utilized for scenic, historic, natural, cultural, or developmental, purposes;

(3) the characteristics which, in the judgment of the appropriate Secretary, make the proposed trail worthy of designation as a national scenic trail;

(4) the current status of land ownership and current and potential use along the designated route;

(5) the estimated cost of acquisition of lands or interest in lands, if any;

(6) the plans for developing and maintaining the trail and the cost thereof;

(7) the proposed Federal administering agency (which, in the case of a national scenic trail wholly or substantially within a national forest, shall be the Department of Agriculture);

(8) the extent to which a State or its political subdivisions and public and private organizations might reasonably be expected to participate in acquiring the necessary lands and in the administration thereof; and

(9) the relative uses of the lands involved, including: the number of anticipated visitor-days for the entire length of, as well as for segments of, such trail; the number of months which such trail, or segments thereof, will be open for recreation purposes; the economic and social benefits which might accrue from alternate land uses; and the estimated man-years of civilian employment and expenditures expected for the purposes of maintenance, supervision and regulation of such trail.

(c) The following routes shall be studied in accordance with the objectives outlined in subsection (b) of this section:

(1) Continental Divide Trail, a three-thousand-one-hundred-mile trail extending from near the Mexican border in southwestern New Mexico northward generally along the Continental Divide to the Canadian border in Glacier National Park.

(2) Potomac Heritage Trail, an eight-hundred-and-twenty-five-mile trail extending generally from the mouth of the Potomac River to its sources in Pennsylvania and West Virginia, including the one-hundred-and-seventy-mile Chesapeake and Ohio Canal towpath.

(3) Old Cattle Trails of the Southwest from the vicinity of San Antonio, Texas, approximately eight hundred miles through Oklahoma via Baxter Springs and Chetopa, Kansas, to Fort Scott, Kansas, including the Chisholm Trail, from the vicinity of San Antonio or Cuero, Texas, approximately eight hundred miles north through Oklahoma to Abilene, Kansas.

(4) Lewis and Clark Trail, from Wood River, Illinois, to the Pacific Ocean in Oregon, following both the outbound and inbound routes of the Lewis and Clark Expedition.

(5) Natchez Trace, from Nashville, Tennessee, approximately six hundred miles to Natchez, Mississippi.

(6) North Country Trail, from the Appalachian Trail in Vermont, approximately three thousand two hundred miles through the States of New York, Pennsylvania, Ohio, Michigan, Wisconsin, and Minnesota, to the Lewis and Clark Trail in North Dakota.

(7) Kittanning Trail from Shirleysburg in Huntingdon County to Kittanning, Armstrong County, Pennsylvania.

(8) Oregon Trail, from Independence, Missouri, approximately two thousand miles to near Fort Vancouver, Washington.

(9) Santa Fe Trail, from Independence, Missouri, approximately eight hundred miles to Santa Fe, New Mexico.

(10) Long Trail, extending two hundred and fifty-five miles from the Massachusetts border northward through Vermont to the Canadian border.

(11) Mormon Trail, extending from Nauvoo, Illinois, to Salt Lake City, Utah, through the States of Iowa, Nebraska, and Wyoming.

(12) Gold Rush Trails in Alaska.

(13) Mormon Battalion Trail, extending two thousand miles from Mount Pisgah, Iowa, through Kansas, Colorado, New Mexico, and Arizona to Los Angeles, California.

(14) El Camino Real from St. Augustine to San Mateo, Florida, approximately 20 miles along the southern boundary of the St. Johns River from Fort Caroline National Memorial to the St. Augustine National Park Monument.

CONNECTING AND SIDE TRAILS

Sec. 6. Connecting or side trails within park, forest, and other recreation areas administered by the Secretary of the Interior or Secretary of Agriculture may be established, designated, and marked as components of a national recreation or national scenic trail. When no Federal land acquisition is involved, connecting or side trails may be located across lands administered by interstate, State, or local governmental agencies with their consent: Provided, That such trails provide additional points of public access to national recreation or scenic trails.

ADMINISTRATION AND DEVELOPMENT

Sec. 7. (a) Pursuant to section 5(a), the appropriate Secretary shall select the rights-of-way for National Scenic Trails and shall publish notice thereof in the Federal Register, together with appropriate maps and descriptions: Provided, That in selecting the rights-of-way full consideration shall be given to minimizing the adverse effects upon the adjacent landowner or user and his operation. Development and management of each segment of the National Trails System shall be designed to harmonize with and complement any established multiple-use plans for that specific area in order to insure continued maximum benefits from the land. The location and width of such rights-of-way across Federal lands under the jurisdiction of another Federal agency shall be by agreement between the head of that agency and the appropriate Secretary. In selecting rights-of-way for trail purposes, the Secretary shall obtain the advice and assistance of the States, local governments, private organizations, and landowners and land users concerned.

(b) After publication of notice in the Federal Register, together with appropriate maps and descriptions, the Secretary charged with the administration of a national scenic trail may relocate segments of a national scenic trail right-of-way, with the concurrence of the head of the Federal agency having jurisdiction over the lands involved, upon a determination that: (i) such a relocation is necessary to preserve the purposes for which the trail was established, or (ii) the relocation is necessary to promote a sound land management program in accordance with established multiple-use principles: Provided, That a substantial relocation of the rights-of-way for such trail shall be by Act of Congress.

(c) National scenic trails may contain campsites, shelters, and related-public-use facilities. Other uses along the trail, which will not substantially interfere with the nature and purposes of the trail, may be permitted by the Secretary charged with the administration of the trail. Reasonable efforts shall be made to provide sufficient access opportunities to such trails and, to the extent practicable, efforts shall be made to avoid activities incompatible with the purposes for which such trails were established. The use of motorized vehicles by the general public along any national scenic trail shall be prohibited and nothing in this Act shall be construed as authorizing the use of motorized vehicles within the natural and historical areas of the national park system, the national wildlife refuge system, the national wilderness preservation system where they are presently prohibited or on other Federal lands where trails are designated as being closed to such use by the appropriate Secretary: Provided, That the Secretary charged with the administration of such trail shall establish regulations which shall authorize the use of motorized vehicles when, in his judgment, such vehicles are necessary to meet emergencies or to enable adjacent landowners or land users to have reasonable access to their lands or timber rights: Provided further, That private lands included in the national recreation or scenic trails by cooperative agreement of a landowner shall not preclude such owner from using motorized vehicles on or across such trails or adjacent lands from time to time in accordance with regulations to be established by the appropriate Secretary. The Secretary of the Interior and the Secretary of Agriculture, in consultation with appropriate governmental agencies and public and private organizations, shall establish a uniform marker, including thereon an appropriate and distinctive symbol for each national recreation and scenic trail. Where the trails cross lands administered by Federal agencies such markers shall be erected at appropriate points along the trails and maintained by the Federal agency administering the trail in accordance with standards established by the appropriate Secretary and where the trails cross non-Federal lands, in accordance with written cooperative agreements, the appropriate Secretary shall provide such uniform markers to cooperating agencies and shall require such agencies to erect and maintain them in accordance with the standards established.

(d) Within the exterior boundaries of areas under their administration that are included in the right-of-way selected for a national recreation or scenic trail, the heads of Federal agencies may use lands for trail purposes and may acquire land or interests in lands by written cooperative agreement, donation, purchase with donated or appropriated funds or exchange: Provided, That not more than twenty-five acres in any one mile may be acquired without the consent of the owner.

(e) Where the lands included in a national scenic trail right-of-way are outside of the exterior boundaries of federally administered areas, the Secretary charged with the administration of such trail shall encourage the States or local governments involved (1) to enter into written cooperative agreements with landowners, private organizations, and individuals to provide the necessary trail right-of-way, or (2) to acquire such lands or interests therein to be utilized as segments of the national scenic trail: Provided, That if the State or local governments fail to enter into such written cooperative agreements or to acquire such lands or interests therein within two years after notice of the selection of the right-of-way is published, the appropriate Secretary may (i) enter into such agreements with landowners, States, local governments, private organizations, and individuals for the use of lands for trail purposes, or (ii) acquire private lands or interests therein by donation, purchase with donated or appropriated funds or exchange in accordance with the provisions of subsection (g) of this section. The lands involved in such rights-of-way should be acquired in fee, if other methods of public control are not sufficient to assure their use for the purpose for which they are acquired: Provided, That if the Secretary charged with the administration of such trail permanently relocates the right-of-way and disposes of all title or interest in the land, the original owner, or his heirs or assigns, shall be offered, by notice given at the former owner's last known address, the right of first refusal at the fair market price.

(f) The Secretary of the Interior, in the exercise of his exchange authority, may accept title to any non-Federal property within the right-of-way and in exchange therefor he may convey to the grantor of such property any federally owned property under his jurisdiction which is located in the State wherein such property is located and which he classifies as suitable for exchange or other disposal. The values of the properties so exchanged either shall be approximately equal, or if they are not approximately equal the values shall be equalized by the payment of cash to the grantor or to the Secretary as the circumstances require. The Secretary of Agriculture, in the exercise of his exchange authority, may utilize authorities and procedures available to him in connection with exchanges of national forest lands.

(g) The appropriate Secretary may utilize condemnation proceedings without the consent of the owner to acquire private lands or the interest therein pursuant to this section only in cases where, in his judgment, all reasonable efforts to acquire such lands or interests therein by negotiation have failed, and in such cases he shall acquire only such title as, in his judgment, is reasonably necessary to provide passage across such lands: Provided, That condemnation proceedings may not be utilized to acquire fee title or lesser interests to more than twenty-five acres in any one mile and when used such authority shall be limited to the most direct or practicable connecting trail right-of-way: Provided further, That condemnation is prohibited with respect to all acquisition of lands or interest in lands for the purposes of the Pacific Crest Trail. Money appropriated for Federal purposes from the land and water conservation fund shall, without prejudice to appropriations from other sources, be available to Federal departments for the acquisition of lands or interests in lands for the purposes of this Act.

(h) The Secretary charged with the administration of a national recreation or scenic trail shall provide for the development and maintenance of such trails within federally administered areas and shall cooperate with and encourage the States to operate, develop, and maintain portions of such trails which are located outside the boundaries of federally administered areas. When deemed to be in the public interest, such Secretary may enter written cooperative agreements with the States or their political subdivisions, landowners, private organizations, or individuals to operate, develop, and maintain any portion of a national scenic trail either within or outside a federally administered area.

Whenever the Secretary of the Interior makes any conveyance of land under any of the public land laws, he may reserve a right-of-way for trails to the extent he deems necessary to carry out the purposes of this Act.

(i) The appropriate Secretary, with the concurrence of the heads of any other Federal agencies administering lands through which a national recreation or scenic trail passes, and after consultation with the States, local governments, and organizations concerned, may issue regulations, which may be revised from time to time, governing the use, protection, management, development, and administration of trails of the national trails system. In order to maintain good conduct on and along the trails located within federally administered areas and to provide for the proper government and protection of such trails, the Secretary of the Interior and the Secretary of Agriculture shall prescribe and publish such uniform regulations as they deem necessary and any person who violates such regulations shall be guilty of a misdemeanor, and may be punished by a fine of not more than $500, or by imprisonment not exceeding six months, or by both such fine and imprisonment.

STATE AND METROPOLITAN AREA TRAILS

Sec. 8. (a) The Secretary of the Interior is directed to encourage States to consider, in their comprehensive statewide outdoor recreation plans and proposals for financial assistance for State and local projects submitted pursuant to the Land and Water Conservation Fund Act, needs and opportunities for establishing park, forest, and other recreation trails on lands owned or administered by States, and recreation trails on lands in or near urban areas. He is further directed, in accordance with the authority contained in the Act of May 28, 1963 (77 Stat. 49), to encourage States, political subdivisions, and private interests, including nonprofit organizations, to establish such trails.

(b) The Secretary of Housing and Urban Development is directed, in administering the program of comprehensive urban planning and assistance under section 701 of the Housing Act of 1954, to encourage the planning of recreation trails in connection with the recreation and transportation planning for metropolitan and other urban areas. He is further directed, in administering the urban open-space program under title VII of the Housing Act of 1961, to encourage such recreation trails.

(c) The Secretary of Agriculture is directed, in accordance with authority vested in him, to encourage States and local agencies and private interests to establish such trails.

(d) Such trails may be designated and suitably marked as parts of the nationwide system of trails by the States, their political subdivisions, or other appropriate administering agencies with the approval of the Secretary of the Interior.

RIGHTS-OF-WAY AND OTHER PROPERTIES

Sec. 9. (a) The Secretary of the Interior or the Secretary of Agriculture as the case may be, may grant easements and rights-of-way upon, over, under, across or along any component of the national trails system in accordance with the laws applicable to the national park system and national forest system, respectively: Provided, That any conditions contained in such easements and rights-of-way shall be related to the policy and purposes of this Act.

(b) The Department of Defense, the Department of Transportation, the Interstate Commerce Commission, the Federal Communications Commission, the Federal Power Commission, and other Federal agencies having jurisdiction or control over or information concerning the use, abandonment, or disposition of roadways, utility rights-of-way, or other properties which may be suitable for the purpose of improving or expanding the national trails system shall cooperate with the Secretary of the Interior and the Secretary of Agriculture in order to assure, to the extent practicable, that any such properties having values suitable for trail purposes may be made available for such use.

AUTHORIZATION OF APPROPRIATIONS

Sec. 10. There are hereby authorized to be appropriated for the acquisition of lands or interests in lands not more than $5,000,000 for the Appalachian National Scenic Trail and not more than $500,000 for the Pacific Crest National Scenic Trail.

APPROVED OCTOBER 2, 1968.

LEGISLATIVE HISTORY:

HOUSE REPORTS; No. 1631 accompanying H.R. 4865 (Comm. On Interior & Insular Affairs) and No. 1891 (Com. Of Conference.)

SENATE REPORT No. 1233 (Comm. On Interior & Insular Affairs).

CONGRESSIONAL RECORD, Vol. 114 (1968):

July 1: Considered and Passed Senate

July 15: Considered and passed House, amended, in lieu of H.R. 4865.

Sept. 18: House agreed to conference report.

Sept. 19: Senate agreed to conference report.

Related Cultural Documents

"AN APPALACHIAN TRAIL: A PROJECT IN REGIONAL PLANNING," BY BENTON MACKAYE

Something has been going on these past few strenuous years which, in the din of war and general upheaval, has been somewhat lost from the public mind. It is the slow quiet development of the recreational camp. It is something neither urban nor rural. It escapes the hecticness of the one, and the loneliness of the other. And it escapes also the common curse of both—the high-powered tension of the economic scramble. All communities face an "economic" problem, but in different ways. The camp faces it through cooperation and mutual helpfulness, the others through competition and mutual fleecing.

We civilized ones also, whether urban or rural, are potentially helpless as canaries in a cage. The ability to cope with nature directly—unshielded by the weakening wall of civilization—is one of the admitted needs of modern times. It is the goal of the "scouting" movement. Not that we want to return to the plights of our Paleolithic ancestors. We want the strength of progress without its puniness. We want its conveniences without its fopperies. The ability to sleep and cook in the open is a good step forward. But "scouting" should not stop there. This is but a feint step from our canary bird existence. It should strike far deeper than this. We should seek the ability not only to cook food but to raise food with less aid—and less hindrance—from the complexities of commerce. And this is becoming daily of increasing practical importance. Scouting, then, has its vital connection with the problem of living.

A New Approach to the Problem of Living

The problem of living is at bottom an economic one. And this alone is bad enough, even in a period of so-called "normalcy." But living has been considerably complicated of late in various ways—by war, by questions of personal liberty, and by "menaces" of one kind or another. There have been created bitter antagonisms. We are undergoing also the bad combination of high prices and unemployment. This situation is world wide—the result of a world-wide war.

It is no purpose of this little article to indulge in coping with any of these big questions. The nearest we come to such effrontery is to suggest more comfortable seats and more fresh air for those who have to consider them. A great professor once said that "optimism is oxygen." Are we getting all the "oxygen" we might for the big tasks before us?

"Let us wait," we are told, "till we solve this cussed labor problem. Then we'll have the leisure to do great things."

But suppose that while we wait the chance for doing them is passed?

It goes without saying that we should work upon the labor problem. Not just the matter of "capital and labor" but the real labor problem—how to reduce the day's drudgery. The toil and chore of life should, as labor saving devices increase, form a diminishing proportion of the average day and year. Leisure and the higher pursuits will thereby come to form an increasing portion of our lives.

But will leisure mean something "higher"? Here is a question indeed. The coming of leisure in itself will create its own problem. As the problem of labor "solves," that of leisure arises. There seems to be no escape from problems. We have neglected to improve the leisure which should be ours as a result of replacing stone and bronze with iron and steam. Very likely we have been cheated out of the bulk of this leisure. The efficiency of modern industry has been placed at 25 percent of its reasonable possibilities. This may be too low or too high. But the leisure that we do succeed in getting—is this developed to an efficiency much higher?

The customary approach to the problem of living relates to work rather than play. Can we increase the efficiency of our working time? Can we solve the problem of labor? If so we can widen the opportunities for leisure. The new approach reverses this mental process. Can we increase the efficiency of our spare time? Can we develop opportunities for leisure as an aid in solving the problem of labor?

An Undeveloped Power—Our Spare Time

How much spare time have we, and how much power does it represent?

The great body of working people—the industrial workers, the farmers, and the housewives have no allotted spare time or "vacations." The business clerk usually gets two weeks' leave, with pay, each year. The U.S. Government clerk gets thirty days. The business man is likely to give himself two weeks or a month. Farmers can get off for a week or more at a time by doubling up on one another's chores. Housewives might do likewise.

As to the industrial worker—in mine or factory—his average "vacation" is all too long. For it is "leave of absence without pay." According to recent official figures the average industrial worker in the United States, during normal times, is employed about four fifths of the time—say 42 weeks in the year. The other ten weeks he is employed in seeking employment.

The proportionate time for true leisure of the average adult American appears, then, to be meagre indeed. But a goodly portion have (or take) about two weeks in the year. The industrial worker during the estimated ten weeks between jobs must of course go on eating and living. His savings may enable him to do this without undue worry. He could, if he felt he could spare the time from job hunting, and if suitable facilities were provided, take two weeks of his ten on a real vacation. In one way or another, therefore, the average adult in this country could devote each year a period of about two weeks in doing the things of his own choice.

Here is enormous undeveloped power—the spare time of our population. Suppose just one percent of it were focused upon one particular job, such as increasing the facilities for the outdoor community life. This would be more than a million people, representing over two million weeks a year. It would be equivalent to 40,000 persons steadily on the job.

A Strategic Camping Base—The Appalachian Skyline

Where might this imposing force lay out its strategic camping ground?

Camping grounds, of course, require wild lands. These in America are fortunately still available. They are in every main region of the country. They are the undeveloped or under-developed areas. Except in the Central States the wild lands now remaining are for the most part among the mountain ranges—the Sierras, the Cascades and the Rocky Mountains of the West and the Appalachian Mountains of the East.

Extensive national playgrounds have been reserved in various parts of the country for use by the people for camping and various kindred purposes. Most of these are in the West where Uncle Sam's public lands were located. They are in the Yosemite, the Yellowstone, and many other National Parks—covering about six million acres in all. Splendid work has been accomplished in fitting these Parks for use. The National Forests, covering about 130 million acres—chiefly in the West—are also equipped for public recreation purposes.

A great public service has been started in these Parks and Forests in the field of outdoor life. They have been called "playgrounds of the people." This they are for the Western people—and for those in the East who can afford time and funds for an extended trip in a Pullman car. But camping grounds to be of the most use to the people should be as near as possible to the center of population. And this is in the East.

It fortunately happens that we have throughout the most densely populated portions of the United States a fairly continuous belt of under-developed lands. These are contained in the several ranges which form the Appalachian chain of mountains. Several National Forests have been purchased in this belt. These mountains, in several ways rivaling the western scenery, are within a day's ride from centers containing more than half the population of the United States. The region spans the climate of New England and the cotton belt; it contains the crops and the people of the North and the South.

The skyline along the top of the main divides and ridges of the Appalachians would overlook a mighty part of the nation's activities. The rugged lands of this skyline would form a camping base strategic in the country's work and play.

Let us assume the existence of a giant standing high on the skyline along these mountain ridges, his head just scraping the floating clouds. What would he see from this skyline as he strode along its length from north to south?

Starting out from Mt. Washington, the highest point in the northeast, his horizon takes in one of the original happy hunting grounds of America—the "Northwoods," a country of pointed firs extending from the lakes and rivers of northern Maine to those of the Adirondacks. Stepping across the Green Mountains and the Berkshires to the Catskills, he gets his first view of the crowded east—a chain of smoky bee-hive cities extending from Boston to Washington and containing a third of the population of the Appalachian drained area. Bridging the Delaware Water Gap and the Susquehanna on the picturesque Allegheny folds across Pennsylvania he notes more smoky columns—the big plants between Scranton and Pittsburgh that get out the basic stuff of modern industry—iron and coal. In relieving contrast he steps across the Potomac near Harpers Ferry and pushes through into the wooded wilderness of the southern Appalachians where he finds preserved much of the primal aspects of the days of Daniel Boone. Here he finds, over on the Monongahela side the black coal of bituminous and the white coal of water power. He proceeds along the great divide of the upper Ohio and sees flowing to waste, sometimes in terrifying floods, waters capable of generating untold hydro-electric energy and of bringing navigation to many a lower stream. He looks over the Natural Bridge and out across the battle fields around Appomattox. He finds himself finally in the midst of the great Carolina hardwood belt. Resting now on the top of Mt. Mitchell, highest point east of the Rockies, he counts up on his big long fingers the opportunities which yet await development along the skyline he has passed.

First he notes the opportunities for recreation. Throughout the Southern Appalachians, throughout the Northwoods, and even through the Alleghenies that wind their way among the smoky industrial towns of Pennsylvania, he recollects vast areas of secluded forests, pastoral lands, and water courses, which, with proper facilities and protection, could be made to serve as the breath of a real life for the toilers in the bee-hive cities along the Atlantic seaboard and elsewhere.

Second, he notes the possibilities for health and recuperation. The oxygen in the mountain air along the Appalachian skyline is a natural resource (and a national resource) that radiates to the heavens its enormous health-giving powers with only a fraction of a percent utilized for human rehabilitation. Here is a resource that could save thousands of lives. The sufferers of tuberculosis, anemia and insanity go through the whole strata of human society. Most of them are helpless, even those economically well off. They occur in the cities and right in the skyline belt. For the farmers, and especially the wives of farmers, are by no means escaping the grinding-down process of our modern life.

Most sanitariums now established are perfectly useless to those afflicted with mental disease, the most terrible, usually, of any disease. Many of these sufferers could be cured. But not merely by "treatment." They need acres not medicine. Thousands of acres of this mountain land should be devoted to them with whole communities planned and equipped for their cure.

Next after the opportunities for recreation and recuperation our giant counts off, as a third big resource, the opportunities in the Appalachian belt for employment on the land. This brings up a need that is becoming urgent—the redistribution of our population, which grows more and more top heavy.

The rural population of the United States, and of the Eastern States adjacent to the Appalachians, has now dipped below the urban. For the whole country has fallen from 60 per cent of the total in 1900 to 49 per cent in 1920: for the Eastern States it has fallen, during this period, from 55 per cent to 45 per cent. Meantime the per capita area of improved farmland has dropped, in the Eastern States, from 3.35 acres to 2.43 acres. This is a shrinkage of nearly 28 percent in 20 years: in the States from Maine to Pennsylvania the shrinkage has been 40 percent. There are in the Appalachian belt probably 25 million acres of grazing and agricultural land awaiting development. Here is room for a whole new rural population. Here is an opportunity—if only the way can be found—for that counter migration from city to country that has so long been prayed for. But our giant in pondering on this resource is discerning enough to know that its utilization is going to depend upon some new deal in our agricultural system. This he knows if he has ever stooped down and gazed in the sunken eyes either of the Carolina "cracker" or of the Green Mountain "hayseed."

Forest land as well as agricultural might prove an opportunity for steady employment in the open. But this again depends upon a new deal. Forestry must replace timber devastation and its consequent haphazard employment. And this the giant knows if he has looked into the rugged face of the homeless "don't care a damn" lumberjack of the Northwoods.

Such are the outlooks—such the opportunities—seen by a discerning spirit from the Appalachian skyline.

Possibilities in the New Approach

Let's put up now to the wise and trained observer the particular question before us. What are the possibilities in the new approach to the problem of living? Would the development of the outdoor community life—as an offset and relief from the various shackles of commercial civilization—be practicable and worthwhile? From the experience of observations and thoughts along the skyline, here is a possible answer:

There are several possible gains from such an approach.

First there would be the "oxygen" that makes for a sensible optimism. Two weeks spent in the real open—right now, this year and next—would be a little real living for thousands of people which they would be sure of getting before they died. They would get a little fun as they went along regardless of problems being "solved." This would not damage the problems and it would help the folks.

Next there would be perspective. Life for two weeks on the mountain top would show up many things about life during the other fifty weeks down below. The latter could be viewed as a whole—away from its heat, and sweat, and irritations. There would be a chance to catch a breath, to study the dynamic forces of nature and the possibilities of shifting to them the burdens now carried on the backs of men. The reposeful study of these forces should provide a broad gauged enlightened approach to the problems of industry. Industry would come to be seen in its true perspective—as a means in life and not as an end in itself.

The actual partaking of the recreative and non-industrial life—systematically by the people and not spasmodically by a few—should emphasize the distinction between it and the industrial life. It should stimulate the quest for enlarging the one and reducing the other. It should put new zest in the labor movement. Life and study of this kind should emphasize the need of going to the roots of industrial questions and of avoiding superficial thinking and rash action. The problems of the farmer, the coal miner, and the lumberjack could be studied intimately and with minimum partiality. Such an approach should bring the poise that goes with understanding.

Finally these would be new clews to constructive solutions. The organization of the cooperative camping life would tend to draw people out of the cities. Coming as visitors they would be loath to return. They would become desirous of settling down in the country—to work in the open as well as play. The various camps would require food. Why not raise food, as well as consume it, on the cooperative plan? Food and farm camps should come about as a natural sequence. Timber also is required. Permanent small-scale operations should be encouraged in the various Appalachian National Forests. The government now claims this as a part of its forest policy. The camping life would stimulate forestry as well as a better agriculture. Employment in both would tend to become enlarged.

How far these tendencies would go the wisest observer of course can not tell. They would have to be worked out step by step. But the tendencies at least would be established. They would be cutting channels leading to constructive achievement in the problem of living: they would be cutting across those now leading to destructive blindness.

A Project for Development

It looks, then, as if it might be worthwhile to devote some energy at least to working out a better utilization of our spare time. The spare time for one per cent of our population would be equivalent, as above reckoned, to the continuous activity of some 40,000 persons. If these people were on the skyline, and kept their eyes open, they would see the things that the giant could see. Indeed this force of 40,000 would be a giant in itself. It could walk the skyline and develop its various opportunities. And this is the job that we propose: a project to develop the opportunities for recreation, recuperation, and employment—in the region of the Appalachian skyline. The project is one for a series of recreational communities throughout the Appalachian chain of mountains from New England to Georgia, these to be connected by a walking trail. Its purpose is to establish a base for a more extensive and systematic development of outdoors community life. It is a project in housing and community architecture.

No scheme is proposed in this particular article for organizing or financing this project. Organizing is a matter of detail to be carefully worked out. Financing depends on local public interest in the various localities affected.

There are four chief features of the Appalachian project:

1. The Trail

The beginnings of an Appalachian trail already exist. They have been established for several years—in various localities along the line. Especially good work in trail building has been accomplished by the Appalachian Mountain Club in the White Mountains of New Hampshire and by the Green Mountain Club in Vermont. The latter association has already built the "Long Trail" for 210 miles thorough the Green Mountains—four fifths of the distance from the Massachusetts line to the Canadian. Here is a project that will logically be extended. What the Green Mountains are to Vermont the Appalachians are to eastern United States. What is suggested, therefore, is a "long trail" over the full length of the Appalachian skyline, from the highest peak in the north to the highest peak in the south—from Mt. Washington to Mt. Mitchell.

The trail should be divided into sections, each consisting preferably of the portion lying in a given State, or subdivision thereof. Each section should be in the immediate charge of a local group of people. Difficulties might arise over the use of private property—especially that amid agricultural lands on the crossovers between ranges. It might be sometimes necessary to obtain a State franchise for the use of rights of way. These matters could readily be adjusted, provided there is sufficient local public interest in the project as a whole. The various sections should be under some sort of general federated control, but no suggestions regarding this form are made in this article.

Not all of the trail within a section could, of course, be built all at once. It would be a matter of several years. As far as possible the work undertaken for any one season should complete some definite usable link—as up or across one peak. Once completed it should be immediately opened for local use and not wait on the completion of other portions. Each portion built should, of course, be rigorously maintained and not allowed to revert to disuse. A trail is as serviceable as its poorest link.

The trail could be made, at each stage of its construction, of immediate strategic value in preventing and fighting forest fires. Lookout stations could be located at intervals along the way. A forest fire service could be organized in each section, which should tie in with the services of the Federal and State Governments. The trail would immediately become a battle line against fire.

A suggestion for the location of the trail and its main branches is shown on the accompanying map. [*Editors' note:* Map not available here.]

2. Shelter Camps

These are the usual accompaniments of the trails which have been built in the White and Green Mountains. They are the trail's equipment for use. They should be located at convenient distances so as to allow a comfortable day's walk between each. They should be equipped always for sleeping and certain of them for serving meals—after the function of the Swiss chalets. Strict regulation is required to assure that equipment is used and not abused. As far as possible the blazing and constructing of the trail and building of camps should be done by volunteer workers. For volunteer "work" is really "play." The spirit of cooperation, as usual in such enterprises, should be stimulated throughout. The enterprise should, of course, be conducted without profit. The trail must be well guarded—against the yegg-man and against the profiteer.

3. Community Groups

These would grow naturally out of the shelter camps and inns. Each would consist of a little community on or near the trail (perhaps on a neighboring lake) where people could live in private domiciles. Such a community might occupy a substantial area—perhaps a hundred acres or more. This should be bought and owned as a part of the project. No separate lots should be sold therefrom. Each camp should be a self-owning community and not a real-estate venture. The use of the separate domiciles, like all other features of the project, should be available without profit.

These community camps should be carefully planned in advance. They should not be allowed to become too populous and thereby defeat the very purpose for which they are created. Greater numbers should be accommodated by more communities, not larger ones. There is room, without crowding, in the Appalachian region for a very large camping population. The location of these community camps would form a main part of the regional planning and architecture.

These communities would be used for various kinds of nonindustrial activity. They might eventually be organized for special purposes—for recreation, for recuperation and for study. Summer schools or seasonal field courses could be established and scientific travel courses organized and accommodated in the different communities along the trail. The community camp should become something more than a mere "playground": it should stimulate every line of outdoor non-industrial endeavor.

4. Food and Farm Camps

These might not be organized at first. They would come as a later development. The farm camp is the natural supplement of the community camp. Here in the same spirit of cooperation and well ordered action the food and crops consumed in the outdoor living would as far as practical be sown and harvested.

Food and farm camps could be established as special communities in adjoining valleys. Or they might be combined with the community camps with the inclusion of surrounding farmlands. Their development could provide tangible opportunity for working out by actual experiment a fundamental matter in the problem of living. It would provide one definite avenue of experiment in getting "back to the land." It would provide an opportunity for those anxious to settle down in the country: it would open up a possible source for new, and needed, employment. Communities of this type are illustrated by the Hudson Guild Farm in New Jersey.

Fuelwood, logs, and lumber are other basic needs of the camps and communities along the trail. These also might be grown and forested as part of the camp activity, rather than bought in the lumber market. The nucleus of such an enterprise has already been started at Camp Tamiment, Pennsylvania, on a lake not far from the route of the proposed Appalachian trail. The camp has been established by a labor group in New York City. They have erected a sawmill on their tract of 2000 acres and have built the bungalows of their community from their own timber.

Farm camps might ultimately be supplemented by permanent forest camps through the acquisition (or lease) of wood and timber tracts. These of course should be handled under a system of forestry so as to have a continuously growing crop of material. The object sought might be accomplished through long-term timber sale contracts with the Federal Government on some of the Appalachian National Forests. Here would be another opportunity for permanent, steady, healthy employment in the open.

Elements of Dramatic Appeal

The results achievable in the camp and scouting life are common knowledge to all who have passed beyond the tenderest age therein. The camp community is a sanctuary and a refuge from the scramble of every-day worldly commercial life. It is in essence a retreat from profit. Cooperation replaces antagonism, trust replaces suspicion, emulation replaces competition. An Appalachian trail, with its camps, communities, and spheres of influence along the skyline, should, with reasonably good management, accomplish these achievements. And they possess within them the elements of a deep dramatic appeal.

Indeed the lure of the scouting life can be made the most formidable enemy of the lure of militarism (a thing with which this country is menaced along with all others). It comes the nearest perhaps, of things thus far projected, to supplying what Professor James once called a "moral equivalent of war." It appeals to the primal instincts of a fighting heroism, of volunteer service and of work in a common cause.

Those instincts are pent up forces in every human and they demand their outlet. This is the avowed object of the Boy Scout and Girl Scout movement, but it should not be limited to juveniles.

The building and protection of an Appalachian trail, with its various communities, interests, and possibilities, would form at least one outlet. Here is a job for 40,000 souls. This trail could be made to be, in a very literal sense, a battle line against fire and flood—and even against disease. Such battles—against the common enemies of man—still lack, it is true, "the punch" of man vs. man. There is but one reason—publicity. Militarism has been made colorful in a world of drab. But the care of the countryside, which the scouting life instills, is vital in any real protection of "home and country." Already basic it can be made spectacular. Here is something to be dramatized.

From Benton MacKaye, "An Appalachian Trail: A Project in Regional Planning," *Journal of the American Institute of Architects* 9 (October 1921): 325–30.

35

Shinkansen—National High-Speed Railways

Japan

DID YOU KNOW . . . ?

➤ Begun in 1959 and completed in 1964, the original Tokaido Shinkansen line was 320 miles (515 kilometers) long, connecting Tokyo and Osaka.

➤ The Shinkansen's average speed is 163 miles per hour (262 kilometers per hour).

➤ On typical days, 225,000 people ride the Shinkansen; on busy days, there can be twice as many.

➤ By 2002, the Shinkansen had 16,347 miles (27,245 kilometers) of rail lines and moved 382 billion passengers and 22 billion tons of freight per year.

➤ Each mile of Japanese rail track is inspected every 10 days.

➤ The Shinkansen is on time 99 percent of the time; if late, it is so by only 12 seconds on average.

➤ When the Shinkansen opened, there were 30 round-trips per day; today there are more than 100.

➤ 260 bullet trains race across Japan daily, covering a distance equal to three times around the globe.

➤ The Shinkansen has a perfect passenger-safety record, with not one death among billions of passengers.

High-speed passenger rail service, introduced in several countries in the latter part of the twentieth century, ranks as one of the most important innovations in the recent history of long-distance transportation. Comfortable, safe, and reliable fast trains that can accommodate large numbers of passengers and travel right into the heart of a city are a boon to the public and to economic life. When, in 1959, the Japanese Diet approved high-speed rail lines, the groundbreaking ceremony occurred an astonishing one week later.

Tsubasa Shinkansen (Bullet Train) in rural Yonezawa, Japan. Courtesy of Shutterstock.

HISTORY

In 1872, Japanese rail history was made with a ceremony dedicating train service—by horse-drawn rail—which began the next day. In 1880 Japan's first rail tunnel, the Osakayama, was completed. The first electric rail was inaugurated in Japan in 1895. Electric locomotives were put in service in 1912, with the EC40 on the Shin-etsu line between Yokokawa and Karuizawa.

Japan pioneered the concept of high-speed lines (defined here as those with a maximum speed of at least 125 miles per hour, or 200 kilometers per hour), which were subsequently built in several European nations and the United States. Before World War II, Japanese railroad planners had begun discussing the creation of an express train between the two major population centers of Tokyo and Osaka.

The desire for a fast train from the Japanese capital of Tokyo to Shimonoseki was first put forward in 1940. The Dangan Ressha (literally, "Bullet Train") was designed to reduce by half the eight hours it took to travel by train from Tokyo to Osaka, the second-largest city in Japan and an important southern nexus for commerce. At the time, steam locomotives were going 95 kilometers per hour, but the bullet trains would be much faster, beginning at 93 miles per hour (150 kilometers per hour), but then quickly ratcheting up to 124 miles per hour (200 kilometers per hour). Construction of the Dangan Ressha was initiated in 1941, but the project was halted when Japan entered World War II.

In the mid-1950s, the notion resurfaced again when Japanese National Railways (JNR) undertook a feasibility study of a super-express train that would take advantage of new technologies. The project was approved in 1958, and construction of the first high-speed railway, the Tokaido Shinkansen, began in

1959. The line, completed in 1964, was 320 miles (515 kilometers) long, connecting Tokyo and Osaka.

In 1949, when JNR was organized, it was originally launched as a government operation. It was not until 1987 that JNR was privatized and split into separate companies: Hokkaido, East, Central, West, Shikoku, Kyushu, and a freight company. The Shinkansen is today run by Japan Railway, formerly part of JNR but now a private consortium.

The Japanese gave the name *Shinkansen* (literally, "New Trunk Line") to the trains; in English, this is often rendered as "bullet train," a tag evoking not only the trains' speed but also the streamlined look of the locomotive. Among the innovations attributed to the Shinkansen are an airplane-like monocoque body structure, low-vibration carriages, lengthy welded rails, prestressed concrete ties, and an automated system for halting trains in the event of an emergency. The line is fitted with snow-melting equipment, allowing the trains to run on time and safely throughout the winter.

Trains on this line were originally designed to run at what were then world-record speeds for passenger rolling stock—up to 130 miles (210 kilometers) per hour. A 1990 survey of scheduled rail passenger service worldwide found that a handful of lines could reach a maximum speed of 190 miles (300 kilometers) per hour, but Japan remained among the leaders: the Shinkansen running between Hiroshima and Kokura, a distance of 119 miles (192 kilometers), had the highest average speed—162.7 miles (261.8 kilometers) per hour. On May 18, 1990, the French TGV (Train de Grande Vitesse) high-speed train reached a record speed of 320.3 miles (515.3 kilometers) per hour.

The building of another Shinkansen line from Osaka southwest to Okayama commenced in 1967. By 1970 work was begun on an extension to Fukuoka on Kyushu Island, the same year the Japanese Diet approved a measure (reproduced here), authorizing the construction of a nationwide Shinkansen network. By the end of the century, Shinkansen lines stretched all the way from Fukuoka in the southwest to Akita in northern Honshu. Lines also extended out to Yamagata and Niigata, and a Shinkansen to Nagano opened shortly before the 1998 Winter Olympic Games were held there. Among new lines are extensions northward to Aomori at the end of Honshu and to Sapporo on Hokkaido Island, and southward on Kyushu from Fukuoka to Kagoshima.

Japanese engineers also explored the feasibility of building even faster trains that use superconducting magnetic-levitation (maglev) technology, which eliminates friction and vibration. By 1999 a Japanese maglev test train reached a record speed of 343 miles (552 kilometers) per hour.

CULTURAL CONTEXT

In 1896 a consumer-oriented innovation was introduced on Japanese trains: color-coding for class of service. The interiors of first-class cars were white, mid-price cars were blue, and economy-class cars were red. Three years later, amenities were added when buffet food service became available in first class. By 1900, sleeping cars were introduced (these perks were discontinued in 1944 under war emergency regulations). That same year, "Tetsudo Shoka" or the "Railway Song" was published, marking the entry of the railroad experience into Japanese folklore. In June 1900 came another milestone: 10 women became employees on the government railway.

The twentieth century saw many improvements, including the first cable car, in 1918, and the first signal with colored lamps. Borrowing technology from Germany, the first automatic ticket machines were introduced in April 1926, with automatic doors following soon thereafter. In 1927, the first subway train went on-line, and in 1929 another import from Germany was put into Japanese service when the first diesel locomotives were used.

True popular acceptance of trains in Japan came in September 1929 when the results of a naming contest were announced and the new trains had affectionate nicknames. The *Fuji* and *Sakura* took their places in history.

Why was Japan so committed to rail? According to Navarra, the country's lack of petroleum reserves, coupled with a determination not to become dependent on outside sources for its energy supply, motivated continuing research into high-speed, low-energy transport. Rail and hybrid cars are part of Japan's strategy to develop innovations that utilize less energy, thereby reducing per-capita costs and lessening dependence on outside resource suppliers.

PLANNING

Following World War II, Japan was determined to develop a prosperous economy using high technology. One tactic for achieving this strategic goal was to send students to technological universities such as the California Institute of Technology and Massachusetts Institute of Technology. One student, Teruo Morita, returned to Japan from MIT and began working with Japan's railways. He also wrote an article about fast trains, which was published in the United States. Many of his colleagues followed the same route—attending technical universities abroad and then bringing knowledge back to Japan, each contributing new science and technology.

Japan had used German innovations early in its train-development history. When, after World War II, Japan began to consider high-speed trains, it sought guidance from France, which had enjoyed such engineering success with its TGV bullet trains. Roger Hutter of the Société Nationale des Chemins de Fers Français (SNCF) had been designated director-general of a company formed in the nineteenth century to build the Channel Tunnel from the French side. Hutter became head of the French Technical Mission to JNR and oversaw the transfer of French technology that was used in the design and construction of the Shinkansen. Roger Hutter was acting under the guidance of Louis Armand, the renowned chairman and modernizer of the French railways. This technology transfer contributed in part to Japan's tremendous forward leap in rail technology.

In moving from planning to reality, Japan cleverly created public interest in the new trains by holding a contest to name them. Nominations and suggestions for names totaled 700,000. As a result, the bullet trains rose to instant celebrity status. Ever since, engineers operating the trains are treated like movie stars, often appearing in advertisements for popular products.

BUILDING

Serious discussion of funding and development of high-speed trains for Japan's railway system began in the 1950s. Legislation was drafted and presented for a vote in 1959. Within one week after approval by the Diet, the first shovel was dug into the earth to begin construction of the Shinkansen.

The Japanese government contributed $640 million to initiate construction. Because a major goal was profitability, the first line to be constructed was the Tokyo-to-Osaka, timed for the start of the Tokyo Olympics in 1964. Just 10 days before the opening ceremony, the project was completed. Since 1975 this route has been an excellent revenue producer, along with the Osaka-to-Fukuoka line. A 1995 study by Japanese historian of technology Hoshimi Uchida reported that the original Tokyo-to-Osaka line has been profitable since its inception. By 2002, there were 16,347 miles (27,245 kilometers) of rail lines in operation in Japan. Passengers numbered 382 billion per year and freight equaled 22 billion tons.

Notwithstanding its profitability, operating such an intensively used high-speed railway creates enormous strains on infrastructure; as a result, about one-third of all costs are for maintenance. This has helped produce an admirable safety record for the operation of the Shinkansen lines: a total absence of fatalities and an impressive record of performance and security. However, regular earthquakes occur in that region. On October 23, 2004, an earthquake derailed a train moving at high speed. However, the 151 passengers aboard the Toki 325 from Tokyo to Niigata were not injured; everyone got off the train and walked back to the Nagaoka station, a trek of about 3.7 miles (six kilometers).

High-technology, high-speed rail lines incur huge construction costs, especially in a country like Japan, which is not only mountainous but consists of a collection of islands. One tunnel on the Tokyo–Niigata line is 14 miles (22.5 kilometers) long. Tunnels account for over half of the line between Okayama and Hakata, which runs under Shimonoseki Strait to Kyushu. Perhaps one of the most remarkable milestones of the Shinkansen was the building of the Seikan Tunnel in 1988, a 32.3-mile stretch that cost $5.6 billion. The greater cost for this tunnel was the 24 years required to complete it because of the enormous land-composition challenges that had to be surmounted. Another amazing achievement is the Tappi Undersea Station, part of the world's lowest railroad tunnel at the time it was completed.

The tunnels created for the Shinkansen not only sped up transport but avoided disasters such as dangerous ferry crossings between Kahodate and Aomori. Many maritime accidents occurred there in what formerly took over four hours across turbulent waters, including the sinking of the *Toya Maru*, which led to the loss of over 1,000 people.

Most of the Shinkansen run on conventional steel rails mounted on concrete sleepers, but the fastest services utilize dedicated tracks to avoid conflict with slower trains. Shinkansen trains run on two different gauges of track—1,067 mm and 1,435 mm—which precludes one part of the system from using the other's trains. However, many of the narrower-gauge routes are being converted to the wider gauge.

An unexpected discovery in the use of the Shinkansen was the loud noise each makes when entering or exiting tunnels. The faster the train's speed, the louder the noise, and engineers found there was a limit to the decibel level of the noise passengers would withstand.

IMPORTANCE IN HISTORY

If Japan was at a stage of rapid expansion following World War II, the early twenty-first century is a period of unprecedented development for China. Among many new technologies being explored is research on very high-speed tube trains that operate in a partial vacuum—hence the term *evacuated tube* because of the partial evacuation of air to reduce resistance. Recognition must be given to the role of General Georges F. Doriot, assistant to the chief of staff of the U.S. Army for research and development in World War II, and founder of the American Research and Development Corporation (ARD), which gave birth to such industry giants as Textron and Digital Equipment Corporation. Under Doriot's leadership, a plan was developed by the Geo-Transport Foundation of New England for high-speed rail in the so-called northeast corridor of the eastern United States (extending from Boston to Washington, DC). While the plan never reached fruition, a few years later the success of the TGV in France proved that such a plan is indeed tenable.

The Shinkansen and TGV are the two great examples of optimizing rail systems within one country. The Trans-Siberian Railway, the Canadian Pacific, and the Union Pacific are excellent examples of transcontinental systems. The next step will be international and intercontinental rail transport. Perhaps the Anglo-French Channel Tunnel will serve as a precedent.

Given the worldwide interest in advancing the technology of train and guideway transport, is it not likely that action will be taken to establish new institutions for this burgeoning field? The *Ecole des Ponts et Chaussees*, officially established in 1775, remains a preeminent center for research and development in the field of highway transport. Should there not be a parallel graduate school for train engineering and management? Is this a topic for consideration by the *Institut Louis Armand*?

The Global Infrastructure Fund (GIF), under the research leadership of Dr. Norio Yamamoto, has been conducting research on a Eurasian transportation network that would link the diverse countries and cultures composing Europe,

Russia, and the central Asian countries of the former Soviet Union. The concept of the GIF was announced in December 1977 by a study task force led by the late Masaki Nakajima, president of Mitsubishi Research Inc., of Tokyo, Japan.

FOR FURTHER REFERENCE

Books and Articles

Krylov, V. V., ed. *Noise and Vibration from High-Speed Trains*. London: Thomas Telford, 2001.

Morita, Teruo. *Intercity High Speed Ground Transportation: History and Prospects*. Master's thesis, June 1984, Massachusetts Insititute of Technology, Center for Transportation and Logistics.

Navarra, John Gabriel. *Supertrains*. Garden City, NY: Doubleday, 1976.

Straszak, A., ed. *The Shinkansen Program: Transportation, Railway, Environmental, Regional and National Development Issues*. Laxenburg, Austria: Institute for Applied Systems Analysis, 1981.

Swartzwelter, Brad. *Faster than Jets: A Solution to America's Long-Term Transportation Problems*. Kingston, WA: Alder Press, 2003.

Taniguchi, Mamoru. *High-Speed Rail in Japan: A Review and Evaluation of the Shinkansen Train*. Berkeley: University of California, Berkeley, Institute of Urban and Regional Development, 1992.

Tuma, J. *The Pictorial Encyclopedia of Transport*. Translated by Alena Emhornova. London: Hamlyn, 1979.

Vranich, Joseph. *Supertrains: Solutions to America's Transportation Gridlock*. New York: St. Martin's Press, 1991.

Internet

For the Global Infrastructure Fund, see http://www.ecdc.net.cn/partners/gif.htm.

For a time line of Japanese railway from 1872 to the present, see http://www.rtri.or/jp/.

To visit the Japan Rail Web site, see http://www.japanrail.com.

For frequently asked question about the Shinkansen, including train types, statistics, dates of construction, services operated, lines used, cars per unit, maximum speed, and colors, see http://www.h2.dion.ne.jp/~dajf/byunbyun/types.htm.

For information about mag-lev trains, see http://www.rtri.or.jp/rtri/future_E.html.

For an interview conducted for the IEEE History Center, Rutgers University, New Brunswick, NJ in 1996, with Yoshihiro Kyotani, an electrical engineer who helped develop the Shinkansen, see http://www.ieee.org/organizations/history_center/oral_histories/transcripts/kyotani.html.

For songs about metros and subways around the world, from Duke Ellington's "Take the A-Train" to the songs composed for Berlin's subway, see http://www.urbanrail.net/metro-song.htm.

Documents of Authorization

LAW FOR THE CONSTRUCTION OF NATION-WIDE HIGH-SPEED RAILWAYS

(Law No. 71, 18 May 1970)

Purpose—Article 1

In view of the importance of the role played by a high-speed transportation system for the comprehensive and extensive development of the land, this Law shall be aimed at the construction of a nation-wide high-speed railway network for the purpose of promoting the growth of the national economy and the enlargement of the people's sphere of life.

Definition—Article 2

The term "high-speed railways" herein shall mean the trunk line railways, on the principal sections of which, operation of trains at high speed over 200 km/h is possible.

Routes of High-Speed Railways—Article 3

The routes of the High-Speed Railways shall be such as will be suited to forming a nation-wide trunk network and link nucleus cities of the nation in a most systematic and efficient way, so that the purpose given in Article 1 will be attained.

Construction and Operation of High-Speed Railways—Article 4

The construction of the High-Speed Railways shall be undertaken by the Japanese National Railways or the Japan Railway Construction Corporation, and the operation thereof, by the Japanese National Railways.

Basic Plan—Article 5

Taking into consideration the trends of transport demand on railways, the emphasized direction of land development and other matters for the effective implementation of the construction of the High-Speed Railways, the Minister of Transport shall, pursuant to the provisions to be set forth by the Government Ordinance, decide upon a basic plan specifying the routes of High-Speed Railways, on which construction is to be started (hereinafter referred to as the "construction line"). (This basic plan is hereinafter referred to as the "basic plan".)

2. The Minister of Transport, in deciding upon the basic plan pursuant to the provision of the preceding Paragraph, shall have previously consulted with the Railway Construction Council. The same shall apply when making changes in the basic plan.

3. The Minister of Transport, when he decides on the basic plan pursuant to Paragraph 1, shall make it public without delay. The same shall apply when making changes in the plan.

Instructions for Investigation on Construction Lines—Article 6

The Minister of Transport, when the basic plan is decided upon as prescribed in the preceding Article, shall instruct the Japanese National Railways or the Japan Railway Construction Corporation to conduct investigations necessary in constructing construction lines. The same shall apply when making changes in the basic plan.

Construction Plan—Article 7

Pursuant to the provisions to be set forth by Government Ordinance, the Minister of Transport shall decide upon a construction plan for the construction of construction lines provided in the basic plan (hereinafter referred to as the "construction plan").

2. The provision of Paragraph 2, Article 5 shall be applied *mutatis mutandis* when the construction plan is to be decided upon or amended.

Instructions for Construction of Construction Lines—Article 8

The Minister of Transport, when the construction plan is decided upon pursuant to the provision of the preceding Article, shall instruct the Japanese National Railways or the Japan Railway Construction Corporation to undertake the construction of the construction lines in accordance with the construction plan. The same shall apply when making changes in the construction plan.

Work Execution Plan—Article 9

The Japanese National Railways or the Japan Railway Construction Corporation, in undertaking the construction of the construction lines under the instructions prescribed in the preceding Article, shall draw up, based on the construction plan, a work execution plan for the construction lines, indicating the name of the lines, sections to be constructed, work methods to be employed and other matters to be stipulated by Ministry of Transport Ordinance and submit it to the Minister of Transport for approval. The same shall apply when making changes in the work execution plan.

2. The work execution plan under the preceding Paragraphs shall be accompanied by drawings indicating the location of the line and other documents to be stipulated by Ministry of Transport Ordinance.

3. The Japan Railway Construction Corporation shall beforehand consult the Japanese National Railways when it draws up or modifies a work execution plan prescribed in Paragraph 1.

4. The Japan Railway Construction Corporation, on receiving the approval prescribed in Paragraph 1, shall submit documents relevant to the work execution plan to the Japanese National Railways.

Designation of Acts Restricted Areas and Cancellation thereof—Article 10

The Minister of Transport may specify areas and designate the same as acts restricted areas when he deems that the restriction of the acts prescribed in Paragraph 1, Article 11 are necessary for the smooth execution of the construction of High-Speed Railways on the land prescribed by Government Ordinance and needed for the construction of High-Speed Railways approved pursuant to Paragraph 1 of the preceding Article.

2. The Minister of Transport, in designating acts restricted areas as provided in the preceding Paragraph, shall beforehand ask the opinion of the Japanese National Railways or the Japan Railway Construction Corporation (hereinafter referred to as the "executor of construction").

3. The Minister of Transport, when he deems it necessary in designating acts restricted areas pursuant to Paragraph 1, may require the executor of construction to submit necessary data.

4. The Minister of Transport, in designating acts restricted areas pursuant to Paragraph 1, shall make public the said acts restricted areas pursuant to the provisions to be set forth by Ministry of Transport Ordinance, and in addition, present the drawing thereof to public inspection.

5. The Minister of Transport, upon completion of construction work on High-Speed Railways in the acts restricted areas designated under the provision of Paragraph 1, shall promptly cancel the designation of the acts restricted areas and make public the cancellation pursuant to the provisions to be set forth by Ministry of Transport Ordinance; the same shall apply when he deems it no longer necessary to keep the area restricted of acts before the construction work is completed.

6. The provisions of Paragraph 2 shall be applied *mutatis mutandis* when canceling the designation of acts restricted areas pursuant to the provision of the preceding Paragraph.

Acts Restricted—Article 11

In the acts restricted areas, designated pursuant to the provision of Paragraph 1 of the preceding Article, no one shall change the form and the constitution of land, or build, reconstruct or add any structure therein; provided that an exception may be made for the acts performed as emergency measures necessitated by an extraordinary disaster or other acts to be prescribed by Government Ordinance.

2. In case a person suffers a loss from the restriction of acts under the provision of the preceding Paragraph, the executor of construction shall compensate for the loss the said person would normally incur.

3. The compensation for the loss referred to in the preceding Paragraph shall be negotiated between the executor of construction and the person sustaining the loss.

4. When no conclusion is reached in the negotiation referred to in the preceding Paragraph, the executor of construction or the person sustaining the loss

may, in accordance with the provisions to be set forth by Government Ordinance, apply to the Land Expropriation Committee for arbitration provided for under Article 94 of the Land Expropriation Law (Law No. 219 of 1951).

Entry into or Temporary Use of Land Belonging to Other Persons—Article 12

The Japanese National Railways or the Japan Railway Construction Corporation, or a person commissioned thereby, may, when necessity compels in performing investigation, survey or work for the construction of High-Speed Railways, enter the land possessed by other persons or temporarily use the land of other persons when no special use is made of the said land, as a yard for placement of materials or as a work place, within the limit of necessity.

2. Those entering the land possessed by other persons by virtue of the provision of the preceding Paragraph shall beforehand notify the occupant to that effect; provided that this shall not apply in cases where it is difficult to give prior notice.

3. In case a person intends to enter the land possessed by other persons pursuant to the provision of Paragraph 1, where there is a building standing or a fence or a railing put up around, the person entering the land shall, in entering, notify the possessor of the said land to that effect beforehand.

4. No one shall enter the land prescribed in the preceding Paragraph before sunrise or after sunset without permission by the possessor of the land.

5. Those entering the land possessed by other persons pursuant to the provision of Paragraph 1 shall carry with them an identification certificate and show it when requested by a person concerned.

6. Those intending to temporarily use as a yard for placement of materials or as a work place, the land belonging to other persons and not in use, pursuant to the provision in Paragraph 1, shall give the occupant and the owner of the land prior notice and ask their opinions.

7. The occupant or the owner of the land shall not, unless he has a just reason, refuse or obstruct the entry into or the temporary use of the land provided in Paragraph 1.

8. The provisions of Paragraph 1 through 4 of the preceding Article shall be applied *mutatis mutandis,* in regard to the compensation for a loss sustained by a person from the entry into or temporary use of land prescribed in Paragraph 1.

9. The form and other matters of the identification certificate provided in Paragraph 5 shall be decided by Ministry of Transport Ordinance.

Financial Measures and Others—Article 13

In view of the importance of the role to be played by High-Speed Railways in the comprehensive and extensive development of the land, in the growth of national economy and in the improvement of the people's life, and in view of the urgent need for construction of High-Speed Railways, the State shall arrange for the provision of aid for the funds needed for the construction thereof and take other necessary measures.

2. In view of the importance of the role which High-Speed Railways are to play in the development and progress of the localities and in the improvement of the life of the inhabitants thereof, local public bodies shall endeavor to take necessary measures with respect to financial aid needed for the construction of High-Speed Railways and lend their good offices for the acquisition of the land needed for the construction thereof.

Exception to Application of the Japanese National Railway Law—Article 14

[Omitted.]

Delegation of Power to Ministry of Transport Ordinance—Article 15

[Omitted.]

Punitive Provisions—Article 16

Any person falling under any of the following items shall be punished with a fine not exceeding ¥100,000:

(1) A person who violates the provision of Paragraph 1, Article 11.
(2) A person who violates the provisions of Paragraph 7, Article 12.

Article 17

When the representative of a juridical person, or a proxy, any of the employees, including servants, of a juridical person or a person, commits an offense of the preceding Article in connection with the business of the juridical person or the person, not only the actual offender but also the juridical person or the person concerned shall be punished with fine prescribed in the same Article.

Article 18

In case the Japanese National Railways or the Japan Railway Construction Corporation fails to seek approval in violation of the provision of Paragraph 1, Article 9, the officer of the Japanese National Railways or of the Japan Railway Construction Corporation having committed the offense shall be punished with a fine not exceeding ¥100,000.

From A. Straszak, ed., *The Shinkansen Program* (Laxenburg, Austria: International Institute for Applied Systems Analysis, 1981).

36
The Trans-Alaska Pipeline

United States and Canada

DID YOU KNOW . . . ?

➤ Construction began in 1974 and was finished in 1977, at a cost of $8 billion.

➤ The project employed more than 20,000 people.

➤ The pipeline is 789 miles (1,270 kilometers) long and 48 inches (122 centimeters) in diameter.

➤ Only half the pipeline is buried; large segments are elevated above ground on supports 50–70 feet (15–21 meters) apart.

➤ Sales of oil leases at Prudhoe Bay brought more than $900 million to Alaska's economy.

➤ The Alaskan oil fields produced 1.5 million barrels of oil daily for 12 years.

➤ In 1989, the *Exxon Valdez* caused the largest oil spill in U.S. history at the time; 1,300 miles of shoreline were contaminated.

➤ The spill released a flood of 257,000 barrels of oil—enough to fill 125 Olympic-size swimming pools.

The discovery of large petroleum deposits at Prudhoe Bay in Alaska's North Slope region in 1968 was an economic bonanza for the state, which reaped more than $900 million from the 1969 sale of North Slope oil leases. Oil proved so valuable to the state's coffers that the personal income tax for state residents was abolished in 1980, and two years later the state government began making annual oil-dividend payments to every man, woman, and child residing in Alaska.

The Trans-Alaska Pipeline stretches into the distance. Courtesy of Corbis.

HISTORY

The Trans-Alaska Pipeline, running from Prudhoe Bay to the ice-free southern Alaska port of Valdez, where the oil could be transferred to tankers, was one of the largest privately financed construction projects ever undertaken. The job employed more than 20,000 people at its peak; the total cost (including the terminal at Valdez) exceeded $8 billion. The result was a steel pipe measuring 789 miles (1,270 kilometers) in length and 48 inches (122 centimeters) in diameter, with a wall thickness ranging from 0.462 inches (1.17 centimeters) to 0.562 inches (1.43 centimeters). Retrieved oil took about six days to traverse the distance to Valdez, in a pipeline designed to carry roughly two million barrels of crude oil daily.

A group of oil companies set up the Alyeska Pipeline Service Company (APSC; the Aleut word *alyeska* means "great land") in 1970 to design, build, and operate what was officially called the Trans-Alaska Pipeline System, or TAPS. Building the pipeline proved an arduous struggle, however, involving both legal and technological problems. Ownership of the land was resolved by the Alaska Native Claims Settlement Act, passed by Congress in late 1971.

Exploitation of the Prudhoe Bay field got off to a slow start because of problems encountered while building a cross-state oil pipeline. Some of the land over which the pipeline would cross was claimed by Native Americans; there was also an ongoing environmental controversy. Not until late 1973 did Congress, in the act presented here, give permission for construction, which was finally completed about four years later.

Dealing with the environmental legal issues was also time-consuming. The proposed pipeline was the subject of the first environmental-impact statement required under the National Environmental Policy Act, which went into effect at the beginning of 1970. In May 1972, the secretary of the interior said he intended to issue a construction permit. A court challenge by environmental groups produced a U.S. Court of Appeals decision that rejected the permit because it would sanction a right-of-way wider than was allowed under the Mineral Leasing Act of 1920. Thus, one purpose of the 1973 legislation authorizing the Trans-Alaska Pipeline was to amend the Mineral Leasing Act to allow greater flexibility regarding the width of the right-of-way.

CULTURAL CONTEXT

As early as 1960, there was talk of great supplies of oil in Alaska. When the rumors turned out to be true and oil was found in Prudhoe Bay in 1968, it proved to be one of the largest discoveries of the twentieth century. Prudhoe Bay produced some 1.5 million barrels a day for almost 12 years. In 1989 it peaked, and yields declined to 350,000 barrels daily.

The discovery also triggered a sizable debate about how to get the oil from the frozen northern site to markets farther south and beyond. Oil companies were eager to participate in any of several solutions under consideration. Ice-breaking tankers on the sea? Huge cargo airplanes? Rail, perhaps an extension of the existing Alaska Railroad? A pipeline? For economic reasons, the pipeline was the first choice.

Many of the technological challenges that confronted the pipeline designers stemmed from the rigors of the Alaskan climate and terrain. The pipeline route ran across mountain ranges; tundra; soft, boggy land known as muskeg; seismic fault lines; and hundreds of streams. It is a huge region, and the completed pipeline would encounter temperatures ranging from 90°F in the summer to −80°F in the winter (roughly 35°C to −60°C). There were also fears that oil leaks might damage the environment. To ensure that the pipeline's welded joints would be sufficiently strong, they were X-rayed and pressure tested. The support structure was built to withstand severe seismic stresses. To allow the pipe to expand and contract with temperature changes, the pipeline was built in a zigzag pattern, which offered the added benefit of letting the pipe flex during an earthquake.

PLANNING

Atlantic Richfield (ARCO), British Petroleum (BP), and Humble Oil, partners in the APSC, applied for a right-of-way permit to build a pipeline in 1969. But there were three obstructions that caused long delays. First, there was disagreement over who owned the land. In 1966 the Native Americans realized they had a legitimate claim on the land; others disagreed, and clarification was sought. The Native Americans began a process of legal intervention, bringing to a halt any action on land all over Alaska. The injunction was still in effect when the APSC applied for its permit, so the company had to wait until 1971, when the Alaska Native Claims Settlement Act was signed. The act gave $1 billion and 44 million acres of land to the Native Americans. But the deal also required that the right-of-way for the proposed pipeline would be free and clear. These decisions resolved the first obstruction.

The same year the APSC applied for its right-of-way, they did not know that another occurrence about the same time would cause long delays and great expense. Ironically, it would also blind them to the very issue they had to solve. In 1969, the United States passed the National Environmental Policy Act (NEPA) heeding a public call to ascertain the appropriate resolution of environmental issues on any federal land. The NEPA act required an environmental-impact study proving that the design proposed in any enterprise would cause no harm. The companies began investigating the environmental impact, and their 12,000-page report filled six volumes. But even this investigation was not sufficient; the findings revealed that the APSC had not studied whether another route, for example, through Canada, might be less environmentally risky. Accordingly, the APSC then had to develop an alternate design; when completed, the environmental impact proved to be about the same. A decision was made to favor the American route, mainly because it would take less time to build. Time was becoming a factor. Not only had delays been time-consuming, but the West was now facing an energy crisis caused by an embargo on Arab-sourced oil. Long lines at the gas pump meant an impatient public. It was decided to build the American pipeline immediately.

One problem began to emerge. Even though the pipeline was now thought of as being in the national interest because it would reduce dependency on foreign oil, the pipeline was delayed again because its design specification did not meet a guideline set forth in the Mineral Leasing Act of 1920. That act restricted the width of rights-of-way, with a maximum boundary of 54 feet (16.5 meters). It required an act of Congress, presented here, to amend the Mineral Leasing Act to permit the Alaska pipeline to be built as specified.

The reader will see from the text presented that there is a detailed focus on the pipeline and the right of way. In title I, section (h), there are four categories of protections, each spelled out in detail. With the guarantees so stipulated, a second act was also agreed upon: title II, the Trans-Alaska Pipeline Authorization Act, also presented here. It is this act that states, in the first section, that "the early development and delivery of oil and gas from Alaska's North Slope to domestic markets is in the national interest because of growing shortages and increasing dependence upon insecure foreign sources." Included in the document of authorization, in title IV, are the "Vessel Construction Standards."

Delays can be costly, especially when considering financing and interest costs. In the case of the Channel Tunnel connecting England and France, it might have been built—with exactly the same specifications—in 1959 for $100 million. By the time the Chunnel was completed in 1993, the cost of construction had risen faster than the rate of inflation, and it cost $15 billion, or 150 times as much arithmetically (not adjusting for the change in the value of money).

BUILDING

After six years of planning and obtaining permissions, the actual building of the pipeline took just over three years, from April 29, 1974, to May 31,

1977. As in so many pipeline projects, the main factor causing delays was the time and difficulty of obtaining permissions from the plethora of regulatory agencies.

There were numerous innovations in the design and construction, some resulting in more delays. Alyeska, the company that would build the pipeline, was working in cooperation with numerous federal agencies, including the U.S. Geological Survey, the Bureau of Land Management, and the Army Corps of Engineers, when they noticed that some permafrost zones were unstable and found that other zones were too close to earthquake-prone territory. The unstable thaw zones were generally south of the Brooks Range, just where earthquake dangers were more apparent. This realization squelched plans to bury the pipeline; instead, a design that kept the pipe above ground in this region was prepared.

Typically, pipelines are buried, but concerns were raised that in permafrost areas the hot oil carried by the pipeline would melt the permafrost and harm the natural ecological balance, not to mention causing the pipe to sag and leak. For this reason, the oil is cooled to about 120°F (50°C) before entering the pipeline. Large segments of the pipeline are elevated above ground, on supports 50–70 feet (15–21 meters) apart; only about half of the pipeline is buried (including segments under many streams, where bridges were too expensive). The aboveground pipe is wrapped with a few inches of fiberglass thermal insulation and covered with aluminum sheet metal. Several miles of underground pipeline are equipped with a cooling system.

Other innovations evolved from the cold-weather engineering research, such as the development of drag reduction by using chemicals. Injection of such an agent was first used in 1979 at Pump Station 1; the procedure was new at the time but is now standard practice. One of the most valuable innovations proved to be the "pigs," a device that cleans and inspects the interior of a petroleum transmission pipeline. "Smart pigs" have onboard computers for greater accuracy, and can detect corrosion in pipes deep inside the earth and measure the interior thickness of the pipe. If the pipe walls are thinner, it means corrosion is causing damage inside the pipe. The pigs proved their value years later, averting a major disaster. Without the smart pigs, the potential disaster is impossible to detect. In the case of the Alaska pipeline, the Atigun section of pipe was deep underground, but in response to data revealed by a smart pig, one section was rerouted and rebuilt. Since then, no other negative results have marred the performance of the project that has delivered as many as 2.03 million barrels of oil per day.

Construction itself was straightforward and efficient. A haul road was built to move 37,500 tons of equipment—reason enough to build the road. Then there was gravel: 73 million cubic yards. Not all the equipment arrived via road; some had to be brought in by air. This led to the construction of 14 airfields, two of which continue to be used. Twenty-nine construction camps were built, both large (3,480 beds) and small (112 beds), to house

Two workers operating oil equipment in connection with the Trans-Alaska Pipeline, 1977. Courtesy of the Library of Congress.

more than 70,000 workers who needed to live on-site.

With the number of contracts on the project totaling over 2,000, the number of laborers needed to build the pipeline was enormous. In contrast to other macro projects, like the Hoover Dam and the Tennessee Valley Authority, where minorities encountered difficulty getting hired, almost 20 percent of the workers on the Alaska pipeline were classified as minorities, including women.

IMPORTANCE IN HISTORY

No consideration of the Alaska pipeline can be concluded without mention of the event that catapulted the project into environmental ignominy for the most glaring omission in environmental-impact statement history. When the *Exxon Valdez* struck rocks on Bligh Reef in Prince William Sound, the gash in the oil tanker's cargo hold released a flood of 257,000 barrels of oil—enough to fill 125 Olympic-size swimming pools. The effects were catastrophic: 1,300 miles (2,000 kilometers) of Alaskan coastline were coated with oil, and 250,000 seabirds, 2,800 sea otters, 300 harbor seals, 250 bald eagles, 22 whales, and billions of salmon and herring eggs were killed.

The disaster pulled together one of the largest cleanup efforts in history. General Thomas McInerney brought to the task a macrosystem engineering approach from his work at Elmendorf Air Force base in Alaska. Known as Command Tactical Information Systems (CTIS), it is not just military but can be applied equally well to civilian disaster rescue and recovery needs by using military hardware and commercial off-the-shelf technology. With this system already in place in Alaska, it was utilized immediately as part of the cleanup. More than 10,000 workers coordinated 1,200 ships in Prince William Sound, all brought together by the CTIS system. Innovations stemming from that experience have advanced the field of telemedicine under the leadership of John Evans of the Medical Defense Performance Review and Massachusetts Institute of Technology Research and Engineering Corporation (MITRE) using CTIS management to improve health care worldwide.

As a result of the oil spill, the Ship Escort/Response Vessel System (SERVS) was mandated by an executive order of the governor of Alaska, requiring that all tankers leaving the facility with a full load of oil must be accompanied by two harbor pilot vessels to guide it safely out of port. The high seas are relatively easy to traverse, but harbors can be dotted with rocks and shifting underwater obstacles. Similar-size ships, such as Cunard's *QE2* cruise ship, have also run aground, that ship doing so near Cape Cod (Massachusetts).

SERVS pulled together a kind of environmental-systems view, a big picture, that should be part of every environmental-impact study. It is ironic that even with all the careful environmental studies performed prior to construction of the Alaska pipeline, an environmental disaster occurred that went well beyond the imagined scope of the environmental-impact statement. The *Exxon Valdez* disaster occurred in part because the government too narrowly defined the environment it was trying to protect; it spent enormous amounts of money to protect the flow of oil without taking into consideration that the flow does not stop at the end of the pipeline. They did not look at the total system.

In the future, it may be possible to take a broader, system-dynamics view of the environment. In the case of the Alaska pipeline, it is clear that the project is a flow system that does not stop once pump stations fill the tanks of ships, but lasts throughout the delivery system. Did the initial interviews and documentation required by the environmental-impact study put too much emphasis on the actual pipeline, its construction, and the land it would traverse, and not enough on the entire project seen as a system?

Another innovation that evolved as a result of the *Exxon Valdez* disaster was a new regulatory entity that brought together all the concerned agencies that would plan, build, manage, and repair the Alaska pipeline. The Joint Pipeline Office (JPO), founded in 1990, is comprised of six state and five federal members:

State of Alaska

Department of Natural Resources
Department of Environmental Conservation
Department of Fish and Game
Department of Labor
Division of Governmental Coordination
Department of Transportation/Public Facilities

Federal

Bureau of Land Management
Department of Transportation/Office of Pipeline Safety
Environmental Protection Agency
U.S. Coast Guard
U.S. Army Corps of Engineers

Does the Alaska pipeline offer a precedent for future pipelines, whether for conveying oil or other products? Yoshihiro Kyotani, who built the world's fastest

test train for Japanese National Railways, a maglev built in Myazaki, envisioned pipes around the world. On November 19, 1991, he gave a lecture before the Washington, DC, Academy of Sciences in which he proposed floating submerged pipelines to transport many things, including freight, in containers on a guideway. Speed could conceivably reach 10,000 miles per hour. The Alaska pipeline may be regarded as a precedent for international and intercontinental pipelines as envisioned by Kyotani.

A similar project began in Libya, commissioned in 1991 by Colonel Muammar Gadhafi of Libya, with a second phase beginning in 1996. When the project is completed it will have more than 2,000 miles of tunnels. Called the Great Man-Made River Project, it will bring water from the huge northern Sahara aquifer and transport the water to Tripoli and other Libyan cities and towns. This multibillion-dollar project includes a tunnel of pipes 13 feet (4 meters) in diameter. More than 12,000 foreign workers are laboring on this project, and Libya expects it will bring sufficient water to make "the desert bloom from Tripoli to Kufra" (Bonner).

Although the magnificent effort that produced the Alaska pipeline contributed to the U.S. economy for several years, it is evident that the United States and its trading partners must now seek resources further afield in order to establish even a semblance of an ever-normal energy supply.

FOR FURTHER INFORMATION

Books and Articles

Bonner, Raymond. "Mysterious Libyan Pipeline Could Be Conduit for Troops." *New York Times*, December 2, 1997. Available at: http://www.fas.org/news/libya/971202-nyt.htm.

Hickel, Walter J. "The Alaskan Pipeline Is Essential." *New York Times*, March 24, 1971.

———. *Crisis in the Commons: The Alaska Solution*. Anchorage: Alaska Pacific University, 2002.

———. *Who Owns America?* Englewood Cliffs, NJ: Prentice Hall, 1971.

Keeble, John. *Out of the Channel: The Exxon Valdez Oil Spill in Prince William Sound*. Cheney: Eastern Washington University Press, 1999.

McInerney, Thomas. "The Command Tactical Information System: Military Software for Macro-engineering Projects." In *Macro-engineering: MIT Brunel Lectures on Global Infrastructure*, edited by Frank P. Davidson, Ernst G. Frankel, and C. Lawrence Meador. Horwood Series in Engineering Science. Chichester, England: Horwood Publishing, 1977.

Rozell, Ned. *Walking My Dog, Jane: From Valdez to Prudhoe Bay along the Trans-Alaska Pipeline*. Pittsburgh, PA: Duquesne University Press, 2000.

Internet

For facts about the pipeline, see http://www.alyeska-pipe.com/pipelinefacts.html.

For frequently asked questions about Trans-Alaska Pipeline history, by the Statewide Library Electronic Doorway: Information Resources for, about, and by Alaskans, see http://sled.alaska.edu/akfaq/aktaps.html.

For L.J. Clifton and B.J. Gallaway's "History of Trans Alaska Pipeline System," a scholarly, well-presented narrative of the project, see http://tapseis.anl.gov/-documents/docs/Section_13_May2.pdf.

For information on the number of contractors, workers, tons of supplies, and other details well laid out and easy to read, see http://www.alyeska-pipe.com/Pipelinefacts/PipelineConstruction.html.

For more on the Arctic National Wildlife Refuge, see http://www.anwr.org/.

To learn more about the Institute of the North and its founder, Walter J. Hickel, see http://www.institutenorth.org/hickel.html.

For more on the U.S. Arctic Research Commission and Mead Treadwell, see http://www.arctic.gov/mtreadwell.htm.

Documents of Authorization

AN ACT

To amend section 28 of the Mineral Leasing Act of 1920, and to authorize a trans Alaska oil pipeline, and for other purposes.

Be it enacted by the Senate and House of Representatives of the United States of America in Congress assembled,

TITLE I

Section 101. Section 28 of the Mineral Leasing Act of 1920 (41 Stat. 449), as amended (30 U.S.C. 185), is further amended to read as follows:

"Grant of Authority

"Sec. 28. (a) Rights-of-way through any Federal lands may be granted by the Secretary of the Interior or appropriate agency head for pipeline purposes for the transportation of oil, natural gas, synthetic liquid or gaseous fuels, or any refined product produced therefrom to any applicant possessing the qualifications provided in section 1 of this Act, as assembled, in accordance with the provisions of this section.

"Definitions

"(b) (1) For the purposes of this section 'Federal lands' means all lands owned by the United States except lands in the National Park System, lands held in trust for an Indian or Indian tribe, and lands on the Outer Continental Shelf. A right-of-way through a Federal reservation shall not be granted if the Secretary or agency head determines that it would be inconsistent with the purposes of the reservation.

"(2) 'Secretary' means the Secretary of the Interior.

"(3) 'Agency head' means the head of any Federal department or independent Federal office or agency, other than the Secretary of the Interior, which has jurisdiction over Federal lands.

"Inter-Agency Coordination

"(c) (1) Where the surface of all of the Federal lands involved in a proposed right-of-way or permit is under the jurisdiction of one Federal agency, the agency head, rather than the Secretary, is authorized to grant or renew the right-of-way or permit for the purposes set forth in this section.

"(2) Where the surface of the Federal lands involved is administered by two or more Federal agencies, the Secretary is authorized, after consultation with the agencies involved, to grant or renew rights-of-way or permits through the Federal lands involved. The Secretary may enter into interagency agreements with all other Federal agencies having jurisdiction over Federal lands for the purpose of avoiding duplication, assigning responsibility, expediting review of rights-of-way or permit applications, issuing joint regulations, and assuring a decision based upon a comprehensive review of all factors involved in any right-of-way or permit application. Each agency head shall administer and enforce the provisions of this section, appropriate regulations, and the terms and conditions of rights-of-way or permits insofar as they involve Federal lands under the agency head's jurisdiction.

"Width Limitations

"(d) The width of a right-of-way shall not exceed fifty feet plus the ground occupied by the pipeline (that is, the pipe and its related facilities) unless the Secretary or agency head finds, and records the reasons for his finding, that in his judgment a wider right-of-way is necessary for operation and maintenance after construction, or to protect the environment or public safety. Related facilities include but are not limited to valves, pump stations, supporting structures, bridges, monitoring and communication devices, surge and storage tanks, terminals, roads, airstrips and campsites, and they need not necessarily be connected or contiguous to the pipe and may be the subjects of separate rights-of-way.

"Temporary Permits

"(e) A right-of-way may be supplemented by such temporary permits for the use of Federal lands in the vicinity of the pipeline as the Secretary or agency head finds are necessary in connection with construction, operation, maintenance, or termination of the pipeline, or to protect the natural environment or public safety.

"Regulatory Authority

"(f) Rights-of-way or permits granted or renewed pursuant to this section shall be subject to regulations promulgated in accord with the provisions of this section and shall be subject to such terms and conditions as the Secretary or agency head may prescribe regarding extent, duration, survey, location, construction, operation, maintenance, use, and termination.

"Pipeline Safety

"(g) The Secretary or agency head shall impose requirements for the operation of the pipeline and related facilities in a manner that will protect the safety of workers and protect the public from sudden ruptures and slow degradation of the pipeline.

"Environmental Protection

"(h) (1) Nothing in this section shall be construed to amend, repeal, modify, or change in any way the requirements of section 102(2)(C) or any other provision of the National Environmental Policy Act of 1969 (Public Law 91–190, 83 Stat. 852).

"(2) The Secretary or agency head, prior to granting a right-of-way or permit pursuant to this section for a new project which may have a significant impact on the environment, shall require the applicant to submit a plan of construction, operation, and rehabilitation for such right-of-way or permit which shall comply with this section. The Secretary or agency head shall issue regulations or impose stipulations which shall include, but shall not be limited to: (A) requirements for restoration, revegetation, and curtailment of erosion of the surface of the land; (B) requirements to insure that activities in connection with the right-of-way or permit will not violate applicable air and water quality standards nor related facility siting standards established by or pursuant to law; (C) requirements designed to control or prevent (i) damage to the environment (including damage to fish and wildlife habitat), (ii) damage to public or private property, and (iii) hazards to public health and safety; and (D) requirements to protect the interests of individuals living in the general area of the right-of-way or permit who rely on the fish, wildlife, and biotic sources of the area for subsistence purposes. Such regulations shall be applicable to every right-of-way or permit granted pursuant to this section, and may be made applicable by the Secretary or agency head to existing rights-of-way or permits, or rights-of-way or permits to be renewed pursuant to this section.

"Disclosure

"(i) If the applicant is a partnership, corporation, association, or other business entity, the Secretary or agency head shall require the applicant to disclose the identity of the participants in the entity. Such disclosure shall include where applicable (1) the name and address of each partner, (2) the name and address of each shareholder owning 3 per centum or more of the shares, together with the number and percentage of any class of voting shares of the entity which such shareholder is authorized to vote, and (3) the name and address of each affiliate of the entity together with, in the case of an affiliate controlled by the entity, the number of shares and the percentage of any class of voting stock of that affiliate owned, directly or indirectly, by that entity, and, in the case of an affiliate which controls that entity, the number of shares and the percentage of any class of voting stock of that entity owned, directly or indirectly, by the affiliate.

"Technical and Financial Capability

"(j) The Secretary or agency head shall grant or renew a right-of-way or permit under this section only when he is satisfied that the applicant has the technical and financial capability to construct, operate, maintain, and terminate the project for which the right-of-way or permit is requested in accordance with the requirements of this section.

"Public Hearings

"(k) The Secretary or agency head by regulation shall establish procedures, including public hearings where appropriate, to give Federal, State, and local government agencies and the public adequate notice and an opportunity to comment upon right-of-way applications filed after the date of enactment of this subsection.

"Reimbursement of Costs

"(l) The applicant for a right-of-way or permit shall reimburse the United States for administrative and other costs incurred in processing the application, and the holder of a right-of-way or permit shall reimburse the United States for the costs incurred in monitoring the construction, operation, maintenance, and termination of any pipeline and related facilities on such right-of-way or permit area and shall pay annually in advance the fair market rental value of the right-of-way or permit, as determined by the Secretary or agency head.

"Bonding

"(m) Where he deems it appropriate the Secretary or agency head may require a holder of a right-of-way or permit to furnish a bond, or other security, satisfactory to the Secretary or agency head to secure all or any of the obligations imposed by the terms and conditions of the right-of-way or permit or by any rule or regulation of the Secretary or agency head.

"Duration of Grant

"(n) Each right-of-way or permit granted or renewed pursuant to this section shall be limited to a reasonable term in light of all circumstances concerning the project, but in no event more than thirty years. In determining the duration of a right-of-way the Secretary or agency head shall, among other things, take into consideration the cost of the facility, its useful life, and any public purpose it serves. The Secretary or agency head shall renew any right-of-way, in accordance with the provisions of this section, so long as the project is in commercial operation and is operated and maintained in accordance with all of the provisions of this section.

"Suspension or Termination of Right-of-Way

"(o) (1) Abandonment of a right-of-way or noncompliance with any provision of this section may be grounds for suspension or termination of the right-of-way if (A) after due notice to the holder of the right of way, (B) a reasonable opportunity to comply with this section, and (C) an appropriate administrative proceeding pursuant to title 5, United States Code, section 554, the Secretary or agency head determines that any such ground exists and that suspension or termination is justified. No administrative proceeding shall be required where the right-of-way by its terms provides that it terminates on the occurrence of a fixed or agreed upon condition, event, or time.

"(2) If the Secretary or agency head determines that an immediate temporary suspension of activities within a right-of-way or permit area is necessary to protect public health or safety or the environment, he may abate such activities prior to an administrative proceeding.

"(3) Deliberate failure of the holder to use the right-of-way for the purpose for which it was granted or renewed for any continuous two-year period shall constitute a rebuttable presumption of abandonment of the right-of-way: *Provided,* That where the failure to use the right-of-way is due to circumstances not within the holder's control the Secretary or agency head is not required to commence proceedings to suspend or terminate the right-of-way.

"Joint Use of Rights-of-Way

"(p) In order to minimize adverse environmental impacts and the proliferation of separate rights-of-way across Federal lands, the utilization of rights-of-way in common shall be required to the extent practical, and each right-of-way or permit shall reserve to the Secretary or agency head the right to grant additional rights-of-way or permits for compatible uses on or adjacent to rights-of-way or permit areas granted pursuant to this section.

"Statutes

"(q) No rights-of-way for the purposes provided for in this section shall be granted or renewed across Federal lands except under and subject to the provisions, limitations, and conditions of this section. Any application for a right-of-way filed under any other law prior to the effective date of this provision may, at the applicant's option, be considered as an application under this section. The Secretary or agency head may require the applicant to submit any additional information he deems necessary to comply with the requirements of this.

"Common Carriers

"(r) (1) Pipelines and related facilities authorized under this section shall be constructed, operated, and maintained as common carriers.

"(2) (A) The owners or operators of pipelines subject to this section shall accept, convey, transport, or purchase without discrimination all oil or gas delivered to the pipeline without regard to whether such oil or gas was produced on Federal or non-Federal lands.

"(B) In the case of oil or gas produced from Federal lands or from the resources on the Federal lands in the vicinity of the pipeline, the Secretary may, after a full hearing with due notice thereof to the interested parties and a proper finding of facts, determine the proportionate amounts to be accepted, conveyed, transported or purchased.

"(3) (A) The common carrier provisions of this section shall not apply to any natural gas pipeline operated by any person subject to regulation under the Natural Gas Act or by any public utility subject to regulation by a State or municipal regulatory agency having jurisdiction to regulate the rates and charges for the sale of natural gas to consumers within the State or municipality.

"(B) Where natural gas not subject to State regulatory or conservation laws governing its purchase by pipelines is offered for sale, each such pipeline shall purchase, without discrimination, any such natural gas produced in the vicinity of the pipeline.

"(4) The Government shall in express terms reserve and shall provide in every lease of oil lands under this Act that the lessee, assignee, or beneficiary, if owner or operator of a controlling interest in any pipeline or of any company operating from pipeline which may be operated accessible to the oil derived from lands under such lease, shall at reasonable rates and without discrimination accept and convey the oil of the Government or of any citizen or company not the owner of any pipeline operating a lease or purchasing gas or oil under the provisions of this Act.

"(5) Whenever the Secretary has reason to believe that any owner or operator subject to this section is not operating any oil or gas pipeline in complete accord with its obligations as a common carrier hereunder, he may request the Attorney General to prosecute an appropriate proceeding before the Interstate Commerce

Commission or Federal power Commission or any appropriate State agency or the United States district court for the district in which the pipeline or any part thereof is located, to enforce such obligation or to impose any penalty provided therefor, or the Secretary may, by proceeding as provided in this section suspend or terminate the said grant of right-of-way for noncompliance with the provisions of this section.

"(6) The Secretary or agency head shall require, prior to granting or renewing a right-of-way, that the applicant submit and disclose all plans, contracts, agreements, or other information or material which he deems necessary to determine whether a right-of-way shall be granted or renewed and the terms and conditions which should be included in the right of way. Such information may include, but is not limited to: (A) conditions for, and agreements among owners or operators, regarding the addition of pumping facilities, looping, or otherwise increasing the pipeline or terminal's throughput capacity in response to actual or anticipated increases in demand; (B) conditions for adding or abandoning intake, offtake, or storage points or facilities; and (C) minimum shipment or purchase tenders.

"Right-of-Way Corridors

"(s) In order to minimize adverse environmental impacts and to prevent the proliferation of separate rights-of-way across Federal lands, the Secretary shall, in consultation with other Federal and State agencies, review the need for a national system of transportation and utility corridors across Federal lands and submit a report of his findings and recommendations to the Congress and the President by July 1, 1975.

"Existing Rights-of-Way

"(t) The Secretary or agency head may ratify and confirm any right-of-way or permit for an oil or gas pipeline or related facility that was granted under any provision of law before the effective date of this subsection, if it is modified by mutual agreement to comply to the extent practical with the provisions of this section. Any action taken by the Secretary or agency head pursuant to this subsection shall not be considered a major Federal action requiring a detailed statement pursuant to section 102(2)(C) of the National Environmental Policy Act of 1970 (Public Law 90-190; 42 U.S.C. 4321).

"Limitations on Export

"(u) Any domestically produced crude oil transported by pipeline over rights-of-way granted pursuant to section 28 of the Mineral Leasing Act of 1920, except such crude oil which is either exchanged in similar quantity for convenience or increased efficiency of transportation with persons or the government of an adjacent foreign state, or which is temporarily exported for convenience or increased efficiency of transportation across parts of an adjacent foreign state and reenters the United States, shall be subject to all of the limitations and licensing requirements of the Export Administration Act of 1969 (Act of December 30, 1969; 83 Stat. 841) and, in addition, before any crude oil subject to this section may be exported under the limitations and licensing requirements and penalty and enforcement provisions of the Export Administration Act of 1969 the President must make and publish an express finding that such exports will not diminish the total quantity or quality of petroleum available to the United States, and are in the national interest and are in accord with the provisions of the Export Administration Act of 1969; *Provided,* That the President shall submit reports to the Congress containing findings made under this section, and after the date of receipt of such report Congress shall have a period of sixty calendar days, thirty days of which Congress must have been in session, to consider whether exports under the terms of this section are in the national interest. If the Congress within this time period passes a concurrent resolution of disapproval stating disagreement with the President's finding concerning the national interest, further exports made pursuant to the aforementioned Presidential findings shall cease.

"State Standards

"(v) The Secretary or agency head shall take into consideration and to the extent practical comply with State standards for right-of-way construction, operation, and maintenance.

"Reports

"(w) (1) The Secretary and other appropriate agency heads shall report to the House and Senate Committees on Interior and Insular Affairs annually on the administration of this section and on the safety and environmental requirements imposed pursuant thereto.

"(2) The Secretary or agency head shall notify the House and Senate Committees on Interior and Insular Affairs promptly upon receipt of an application for a right-of-way for a pipeline twenty-four inches or more in diameter, and no right-of-way for such a pipeline shall be granted until sixty days (not counting days on which the House of Representatives or the Senate has adjourned for more than three days) after a notice of intention to grant the right-of-way, together with the Secretary's or agency head's detailed findings as to terms and conditions he proposes to impose, has been submitted to such committees, unless each committee by resolution waives the waiting period.

"(3) Periodically, but at least once a year, the Secretary of the Department of Transportation shall cause the examination of all pipelines and associated facilities on Federal lands and shall cause the prompt reporting of any potential leaks or safety problems.

"(4) The Secretary of the Department of Transportation shall report annually to the President, the Congress, the Secretary of the Interior, and the Interstate Commerce Commission any potential dangers of or actual explosions, or potential or actual spillage on Federal lands and shall include in such report a statement of corrective action taken to prevent such explosion or spillage.

"Liability

"(x) (1) The Secretary or agency head shall promulgate regulations and may impose stipulations specifying the extent to which holders of rights-of-way and permits under this Act shall be liable to the United States for damage or injury incurred by the United States in connection with the right-of-way or permit. Where the right-of-way or permit involved lands which are under the exclusive jurisdiction of the Federal Government, the Secretary or agency head shall promulgate regulations specifying the extent to which holders shall be liable to third parties for injuries incurred in connection with the right-of-way or permit.

"(2) The Secretary or agency head may, by regulation or stipulation, impose a standard of strict liability to govern activities taking place on a right-of-way or permit area which the Secretary or agency head determines, in his discretion, to present a foreseeable hazard or risk of danger to the United States.

"(3) Regulations and stipulations pursuant to this subsection shall not impose strict liability for damage or injury resulting from (A) an act of war, or (B) negligence of the United States.

"(4) Any regulation or stipulation imposing liability without fault shall include a maximum limitation on damages commensurate with the foreseeable risks or hazards presented. Any liability for damage or injury in excess of this amount shall be determined by ordinary rules of negligence.

"(5) The regulations and stipulations shall also specify the extent to which such holders shall indemnify or hold harmless the United States for liability, damage, or claims arising in connection with the right-of-way or permit.

"(6) Any regulation or stipulation promulgated or imposed pursuant to this section shall provide that all owners of any interest in, and all affiliates or subsidiaries of any holder of, a right-of-way or permit shall be liable to the United States in the event that a claim for damage or injury cannot be collected from the holder.

"(7) In any case where liability without fault is imposed pursuant to this subsection and the damages involved were caused by the negligence of a third party, the rules of subrogation shall apply in accordance with the law of the jurisdiction where the damages occurred.

"Antitrust Laws

"(y) The grant of a right-of-way or permit pursuant to this section shall grant no immunity from the operation of the Federal antitrust laws."

TITLE II

SHORT TITLE

SEC. 201. This title may be cited as the "Trans-Alaska Pipeline Authorization Act".

CONGRESSIONAL FINDINGS

SEC. 202. The Congress finds and declares that:

(a) The early development and delivery of oil and gas from Alaska's North Slope to domestic markets is in the national interest because of growing domestic shortages and increasing dependence upon insecure foreign sources.

(b) The Department of the Interior and other Federal agencies, have, over a long period of time, conducted extensive studies of the technical aspects and of the environmental, social, and economic impacts of the proposed trans-Alaska oil pipeline, including consideration of a trans-Canada pipeline.

(c) The earliest possible construction of a trans-Alaska oil pipeline from the North Slope of Alaska to Port Valdez in that State will make the extensive proven and potential reserves of low-sulfur oil available for domestic use and will best serve the national interest.

(d) A supplemental pipeline to connect the North Slope with a trans-Canada pipeline may be needed later and it should be studied now, but it should not be regarded as an alternative for a trans-Alaska pipeline that does not traverse a foreign country.

CONGRESSIONAL AUTHORIZATION

SEC. 203. (a) The purpose of this title is to insure that, because of the extensive governmental studies already made of this project and the national interest in early delivery of North Slope oil to domestic markets, the trans-Alaska oil pipeline be constructed promptly without further administrative or judicial delay or impediment. To accomplish this purpose it is the intent of the Congress to exercise its constitutional powers to the fullest extent in the authorizations and directions herein made and in limiting judicial review of the actions taken pursuant thereto.

(b) The Congress hereby authorizes and directs the Secretary of the Interior and other appropriate Federal officers and agencies to issue and take all necessary action to administer and enforce rights-of-way, permits, leases, and other authorizations that are necessary for or related to the construction, operation, and maintenance of the trans-Alaska oil pipeline system, including roads and airstrips, as that system is generally described in the Final Environmental Impact Statement issued by the Department of the Interior on March 20, 1972. The route of the pipeline may be modified by the Secretary to provide during construction greater environmental protection.

(c) Rights-of-way, permits, leases, and other authorizations issued pursuant to this title by the Secretary shall be subject to the provisions of section 28 of the Mineral Leasing Act of 1920, as amended by title I of this Act (except the provisions of subsections (h)(1), (k), (q), (w)(2), and (x)); all authorizations issued by the Secretary and other Federal officers and agencies pursuant to this title shall include the terms and conditions required, and may include the terms and conditions permitted, by the provisions of law that would otherwise be applicable if this title had not been enacted, and they may waive any procedural requirements of law or regulation which they deem desirable to waive in order to accomplish the purposes of this title. The direction contained in section 203(b) shall supersede the provisions of any law or regulation relating to an administrative determination as to whether the authorizations for construction of the trans-Alaska oil pipeline shall be issued.

(d) The actions taken pursuant to this title which relate to the construction and completion of the pipeline system, and to the applications filed in connection therewith necessary to the pipeline's operation at full capacity, as described in the Final Environmental Impact Statement of the Department of the Interior, shall be taken without further action under the National Environmental Policy Act of 1969; and the actions of the Federal officers concerning the issuance of the necessary rights-of-way, permits, leases, and other authorizations for construction and initial operation at full capacity of said pipeline system shall not be subject to judicial review under any law except that claims alleging the invalidity of this section may be brought within sixty days following its enactment, and claims alleging that an action will deny rights under the Constitution of the United States, or that the action is beyond the scope of authority conferred by this title, may be brought within sixty days following the date of such action. A claim shall be barred unless a complaint is filed within the time specified. Any such complain shall be filed in a United States district court, and such court shall have exclusive jurisdiction to determine such proceeding in accordance with the procedures hereinafter provided, and no other court of the United States, of any State, territory, or possession of the United States, or of the District of Columbia, shall have jurisdiction of any such claim whether in a proceeding instituted prior to or on or after the date of the enactment of this Act. Any such proceeding shall be assigned for hearing at the earliest possible date, shall take precedence over all other matters pending on the docket of the district court at that time, and shall be expedited in every way by such court. Such court shall not have jurisdiction to grant any injunctive relief against the issuance of any right-of-way, permit, lease, or other authorization pursuant to this section except in conjunction with a final judgment entered in a case involving a claim filed pursuant to this section. Any review of an interlocutory or final judgment, decree, or order of such district court may be had only upon direct appeal to the Supreme Court of the United States.

(e) The Secretary of the Interior and the other Federal officers and agencies are authorized at any time when necessary to protect the public interest, pursuant to the authority of this section and in accordance with its provisions, to amend or modify any right-of-way, permit, lease, or other authorization issued under this title.

LIABILITY

SEC. 204. (a)(1) Except when the holder of the pipeline right-of-way granted pursuant to this title can prove that damages in connection with or resulting from activities along or in the vicinity of the proposed trans-Alaskan pipeline right-of-way were caused by an act of war or negligence of the United States, other government entity, or the damaged party, such holder shall be strictly liable to all damaged parties, public or private, without regard to fault for such damages, and without regard to ownership of any affected lands, structures, fish, wildlife, or biotic or other natural resources relied upon by Alaska Natives, Native organizations, or others for subsistence or economic purposes. Claims for such injury or damages may be determined by arbitration or judicial proceedings.

(2) Liability under paragraph (1) of this subsection shall be limited to $50,000,000 for any one incident, and the holders of the right-of-way or permit shall be liable for any claim allowed in proportion to their ownership interest in the right-of-way or permit. Liability of such holders for damages in excess of $50,000,000 shall be in accord with ordinary rules of negligence.

(3) In any case where liability without fault is imposed pursuant to this subsection and the damages involved were caused by negligence of a third party, the rules of subrogation shall apply in accordance with the law of the jurisdiction where the damage occurred.

(4) Upon order of the Secretary, the holder of a right-of-way or permit shall provide emergency subsistence and other aid to an affected Alaska native, Native organization, or other person pending expeditious filing of, and determination of, a claim under this subsection.

(5) Where the State of Alaska is the holder of a right-of-way or permit under this title, the State shall not be subject to the provisions of subsection 204(a), but the holder of the permit or right-of-way for the trans-Alaska pipeline shall be subject to that subsection with respect to facilities constructed or activities conducted under rights-of-way or permits issued to the State to the extent that such holder engages in the construction, operation, maintenance, and termination of facilities, or in other activities under rights-of-way or permits issued to the State.

(b) If any area within or without the right-of-way or permit area granted under this title is polluted by any activities conducted by or on behalf of the holder to whom such right-of-way or permit was granted, and such pollution damages or threatens to damage aquatic life, wildlife, or public or private property, the control and total removal of the pollutant shall be at the expense of such holder, including any administrative and other costs incurred by the Secretary or any other Federal officer or agency. Upon failure of such holder to adequately control and remove such pollutant, the Secretary, in cooperation with other Federal, State, or local agencies, or in cooperation with such holder, or both, shall have the right to accomplish the control and removal at the expense of such holder.

(c) (1) Notwithstanding the provisions of any other law, if oil that has been transported through the trans-Alaska pipeline is loaded on a vessel at the terminal facilities of the pipeline, the owner and operator of the vessel (jointly and severally) and the Trans-Alaska Pipeline Liability Fund established by this subsection, shall be strictly liable without regard to fault in accordance with the provisions of this subsection for all damages, including clean-up costs, sustained by any person or entity, public or private, including residents of Canada, as the result of discharges of oil from such vessel.

(2) Strict liability shall not be imposed under this subsection if the owner or operator of the vessel, or the Fund, can prove that the damages were caused by an act of war or by the negligence of the United States or other governmental agency. Strict liability shall not be imposed under this subsection with respect to the claim of a damaged party if the owner or operator of the vessel, or the Fund, can prove that the damage was caused by the negligence of such party.

(3) Strict liability for all claims arising out of any one incident shall not exceed $100,000,000. The owner and operator of the vessel shall be jointly and severally liable for the first $14,000,000 of such claims that are allowed. Financial responsibility for $14,000,000 shall be demonstrated in accordance with the provisions of section 311(p) of the Federal Water Pollution Control Act, as amended (33 U.S.C. 1321(p)) before the oil is loaded. The Fund shall be liable for the balance of the claims that are allowed up to $100,000,000. If the total claims allowed exceed $100,000,000, they shall be reduced proportionately. The unpaid portion of any claim may be asserted and adjudicated under other applicable Federal or state law.

(4) The Trans-Alaska Pipeline Liability Fund is hereby established as a nonprofit corporate entity that may sue and be sued in its own name. The Fund shall be administered by the holders of the trans-Alaska pipeline right-of-way under regulations prescribed by the Secretary. The Fund shall be subject to an annual audit by the Comptroller General, and a copy of the audit shall be submitted to the Congress.

(5) The operator of the pipeline shall collect from the owner of the oil at the time it is loaded on the vessel a fee of five cents per barrel. The collection shall cease when $100,000,000 has been accumulated in the Fund, and it shall be resumed when the accumulation in the Fund falls below $100,000,000.

(6) The collections under paragraph (5) shall be delivered to the Fund. Costs of administration shall be paid from the money paid to the Fund, and all sums not needed for administration and the satisfaction of claims shall be invested prudently in income-producing securities approved by the Secretary. Income from such securities shall be added to the principal of the Fund.

(7) The provisions of this subsection shall apply only to vessels engaged in transportation between the terminal facilities of the pipeline and ports under the

jurisdiction of the United States. Strict liability under this subsection shall cease when the oil has first been brought ashore at a port under the jurisdiction of the United States.

(8) In any case where liability without regard to fault is imposed pursuant to this subsection and the damages involved were caused by the unseaworthiness of the vessel or by negligence, the owner and operator of the vessel, and the Fund, as the case may be, shall be subrogated under applicable State and Federal laws to the rights under said laws of any person entitled to recovery hereunder. If any subrogee brings an action based on unseaworthiness of the vessel or negligence of its owner or operator, it may recover from any affiliate of the owner or operator, if the respective owner or operator fails to satisfy any claim by the subrogee allowed under this paragraph.

(9) This subsection shall not be interpreted to preempt the field of strict liability or to preclude any State from imposing additional requirements.

(10) If the Fund is unable to satisfy a claim asserted and finally determined under this subsection, the Fund may borrow the money needed to satisfy the claim from any commercial credit source, at the lowest available rate of interest, subject to approval of the Secretary.

(11) For purposes of this subsection only, the term "affiliate" includes—

(A) Any person owned or effectively controlled by the vessel owner or operator; or

(B) Any person that effectively controls or has the power effectively to control the vessel owner or operator by—

(i) stock interest, or

(ii) representation on a board of directors or similar body, or

(iii) contract or other agreement with other stockholders, or

(iv) otherwise; or

(C) Any person which is under common ownership or control with the vessel owner or operator.

(12) The term "person" means an individual, a corporation, a partnership, an association, a joint-stock company, a business trust, or an unincorporated organization.

ANTITRUST LAWS

SEC. 205. The grant of a right-of-way, permit, lease, or other authorization pursuant to this title shall grant no immunity from the operation of the Federal anti-trust laws.

ROADS AND AIRPORTS

SEC. 206. A right-of-way, permit, lease, or other authorization granted under section 203(b) for a road or airstrip as a related facility of the trans-Alaska pipeline may provide for the construction of a public road or airstrip.

TITLE III

NEGOTIATIONS WITH CANADA

SEC. 301. The President of the United States is authorized and requested to enter into negotiations with the Government of Canada to determine—

(a) the willingness of the Government of Canada to permit the construction of pipelines or other transportation systems across Canadian territory for the transport of natural gas and oil from Alaska's North Slope to markets in the United States, including the use of tankers by way of the Northwest Passage;

(b) the need for intergovernmental understandings, agreements, or treaties to protect the interest of the Governments of Canada and the United States and any party or parties involved with the construction, operation, and maintenance of pipelines or other transportation systems for the transport of such natural gas or oil;

(c) the terms and conditions under which pipelines or other transportation systems could be constructed across Canadian territory;

(d) the desirability of undertaking joint studies and investigations designed to insure protection of the environment, reduce legal and regulatory uncertainty, and insure that the respective energy requirements of the people of Canada and of the United States are adequately met;

(e) the quantity of such oil and natural gas from the North Slope of Alaska for which the Government of Canada would guarantee transit; and

(f) the feasibility, consistent with the needs of other sections of the United States, of acquiring additional energy from other sources that would make unnecessary the shipment of oil from the Alaska pipeline by tanker into the Puget Sound area.

The President shall report to the House and Senate Committees on Interior and Insular Affairs the actions taken, the progress achieved, the areas of disagreement, and the matters about which more information is needed, together with his recommendations for further action.

SEC. 302. (a) The Secretary of the Interior is authorized and directed to investigate the feasibility of one or more oil or gas pipelines from the North Slope of Alaska to connect with a pipeline through Canada that will deliver oil or gas to United States markets.

(b) All costs associated with making the investigations authorized by subsection (a) shall be charged to any future applicant who is granted a right-of-way for one of the routes studied. The Secretary shall submit to the House and Senate Committees on Interior and Insular Affairs periodic reports of his investigation, and the final report of the Secretary shall be submitted within two years from the date of this Act.

SEC. 303. Nothing in this title shall limit the authority of the Secretary of the Interior or any other Federal official to grant a gas or oil pipeline right-of-way or permit which he is otherwise authorized by law to grant.

TITLE IV—MISCELLANEOUS

VESSEL CONSTRUCTION STANDARDS

SEC. 401. Section 4417a of the Revised Statutes of the United States (46 U.S.C. 391a), as amended by the Ports and Waterways Safety Act of 1972 (86 Stat. 424, Public Law 92-340), is hereby amended as follows:

"(C) Rules and regulations published pursuant to subsection (7)(A) shall be effective not earlier than January 1, 1974, with respect to foreign vessels and the United States-flag vessels operating in the foreign trade, unless the Secretary shall earlier establish rules and regulations consonant with international treaty, convention, or agreement, which generally address the regulation of similar topics for the protection of the marine environment. In absence of the promulgation of such rules and regulations consonant with international treaty, convention, or agreement, the Secretary shall establish an effective date not later than January 1, 1976, with respect to foreign vessels and United States-flag vessels operating in the foreign trade, for rules and regulations previously published pursuant to this subsection (7) which he then deems appropriate. Rules and regulations published pursuant to subsection (7)(A) shall be effective not later than June 30, 1974, with respect to United States-flag vessels engaged in the coast-wise trade.".

VESSEL TRAFFIC CONTROL

SEC. 402. The Secretary of the Department in which the Coast Guard is operating is hereby directed to establish a vessel traffic control system for Prince William Sound and Valdez, Alaska, pursuant to authority contained in title I of the Ports and Waterways Safety Act of 1972 (86 Stat. 424, Public Law 92-340).

CIVIL RIGHTS

SEC. 403. The Secretary of the Interior shall take such affirmative action as he deems necessary to assure that no person shall, on the grounds of race, creed, color, national origin, or sex, be excluded from receiving, or participating in any activity conducted under, any permit, right-of-way, public land order, or other Federal authorization granted or issued under title II. The Secretary of the Interior shall promulgate such rules as he deems necessary to carry out the purposes of this subsection and may enforce this subsection, and any rules promulgated under this subsection, through agency and department provisions and rules which shall be similar to those established and in effect under title VI of the Civil Rights Act of 1964.

CONFIRMATION OF THE DIRECTOR OF THE ENERGY POLICY OFFICE

SEC. 405. The head of the Mining Enforcement and Safety Administration established pursuant to Order Numbered 2953 of the Secretary of the Interior issued in accordance with the authority provided by section 2 of Reorganization Plan Numbered 3 of 1950 (64 Stat. 1262) shall be appointed by the President, by and with the advice and consent of the Senate: *Provided,* That if any individual who is serving in this office on the date of enactment of this Act is nominated for such position, he may continue to act unless and until such nomination shall be disapproved by the Senate.

EXEMPTION OF FIRST SALE OF CRUDE OIL AND NATURAL GAS OF CERTAIN LEASES FROM PRICE RESTRAINTS AND ALLOCATION PROGRAMS

SEC. 406. (a) The first sale of crude oil and natural gas liquids produced from any lease whose average daily production of such substances for the preceding calendar month does not exceed ten barrels per well shall not be subject to price restraints established pursuant to the Economic Stabilization Act of 1970, as amended, or to any allocation program for fuels or petroleum established pursuant to that Act or to any Federal law for the allocation of fuels or petroleum.

(b) To qualify for the exemption under this section, a lease must be operating at the maximum feasible rate of production and in accord with recognized conservation practices.

(c) The agency designated by the President or by law to implement any such fuels or petroleum allocation program is authorized to conduct inspections to insure compliance with this section and shall promulgate and cause to be published regulations implementing the provisions of this section.

ADVANCE PAYMENTS TO ALASKA NATIVES

SEC. 407. (a) In view of the delay in construction of a pipeline to transport North Slope crude oil, the sum of $5,000,000 is authorized to be appropriated from the United States Treasury into the Alaska Native Fund every six months of each fiscal year beginning with the fiscal year ending June 30, 1976, as advance payments chargeable against the revenues to be paid under section 9 of the Alaska Native Claims Settlement Act, until such time as the delivery of North Slope crude oil to a pipeline is commenced.

(b) Section 9 of the Alaskan native Claims Settlement Act is amended by striking the language in subsection (g) thereof and substituting the following language: "The payments required by this section shall continue only until a sum of $500,000,000 has been paid into the Alaska Native Fund less the total of advance payments paid into the Alaska Native Fund pursuant to section 407 of the Trans-Alaska Pipeline Authorization Act. Thereafter, payments which would otherwise go into the Alaska Native Fund will be made to the United States Treasury as reimbursement for the advance payments authorized by section 407 of the Trans-Alaskan Pipeline Authorization Act. The provisions of this section shall no longer apply, and the reservation required in patents under this section shall be of no further force and effect, after a total sum of $500,000,000 has been paid to the Alaska Native Fund and to the United States Treasury pursuant to this subsection.".

FEDERAL TRADE COMMISSION AUTHORITY

SEC. 408. (a) (1) The Congress hereby finds that the investigative and law enforcement responsibilities of the Federal Trade Commission have been restricted and hampered because of inadequate legal authority to enforce subpoenas and to seek preliminary injunctive relief to avoid unfair competitive practices.

(2) The Congress further finds that as a result of this inadequate legal authority significant delays have occurred in a major investigation into the legality of the structure, conduct, and activities of the petroleum industry, as well as in other major investigations designed to protect the public interest.

(b) It is the purpose of this Act to grant the Federal Trade Commission the requisite authority to insure prompt enforcement of the laws the Commission administers by granting statutory authority to directly enforce subpoenas issued by the Commission and to seek preliminary injunctive relief to avoid unfair competitive practices.

(c) Section 5(1) of the Federal Trade Commission Act (15 U.S.C. 45(1)) is amended by striking subsection (1) and inserting in lieu thereof:

"(1) Any person, partnership, or corporation who violates an order of the Commission after it has become final, and while such order is in effect, shall forfeit and pay to the United States a civil penalty of not more than $10,000 for each violation, which shall accrue to the United States and may be recovered in a civil action brought by the Attorney General of the United States. Each separate violation of such an order shall be a separate offense, except that in the case of a violation through continuing failure to obey or neglect to obey a final order of the Commission, each day of continuance of such failure or neglect shall be deemed a separate offense. In such actions, the United States district courts are empowered to grant mandatory injunctions and such other and further equitable relief as they deem appropriate in the enforcement of such final orders of the Commission."

(d) Section 5 of the Federal Trade Commission Act (15 U.S.C. 45) is amended by adding at the end thereof the following new subsection:

"(m) Whenever in any civil proceeding involving this Act the Commission is authorized or required to appear in a court of the United States, or to be represented therein by the Attorney General of the United States, the Commission may elect to appear in its own name by any of its attorneys designated by it for such purpose, after formally notifying and consulting with and giving the Attorney General 10 days to take the action proposed by the Commission."

(e) Section 6 of the Federal Trade Commission Act (15 U.S.C. 46), is amended by adding to the end thereof the following proviso:

"*Provided,* That the exception of 'banks and common carriers subject to the Act to regulate commerce' from the Commission's powers defined in clauses

(a) and (b) of this section, shall not be construed to limit the Commission's authority to gather and compile information, to investigate, or to require reports or answers from, any such corporation to the extent that such action is necessary to the investigation of any corporation, group of corporations, or industry which is not engaged or is engaged only incidentally in banking or in business as a common carrier subject to the Act to regulate commerce."

(f) Section 13 of the Federal Trade Commission Act (15 U.S.C. 53) is amended by redesignating "(b)" and "(c)" and inserting the following new subsection:

"(b) Whenever the Commission has reason to believe—

"(1) that any person, partnership, or corporation is violating, or is about to violate, any provision of law enforced by the Federal Trade Commission, and

"(2) that the enjoining thereof pending the issuance of a complaint by the Commission and until such complaint is dismissed by the Commission or set aside by the court on review, or until the order of the Commission made thereon has become final, would be in the interest of the public—

"the Commission by and of its attorney designated by it for such purpose may bring suit in a district court of the United States to enjoin any such act or practice. Upon a proper showing that, weighing the equities and considering the Commission's likelihood of ultimate success, such action would be in the public interest, and after notice to the defendant, a temporary restraining order or a preliminary injunction may be granted without bond: *Provided, however,* That if a complaint is not filed within such period (not exceeding 20 days) as may be specified by the court after issuance of the temporary restraining order or preliminary injunction, the order or injunction shall be dissolved by the court and be of no further force and effect: *Provided further,* That in proper cases the Commission may see, and after proper proof, the court may issue, a permanent injunction. Any such suit shall be brought in the district in which such person, partnership, or corporation resides or transacts business."

(g) Section 16 of the Federal Trade Commission [Act (15 U.S.C. 16) is amended to read as follows:

"SEC. 16. Whenever the Federal Trade Commission has reason to believe that any person, partnership, or corporation is liable to a penalty under section 14 or under subsection (1) of section 5 of this act, it shall—

"(a) certify the facts to the Attorney General, whose duty it shall be to cause appropriate proceedings to be brought for the enforcement of the provisions of such section or subsection; or

"(b) after compliance with the requirements with section 5(m), itself cause such appropriate proceedings to be brought."

SEC. 409. (a) Section 3502 of title 44, United States Code, is amended by inserting in the first paragraph defining "Federal agency" after the words "the General Accounting Office" and before the words "nor the governments" the words "independent Federal regulatory agencies,"

(b) Chapter 35 of title 44, United States Code, is amended by adding after section 3511 the following new section:

"§ 3512. Information for independent regulatory agencies

"(a) The Comptroller General of the United States shall review the collection of information required by independent Federal regulatory agencies described in section 3502 of this chapter to assure that information required by such agencies is obtained with a minimum burden upon business enterprises, especially small business enterprises, and other persons required to furnish the information. Unnecessary duplication of efforts in obtaining information already filed with other Federal agencies or departments through the use of reports, questionnaires, and other methods shall be eliminated as rapidly as practicable. Information collected and tabulated by an independent regulatory agency shall, as far as is expedient, be tabulated in a manner to maximize the usefulness of the information to other Federal agencies and the public.

"(b) In carrying out the policy of this section, the Comptroller General shall review all existing information gathering practices of independent regulatory agencies as well as requests for additional information with a view toward—

"(1) avoiding duplication of effort by independent regulatory agencies, and

"(2) minimizing the compliance burden on business enterprises and other persons.

"(c) In complying with this section, an independent regulatory agency shall not conduct or sponsor the collection of information upon an identical item from ten or more persons, other than Federal employees, unless, in advance of adoption or revision of any plans or forms to be used in the collection—

"(1) the agency submitted to the Comptroller General the plans or forms, together with the copies of pertinent regulations and of other related materials as the Comptroller General has specified; and

"(2) the Comptroller General has advised that the information is not presently available to the independent agency from another source within the Federal Government and has determined that the proposed plans or forms are consistent with the provision of this section. The Comptroller General shall maintain facilities for carrying out the purposes of this section and shall render such advice to the requestive independent regulatory agency within forty-five days.

"(d) While the Comptroller General shall determine the availability from other Federal sources of the information sought and the appropriateness of the forms for the collection of such information, the independent regulatory agency shall make the final determination as to the necessity of the information in carrying out its statutory responsibilities and whether to collect such information. If no advice is received form the Comptroller General within forty-five days, the independent regulatory agency may immediately proceed to obtain such information.

"(e) Section 3508(a) of this chapter dealing with unlawful disclosure of information shall apply to the use of information by independent regulatory agencies.

"(f) The Comptroller General may promulgate rules and regulations necessary to carry out this chapter."

EQUITABLE ALLOCATION OF NORTH SLOPE CRUDE OIL

SEC. 410. The Congress declares that the crude oil on the North Slope of Alaska is an important part of the Nation's oil resources, and that the benefits of such crude oil should be equitably shared, directly or indirectly, by all regions of the country. The President shall use any authority he may have to insure an equitable allocation of available North Slope and other crude oil resources and petroleum products among all regions and all of the several States.

SEPARABILITY

SEC. 411. If any provision of this Act or the applicability thereof is held invalid the remainder of this Act shall not be affected thereby.

Approved November 16, 1973.

From Public Law 93-153. Nov. 16, 1973 87 Stat.

Construction of the Itaipú Hydroelectric Power Station. © Alamy.

37
The Itaipú Hydroelectric Power Project

Brazil and Paraguay

DID YOU KNOW . . . ?

- Construction began in 1984, and the 20th unit went on-line in 2002 at a total cost of $20 billion to date.
- Itaipú provides 78 percent of Paraguay's energy and 25 percent of Brazil's.
- It ranks number one in generating capacity (12,600 megawatts) and annual production (93,428 gigawatt-hours annually).
- In comparison, at its completion the Three Gorges dam of China is expected to generate 18,200 megawatts of power when completed in 2009, 50 percent more than Itaipú.
- The dam is taller than a 55-story building and five miles (8 kilometers) wide. The dam's reservoir took 14 days to fill.
- The iron and steel used would build 380 Eiffel Towers.
- Each generating unit could handle the flow of the entire Seine River in France.
- 8.5 times more earth was dug for Itaipú than for the Channel Tunnel, and 15 times more concrete was used.
- It is designated as one of the seven wonders of the modern world.
- The Brazil Symphony Orchestra gave a concert inside one of the generators to celebrate the project.

Located on the Paraná River between Brazil and Paraguay and built by the two countries as a joint undertaking, the Itaipú hydroelectric power plant is one of the largest hydroelectric projects in history. (*Itaipú* means "the singing stone" in Guarani, the local Indian language.) In addition to producing most of Paraguay's electricity and one-quarter of Brazil's, it supplies water for homes, farms, and industry. At the beginning of the twenty-first century, it ranked number-one in the world in generating capacity (12,600 megawatts) and annual production (93,428 gigawatt-hours in 2000). Construction of the mammoth facility, at a total cost estimated at $20 billion, required the involvement of foreign investors as well as the two national governments and regional administrative units, necessitated the resettling of half a million people, and had far-reaching effects on local ecosystems.

HISTORY

Drought plagued Brazil and Paraguay for many years. As the need for water for drinking, irrigation, and industrial use became more pronounced, a stable supply of water became an urgent problem for both countries.

Authoritarian rule was instituted in Brazil and Paraguay in 1964 and 1966, respectively. In 1966 the military rulers of Brazil unilaterally took control of the Sete Quedas waterfalls on the Paraná River—a natural resource that had been claimed by Paraguay. On June 22, 1966, the Act of Iguaçu was signed, marking agreement between General Humberto Castelo Branco of Brazil and General Alfredo Stroessner of Paraguay to cooperatively study and evaluate the resources of the Paraná River pursuant to creating a stable supply of hydroelectric power for both countries. It was agreed that the electricity generated by Itaipú would be divided equally between the two countries. However, it was stipulated that Paraguay would pay off extensive Brazilian financial assistance within 50 years by selling its surplus electricity, at cost, only to Brazil.

In 1973, the Itaipú Treaty was signed. It set forth guidelines for the control and development of water resources and implementation of the Itaipú project. It also provided for the creation of a binational entity, Itaipú Binacional. That treaty is reproduced here, along with supplementary diplomatic correspondence between the signatories.

A year later, in May 1974, the entity called Itaipú Binacional was formally created. Owned in equal proportions by ELETROBRÁS of Brazil and ANDE of Paraguay, Itaipú was given responsibility for planning, building, and operating a power-generating facility (whose tentative specifications were listed in the treaty). The treaty also reiterated that the electricity generated by Itaipú would be divided equally between the two countries, and that excess not used by one country would first be made available to the other country at a fair market price.

CULTURAL CONTEXT

The current ability of South American countries to work cooperatively is noteworthy. How did this evolve? Early independence from colonial rule by Spain, coupled with native languages and cultural traditions that are related (Spanish and Portuguese), may be the successful combination. While the Hoover Dam was also a cooperative undertaking between the numerous companies participating in the project, Itaipú went further because it involved two countries. In 1978, Argentina also became involved.

The need for a stable water supply is critical for the economic growth of the Latin American countries, and for the well-being of their populations. Itaipú accounts for about 25 percent of power production in Brazil and 80 percent in Paraguay, and, as such, it plays an important role in the economies of both nations.

Itaipú epitomizes the efforts of two neighboring countries, Brazil and Paraguay, to develop their common energy resources for mutual benefit. It testifies to the spirit of international cooperation that, from the beginning, pervaded all levels of Brazilian and Paraguayan government officials as well as workers and employees who were engaged in this power-development project (Tihlo).

PLANNING

Studies were conducted by the Brazilian-Paraguayan Joint Technical Commission, established in February 1967, that found that (1) Brazil and Paraguay both wanted free navigation on the international rivers of the River Plate Basin, and (2) Brazil and Paraguay both wanted to use the Paraná River for hydroelectric power.

Itaipú Binacional is a model of cooperative planning with several stipulations, including the following:

1. All meeting records and notes would be in both Spanish and Portuguese.
2. There would be two central headquarters, Brasília and Asunción.
3. There would be equal numbers of nationals on the Governing Council and Executive Directorate.
4. Each country was to have an equal share of the electricity generated, and would have a right to acquire the energy not used by the other partner for its own consumption.
5. Each country would pay royalties in equal amounts.

The agreement presented here recommends, in article XII, financial advantages that would make the construction and operation economically attractive: no taxes on the new joint entity and no taxes on the electrical services to be produced. No taxes would be levied on materials and equipment, even if obtained from a third country.

Article XVIII delineates many aspects of the project that were to influence its development, providing a list of diplomatic, administrative, and governmental considerations that must accompany a macro project: "(a) Diplomatic and consular; (b) Administrative and financial; (c) Employment and social security; (d) Tax and customs; (e) Passage across the international frontier; (f) Urban and housing; (g) Policy and security; (h) Control of access."

Considerable attention was paid to cultural artifacts and environmental treasures. Before building, stakeholders were surveyed and consulted about environmental measures to be enacted that would keep valuable resources and artifacts safe. However, unlike other more current macro projects, no environmental-impact statement was required prior to receiving permission to build.

BUILDING

Named after the island in the Paraná River that became the site of the project's main dam, Itaipú is a complex of spillways and dams stretching 4.8 miles (7.7 kilometers) across the river. The centerpiece of Itaipú is a hollow-concrete gravity dam nearly 3,300 feet (1 kilometer) long and 643 feet (196 meters) high. The plant produces electricity with huge generating units, each rated at 700 megawatts. The first was commissioned in 1984, and the 18th went on-line in April 1991. Two new units came on-line in 2002, raising the total installed capacity to 14,000 megawatts. Stored water is held in a reservoir over 100 miles (170 kilometers) long with an area of about 520 square miles (1,350 square kilometers).

More than 30,000 workers were employed on Itaipú over the seven years it took to build. They were governed by Article XXI, which suggests that employees of a third nationality be treated in accordance with Brazilian or Paraguayan law depending upon the worker's assignment.

It was the first time a river as big as the Paraná was fully diverted; just carving out the diversion channel, which measured 1.3 miles (2 kilometers) long, 300 feet (90 meters) deep, and 490 feet (150 meters) wide, took almost three years.

Financing for the project was complex. It was basically funded by loans obtained from several sources, with the Brazilian government as guarantor. As of December 1998, the cost of the undertaking amounted to $11.7 billion. After Argentina became part of the deal in 1979, discussions ensued as to the possibility of Argentina purchasing electricity from Brazil and Paraguay. Discussions led to linking Brazil and Argentina's electrical systems in 1983.

Farm animals were a prominent consideration in the planning and building of the dam. One purpose of the project was to provide drinking water for cattle. When the dam was built and the reservoir filled, exotic wildlife was protected by forming seven separate wildlife habitat areas where native fauna and flora could be preserved. New roads also brought tourists to view the animals in Iguaçu National Park.

Still there was major damage to wildlife. In its early stages the construction project threatened to wreak serious environmental consequences, wiping out endangered species and forest habitats. When the reservoir was completely filled, it came at the loss of valuable rainforest and the devastation of large populations of native parrots whose tree homes were destroyed. By 1974 it was reported that almost 85 percent of the forest on the Paraguayan side of the Paraná River had disappeared. Later, stepped-up environmental protection efforts by both Brazil and Paraguay, including reforestation and transplantation programs and the creation of several nature reserves, saved many species from extinction.

IMPORTANCE IN HISTORY

Itaipú could be a topic in the development of the waters of South America. Can the Amazon River be used on a worldwide basis? If an ever-normal water supply became an agreed objective of the world community, the immense freshwater resources of the Amazon River would surely be assessed (after the necessary environmental studies). Use of the Amazon was recommended by J. Vincent Harrington at a meeting presided over by Sir Robert G. A. Jackson, the first person appointed second-in-charge after the formation of the United Nations.

Because Itaipú Binacional is a single entity endowed by two governments, with the power to own, build, and operate on behalf of both countries the hydroelectric power plant on the river that is owned in condominium by the two countries, Itaipú Binacional has special legal standing under international law.

The Itaipú project is unique because of its social and economic relationships with the local communities. About $13 million is paid on a monthly basis by Itaipú to each of the partner countries as royalties for use of the Paraná River waters to produce electric power. In Brazil, about 38 percent of that amount is distributed by the federal government to the municipalities, in proportion to the areas of their respective territories that have been lost to the reservoir. In this way, social benefits of the project are amplified. As a result of Itaipú, the population living near the reservoir receives an income comparable to what would otherwise have been obtained from economic developments in the flooded areas. This income has played a pivotal role in the social and economic development of the region served by the power plant and its reservoir.

How many dams have inspired works of art? Itaipú is the subject of music created by composer Philip Glass. When commissioned to develop a new work for the Atlanta Symphony Orchestra, the artist was looking at ways to contrast nature and technology. Glass visited Itaipú in 1988 with friends, and when he saw the massive structure, more than 55 stories high, he was awestruck, comparing it to the Egyptian pyramids. Later a friend told Glass the meaning of the name *Itaipú*—"the singing stone." Glass brought his inspiration forward in great detail, using Guarani language and poetry in the libretto of his work of four movements, which was performed on November 2, 1989, in Atlanta.

FOR FURTHER REFERENCE

Books and Articles

Dryzek, John S., and David Schlossberg. *Debating the Earth: The Environmental Politics Reader*. Oxford: Oxford University Press, 1998.

Keck, Margaret E. "Amazônia and Environmental Politics." In Tulchin, Joseph S. and Heather A. Golding, eds. *Environment and Security in the Amazon Basin*. Washington, DC: Woodrow Wilson International Center for Scholars, 2002, 47. Provides a discussion of Robert Panero's and Herman Kahn's plan to link all rivers to make "great lakes" in South America.

Kohlhepp, Gerd. *Itaipú: Basic Geopolitical and Energy Situation; Socio-economic and Ecological Consequences of the Itaipú Dam and Reservoir on the Rio Paraná*. Braunschweig, Germany: F. Vieweg, 1987.

Krech, Shepard. *The Ecological Indian: Myth and History*. New York: Norton, 1999.

Mufson, Steven. "The Yangtze Dam: Feat or Folly?" *Washington Post*, Nov. 9, 1997, A-01. The article compares the Three Gorges dam to Itaipú.

Reaka-Kudla, Marjorie, Don Wilson, and Edward O. Wilson. *Biodiversity II: Understanding and Protecting our Biological Resources*. Washington, DC: Joseph Henry Press, 1997.

Redman, Charles. *Human Impact on Ancient Environments*. Tucson: University of Arizona Press, 1999.

Internet

For information about the Itaipú project, see http://www.itaipu.gov.br/, http://ce. eng.usf.edu/pharos/wonders/Modern/itaipu.html, and http://www.sovereign-publications.com/itaipu.htm.

For an informational article by Altino Ventura Fihlo, "Itaipú: A Binational Hydroelectric Power Plant, Its Benefits and Regional Context," see http://www.dams.org/ kbase/submissions/showsub.php?rec=ins237.

For a snapshot of key facts, see http://www.pbs.org/wgbh/buildingbig/wonder/structure/ itaipu.html.

For comments on the project from the International Research Institute for Climate and Society, see http://iri.columbia.edu/.

For more on Philip Glass's composition *Itaipú*, see http://www.philipglass.com/html/ compositions/itaipu.html.

For an overview of dams and the environment, see http://www.unep.org/dams.

For an online copy of the authorizing document in this chapter, see http://www. internationalwaterlaw.org/RegionalDocs/Parana2.htm.

Music

Glass, Philip. *Itaipú*. Libretto translated from Guarani Indian text by Daniela Thomas, copyright 1988 Dunvagen Music Publishers, Inc. New York: Sony 46352, 1993.

Documents of Authorization—I

TREATY BETWEEN THE FEDERATIVE REPUBLIC OF BRAZIL AND THE REPUBLIC OF PARAGUAY CONCERNING THE HYDROELECTRIC UTILIZATION OF THE WATER RESOURCES OF THE PARANÁ RIVER OWNED IN CONDOMINIUM BY THE TWO COUNTRIES, FROM AND INCLUDING THE SALTO GRANDE DE SETE QUEDAS OR SALTO DEL GUIRÁ, TO THE MOUTH OF THE IGUASSU RIVER

The President of the Federative Republic of Brazil, General of the Army Emilio Garrastazu Médici, and the President of the Republic of Paraguay, General of the Army Alfredo Stroessner,

Considering the spirit of cordiality between the two countries and the ties of fraternal friendship that unite them;

Their common interest in the hydroelectric utilization of the water resources of the Paraná River owned in condominium by the two countries, from and including the Salto Grande de Sete Quedas or Salto del Guairá to the mouth of the Iguassu River;

The provisions of the Final Act signed at Foz do Iguaçu on 22 June 1966, concerning the division into equal parts between the two countries of such electrical energy as may be produced by the differences in level of the Paraná River on the stretch referred to above;

The provisions of article VI of the Treaty of the River Plate Basin;

The stipulations in the Declaration of 3 June 1971 of Asunción on the utilization of international rivers;

The studies of the Brazilian-Paraguayan Joint Technical Commission established on 12 February 1967;

The identical positions traditionally held by the countries concerning free navigation on the international rivers of the River Plate Basin,

Have resolved to conclude a treaty and, to that end, have designated as their plenipotentiaries:

The President of the Federative Republic of Brazil:

Ambassador Mário Gibson Barboza, Minister for Foreign Affairs;

The President of the Republic of Paraguay:

Dr. Raúl Sapena Pastor, Minister for Foreign Affairs,

who, having exchanged their full powers, found in good and due form,

Have agreed as follows:

Article I. The High Contracting parties agree to utilize for hydroelectric purposes, jointly and in accordance with the provisions of this Treaty and the annexes thereto, the water resources of the Paraná River owned in condominium by the two countries, from and including the Salto Grande de Sete Quedas, or Salto del Guairá, to the mouth of the Iguassu River.

Article II. For the purposes of this Treaty:

(a) "Brazil" means the Federative Republic of Brazil;

(b) "Paraguay" means the Republic of Paraguay;

(c) "Commission" means the Brazilian-Paraguayan Joint Technical Commission established on 12 February 1967;

(d) "ELETROBRÁS" means Centrais Elétricas Brasileriras S.A.— ELETROBRÁS, of Brazil or such legal entity as may succeed it;

(e) "ANDE" means the Administración Nacional de Electricidad of Paraguay, or such legal entity as may succeed it;

(f) "ITAIPU" means the binational entity created by this Treaty.

Article III. The High Contracting Parties shall create, with equal rights and obligations, a binational entity known as Itaipu, with a view to undertaking the hydroelectric utilization referred to in article I.

1. ITAIPU shall be constituted by ELETROBRÁS and ANDE, each having equal participation in the capital, and shall be governed by the rules laid down in this Treaty, in the Statute which constitutes annex A thereto and in the other annexes.

2. The meeting records, resolutions, reports or other official documents of the administrative organs of ITAIPU shall be written in Portuguese and in Spanish.

Article IV. ITAIPU shall have headquarters at Brasília, capital of the Federative Republic of Brazil, and at Asunción, capital of the Republic of Paraguay.

1. ITAIPU shall be administered by a Governing Council and an Executive Directorate composed of equal numbers of nationals of the two countries.

2. The meeting records, resolutions, reports or other official documents of the administrative organs of ITAIPU shall be written in Portuguese and in Spanish.

Article V. The High Contracting Parties authorize ITAIPU to undertake, during the period of validity of this Treaty, the hydroelectric utilization of the stretch of the Paraná River referred to in article I.

Article VI. The following shall be part of this Treaty:

(a) The Statute of the binational entity known as ITAIPU (annex A);

(b) The general description of the facilities for the production of electrical energy and the auxiliary works, with any modifications that may prove necessary (annex B);

(c)   The financial bases of ITAIPU and the conditions for the provision of its electrical services (annex C).

Article VII. The facilities for the production of electrical energy and the auxiliary works shall not produce any change in the boundaries between the two countries established in the treaties now in effect.

1.   The facilities and works set up pursuant to this Treaty shall not confer upon either of the High Contracting Parties the right of ownership or of jurisdiction over any part of the other's territory.
2.   The authorities declared competent by each of the High Contracting Parties shall establish, as appropriate and by such procedures as they deem proper, a suitable signal system in the works to be constructed, for the practical purposes of exercising jurisdiction and control.

Article VIII. The resources needed to constitute ITAIPU's capital shall be furnished to ELECTROBRÁS and ANDE respectively by the Brazilian Treasury and by the Paraguayan Treasury or by the financing institutions designated by the Governments.

Either of the High Contracting Parties may, with the consent of the other, advance to it the funds to constitute the capital, on conditions established by agreement.

Article IX. Such resources in addition to those mentioned in article VIII as are needed for studies, construction and operation of the power station and of the auxiliary works and facilities shall be provided by the High Contracting Parties or obtained by ITAIPU through credit operations.

Article X. The High Contracting Parties shall jointly or separately, directly or indirectly, and in such manner as they may agree upon, give to ITAIPU, at its request, a guarantee for any credit operations it may carry out. They shall ensure in the same manner the exchange transactions necessary for the payment of the obligations assumed by ITAIPU.

Article XI. In so far as possible and under comparable conditions, the skilled and unskilled manpower, equipment and materials available in the two countries shall be utilized in an equitable manner.

1.   The High Contracting Parties shall adopt all the necessary measures to enable their nationals to work, without distinction, on projects related to the purpose of this Treaty, carried out in the territory of either Party.
2.   The provisions of this article shall not apply to the conditions agreed upon with financing institutions concerning the engagement of skilled personnel or the acquisition of equipment or materials. The provisions of this article shall also be inapplicable if technological conditions so require.

Article XII. The High Contracting Parties shall adopt the following standards with respect to taxation:

(a)   They shall not impose taxes, charges or compulsory loans of any nature on ITAIPU and the electrical services provided by it;

(b)   They shall not impose taxes, charges or compulsory loans of any nature on such materials and equipment as ITAIPU may acquire in either country or import from a third country for use in the construction of the power station, its accessories and supplementary works or for incorporation in the power station, its accessories and supplementary works. Similarly, they shall not impose taxes, charges or compulsory loans of any nature affecting operations which are related to such materials and equipment and to which ITAIPU is party;

(c)   They shall not impose taxes, charges or compulsory loans of any nature on ITAIPU's profits and on its payments and remittances to any individual or corporate body provided that payment of such taxes, charges and compulsory loans is the legal responsibility of ITAIPU;

(d)   They shall impose no restriction of taxation on any movement of ITAIPU's funds resulting from the implementation of this Treaty;

(e)   They shall impose no restrictions of any nature on the transit or storage of the materials and equipment referred to in sub-paragraph (b) of this article;

(f)   The materials and equipment referred to in sub-paragraph (b) of this article shall be admitted into the territories of the two countries.

Article XIII. The energy produced by the hydroelectric utilization scheme referred to in article I shall be divided into equal parts between the two countries and each one shall have the right to acquire, in the manner laid down in article XIV, the energy not utilized by the other country for its own consumption.

The High Contracting Parties pledge to acquire, jointly or separately, in such manner as they may agree upon, the total amount of installed power.

Article XIV. The electrical services of ITAIPU shall be acquired by ELETROBRÁS and ANDE, which may also acquire them through such Brazilian or Paraguayan enterprises or entities as they may designate.

Article XV. Annex C contains the financial bases of ITAIPU and the conditions for the provision of its electrical services.

1.   ITAIPU shall pay royalties to the High Contracting Parties in equal amounts, for the utilization of the hydraulic potentials.

2.   ITAIPU shall include in its cost of services the amount needed to pay profits.

3.   ITAIPU shall also include in its cost of services the amount needed to compensate the High Contracting Party ceding energy to the other.

4. The real value of the amount in United States dollars intended for the payment of royalties, profits and compensation, laid down in annex C, shall be kept constant, and to that end, the said amount shall follow the fluctuations in the value of the United States dollar in terms of its equivalent in gold of the weight and fineness in effect on the date of the exchange of the instruments of ratification of this Treaty.

5. The said value of the United States dollar in terms of weight and fineness of gold may be replaced if the official parity of the dollar ceases to be tied to gold.

Article XVI. The High Contracting Parties express their determination to bring about all the conditions that will make it possible for the first generating unit to start operating within eight years after the ratification of this Treaty.

Article XVII. The High Contracting Parties undertake to declare to be areas of public utility the areas necessary for the hydroelectric utilization facility and auxiliary works and their operation, and take in their respective areas of sovereignty any administrative or judicial actions designed for expropriating land or land improvements or for establishing easements over the same.

1. ITAIPU shall be responsible for delimiting such areas, subject to approval by the High Contracting Parties.
2. ITAIPU shall be responsible for paying for the expropriation of the delimited areas.
3. Persons providing services to ITAIPU and goods consigned to ITAIPU or to individuals or bodies corporate under contract to it shall have free passage in the delimited areas.

Article XVIII. The High Contracting Parties may, by means of additional protocols or unilateral acts, adopt any measures necessary for the implementation of this Treaty, particularly those relating to the following aspects:

(a) Diplomatic and consular;
(b) Administrative and financial;
(c) Employment and social security;
(d) Tax and customs;
(e) Passage across the international frontier;
(f) Urban and housing;
(g) Police and security;
(h) Control of access to the areas delimited in accordance with article XVII.

Article XIV. The competent jurisdictions for ITAIPU, with respect to individuals or bodies corporate domiciled or headquartered in Brazil or Paraguay, shall be those of Brasília and Asunción respectively. For that purpose, each High Contracting Party shall apply its own laws, taking account of the provisions of this Treaty and the annexes thereto.

In connection with individuals or bodies corporate domiciled or headquartered outside Brazil or Paraguay, ITAIPU shall establish by agreement the clauses that will govern contractual relations with regard to works and supplies.

Article XX. The High Contracting Parties shall adopt, by means of an additional protocol to be signed within 90 days after the date of the exchange of the instruments of ratification of this Treaty, the legal rules applicable to the employment and social security relations of workers engaged by ITAIPU.

Article XXI. The civil and/or penal responsibility of the councilors, directors, deputy directors and other Brazilian and Paraguayan employees of ITAIPU for acts harmful to the latter's interest shall be investigated and judged in accordance with the provisions of the respective national laws.

In connection with employees of a third nationality, proceedings shall be in accordance with Brazilian or Paraguayan national law, depending on whether the employees are assigned to Brazil or to Paraguay.

Article XXII. Any disagreement over the interpretation or implementation of this Treaty and the annexes thereto shall be settled through the usual diplomatic channels, with no resultant delay or interruption in the construction and/or operation of the hydroelectric utilization scheme and of its auxiliary works and facilities.

Article XXIII. The Brazilian-Paraguayan Joint Technical Commission established on 12 February 1967 with a view to carrying out the studies referred to in the preamble to this Treaty shall remain in existence until it submits to the High Contracting Parties the final report on the assignment entrusted to it.

Article XXIV. This treaty shall be ratified, and the respective instruments shall be exchanged as soon as possible at the city of Asunción.

Article XXV. This Treaty shall enter into force on the date of the exchange of the instruments of ratification and shall remain in force until the High Contracting Parties, by a new agreement, adopt such decision as they may deem appropriate.

IN WITNESS WHEREOF the aforesaid plenipotentiaries have signed this Treaty, in duplicate in the Portuguese and Spanish languages, both texts being equally authentic.

Done at the City of Brasilia on 26 April 1973.
MÁRIO GIBSON BARBOZA
RAÚL SAPENA PASTOR

ANNEX A

STATUTE OF "ITAIPU"

Chapter 1. DENOMINATION AND PURPOSE

Article 1. ITAIPU is a binational entity established under article III of the Treaty signed by Brazil and Paraguay on 26 April 1973, the parties constituting it being:

(a) Centrais Elétricas Brasileiras S.A.—ELETROBRÁS, a Brazilian limited-liability company of mixed economy;
(b) The Administración Nacional de Electricidad—ANDE, a Paraguayan autarch entity.

Article 2. The purpose of ITAIPU is the hydroelectric utilization of the water resources of the Paraná River owned in condominium by the two countries, from and including the Salto Grande de Sete Quedas, or Salto del Guairá, to the mouth of the Iguassu River.

Article 3. ITAIPU shall be governed by the rules laid down in the Treaty of 26 April 1973 in this Statute and in the other annexes.

Article 4. ITAIPU shall, in accordance with the provisions of the Treaty and the annexed thereto, have the juridical, financial and administrative capacity and the technical responsibility to study, plan, direct and execute the works for which it was established, bring thence into service and operate them, for which purposes it may require rights and undertake obligations.

Article 5. ITAIPU shall have headquarters at Brasilia, capital of the Federative Republic of Brazil, and at Asunción, capital of the Republic of Paraguay.

Chapter II. CAPITAL

Article 6. ITAIPU shall have a capital equivalent to US 100,000,000.00 dollars (one hundred million United States dollars), belonging to ELETROBRÁS and ANDE in equal and non-transferable parts.

The capital shall be kept at a constant value in accordance with the provisions of article XV, paragraph 4 of the Treaty.

Chapter III. ADMINISTRATION

Article 7. ITAIPU shall have as its administrative organs a Governing Council and an Executive Directorate.

Article 8. The Governing Council shall be composed of 12 Councilors appointed as follows:

(a) Six by the Brazilian Government, including one designated by the Ministry of Foreign Affairs and two by ELETROBRÁS;
(b) Six by the Paraguayan Government, including one designated by the Ministry of Foreign Affairs and two by ANDE.

1. In addition, the Director-General and Deputy Director-General provided for in article 12 shall be participating but non-voting members of the Council.

2. The meetings of the Council shall be presided over alternately by a Brazilian and a Paraguayan Councilor and, on a rotating basis, by all members of the Council.

3. The Council shall appoint two Secretaries, one Brazilian and the other Paraguayan, who shall be responsible, *inter alia,* for certifying documents of ITAIPU in Portuguese and in Spanish respectively.

Article 9. The Governing Council shall be responsible for implementing and seeing to the implementation of the Treaty and the annexes thereto and for determining:

(a) The fundamental administrative guidelines of ITAIPU;
(b) The rules of procedure;
(c) The plan of organization of basic services;
(d) Actions resulting in any transfer of ITAIPU's patrimony, after consultation with ELETROBRÁS and ANDE;
(e) Reevaluations of assets and liabilities, after consultation with ELETROBRÁS and ANDE, taking account of the provisions of article XV, paragraph 4 of the Treaty;
(f) The conditions for the provision of electrical service;
(g) Proposals of the Executive Directorate relating to obligations and loans;
(h) The proposed budget for each financial year and the revisions thereof, submitted by the Executive Directorate.

1. The Governing Council shall consider the annual report, balance sheet and statement of account, drawn up by the Executive Directorate and shall submit them, together with its comments, to ELETROBRÁS and ANDE in accordance with the provisions of article 24 of this Statute.

2. The Governing Council shall take note of the progress of ITAIPU's affairs by means of the statements which shall normally be made by the Director-General or other statements which the Council may request through him.

Article 10. The Governing Council shall meet ordinarily every two months and, in extraordinary circumstances, when convened through the secretaries, by the Director-General or by half of the Councilors minus one.

The Governing Council may take valid decisions only when a majority of the Councilors of each country are present and with parity of votes equal to the smaller of the two national representations present.

Article 11. Councilors shall have a term of office of four years and may be reappointed.

1. The Governments may at any time replace the Councilors appointed by them.

2.   When a post of Councilor becomes definitively vacant, the Government concerned shall appoint a replacement, who shall serve for the remainder of his predecessor's term of office.

Article 12. The Executive Directorate, constituted by nationals of both countries in equal number, shall consist of the Director-General and the Technical, Legal, Administrative, Financial and Coordinating Directors.

1.   For each Director there shall be a Brazilian or Paraguayan Deputy Director of nationality other than that of the Director.
2.   The Directors and Deputy Directors shall be appointed by the respective Governments on the proposal of ELETROBRÁS or ANDE as appropriate.
3.   The Directors and Deputy Directors shall have a term of office of five years and may be reappointed.
4.   The Governments may at any time replace the Directors and Deputy Directors appointed by them.
5.   In the event of the absence or temporary incapacity of a Director, ELETROBRÁS or ANDE, as appropriate, shall designate a replacement from among the other Directors, who shall also be entitled to the vote of the Director he is replacing.
6.   When a post of Director becomes definitively vacant, ELECTROBRÁS or ANDE, as appropriate, shall propose a replacement, who, after being appointed, shall serve for the remainder of his predecessor's term of office.

Article 13. The responsibilities and duties of the Executive Directorate shall be as follows:

(a)   To implement the Treaty and the annexes thereto and the decisions of the Governing Council;
(b)   To implement and see to the implementation of the rules of procedure;
(c)   To carry out the administrative actions necessary for the conduct of the entity's business;
(d)   To propose fundamental administrative guidelines to the Governing Council;
(e)   To propose rules for personnel management to the Governing Council;
(f)   To prepare and submit to the Governing Council during each financial year the proposed budget for the following year and any revisions thereto;
(g)   To prepare and submit to the Governing Council the annual report, balance sheet and statement of account for the preceding financial year;
(h)   To implement the rules and conditions for the provision of electrical services;

(i)   To create and install, where appropriate, such technical and/or administrative offices as it may deem necessary.

Article 14. The Executive Directorate shall meet ordinarily at least twice a month and, in extraordinary circumstances, when convened by the Director-General or at the request of one of the Directors to the Director-General.

1.   Resolutions of the Executive Directorate shall be adopted by a majority of votes; the Director-General shall have the casting vote.
2.   The Executive Directorate shall establish itself at such place as it may deem most suitable for the exercise of its functions.

Article 15. ITAIPU shall be able to undertake obligations or issue powers of attorney only with the signatures of both the Director-General and another Director.

Article 16. The honoraria of members of the Council, Directors and Deputy Directors shall be fixed annually by agreement between ELETROBRÁS and ANDE.

Article 17. The Director-General shall be responsible for coordinating, organizing and managing ITAIPU's activities and shall represent it, at law and elsewhere, and be responsible for carrying out all the ordinary administrative actions necessary for the functioning of the entity, with the exception of those which are the responsibility of the Governing Council and the Executive Directorate. In addition, he shall be responsible for engaging and dismissing personnel.

Article 18. The Technical Director shall be responsible for managing the project, constructing the works and operating the facilities.

Article 19. The Legal Director shall be responsible for managing the entity's legal affairs.

Article 20. The Administrative Director shall be responsible for personnel management and for supervising general services.

Article 21. The Financial Director shall be responsible for implementing economic and financial policy and policy relating to supplies and purchases.

Article 22. The Coordinating Director shall be responsible for handling administrative matters vis-à-vis the authorities of the two countries.

Article 23. The Deputy Directors shall have such responsibilities as the Directors by agreement with them, may delegate to them.

1.   The Deputy Directors shall keep informed about the business of their respective departments and shall report on the progress of the matters entrusted to them.
2.   The Deputy Directors shall attend the meetings of the Executive Directorate and may participate in the discussions without a vote.

Chapter IV. FINANCIAL YEAR

Article 24. The financial year shall end on 31 December of each year.

1. ITAIPU shall submit the annual report, balance sheet and statement of account for the preceding financial year at any time up to 30 April of each year for decision by ELETROBRÁS and ANDE.
2. ITAIPU shall adopt the currency of the United States of America as a standard for its accounting operations. The said standard may be replaced with another by agreement between the two Governments.

Chapter V. GENERAL PROVISIONS

Article 25. ITAIPU shall assume as part of the capitalization by ELETRO-BRÁS and ANDE, the expenses incurred by the said enterprises, prior to the establishment of the entity, in connection with the following:

(a) Studies resulting from the Co-operation Agreement, signed 10 April 1970;
(b) Preliminary works and services relating to the construction of the hydroelectric utilization scheme.

Article 26. Councilors, directors, deputy directors and other employees may not exercise management, administrative or consultative functions in enterprises that supply or contract for any materials and services utilized by ITAIPU.

Article 27. Brazilian or Paraguayan public officials and employees of autarchic entities and of mixed-economy companies may perform services for ITAIPU without forfeiting their original connection or any pension and/ or social security benefits, due regard being had for the respective national legislation.

Article 28. The rules of procedure of ITAIPU, referred to in article 9, shall be submitted by the Executive Directorate to the Governing Council for approval and shall deal, *inter alia,* with the following matters: the accounting and financial régime; the régime for obtaining bids and awarding and concluding contracts for services and works and the acquisition of property; rules for the exercise of their functions by the members of the Governing Council and the Executive Directorate.

Article 29. Cases for which no provision is made in this Statute and which cannot be resolved by the Governing Council shall be settled by the two Governments after consultation with ELETROBRÁS and ANDE.

ANNEX B

GENERAL DESCRIPTION OF THE FACILITIES FOR THE PRODUCTION OF ELECTRICAL ENERGY AND THE AUXILIARY WORKS

I. PURPOSE

The purpose of this annex is to describe and identify the main features of the project for the hydroelectric utilization of the Paraná River at the site called ITAIPU, hereinafter referred to as "the Project".

This annex was prepared on the basis of the "Preliminary Report" submitted to the Governments of Brazil and Paraguay by the Brazilian-Paraguayan Joint Technical Commission on 12 January 1973.

The works described in this annex may be modified or expanded, including their altitudes and dimensions if that should prove necessary during the implantation stage for technical reasons. In addition, if, for similar reasons, it proves necessary to reduce substantially the altitude of the dam crest, consideration will be given to the advisability of building an additional hydroelectric utilization scheme upstream, pursuant to the provisions of the aforementioned "Preliminary Report".

II. GENERAL DESCRIPTION

1. *Location.* The Project will be located on the Paraná River approximately 14 kilometers upstream from the international bridge joining Foz do Iguaçu, in Brazil, to Puerto Presidente Stroessner, in Paraguay.
2. *General arrangement.* The Project will consist of a main gravity dam, in concrete, across the Paraná River, with a powerhouse at its foot and of lateral rock-fill dams and earth dykes on each bank of the river. The lateral dam on the right bank will include the structure of the spillway, together with its gates.

The general direction of the works will be east-west along a broken-line axis with a total development of 8.5 kilometers. The normal maximum water level in the reservoir was established at about 220 meters above sea level. The reservoir will flood an area of approximately 1,400 km² (800 km² in Brazil and 600 km² in Paraguay) and will extend approximately 200 kilometers upstream up to and including the Salto Grande de Sete Quedas, or Salto del Guairá.

III. MAIN COMPONENTS OF THE PROJECT

The Project includes the following main components, listed in sequence, beginning with the right bank:

1. *Right lateral dyke.* An earth dyke with the crest at an altitude of 225 meters, 700 meters long and with a volume of 103,000 m³.

2. *Spillway.* A concrete spillway with 14 gates, 380 meters long, capable of spilling up to 58,000 m³ per second, with an access channel dug upstream from the spillway. A channel lined with concrete will transmit the overflow from the spillway to the Paraná River approximately 1,500 meters downstream from the main dam.

3. *Right lateral dam.* A rock-fill dam with a crest at an altitude of 225 meters, 800 meters long and with a volume of 3,514,000 m³ joining the spillway to the main dam.

4. *Main dam and water intake.* The main dam will be a gravity structure, in solid concrete, with a crest at an altitude of 224 meters, 1,400 meters long and with a volume of 6,800,000 m³ to be built across the Paraná River and the channel on the left bank which will be dug in order to divert the river temporarily. The dam will have 14 water-intake openings equipped with gates. Each intake will give access, via a pressure pipe, to a turbine in the powerhouse.

5. *Powerhouse.* The powerhouse will be located at the foot of the main dam and will be 900 meters long and contain 14 generating units of 765 megawatts each. Four of the units will be situated in that part of the dam and intake to be built across the diversion channel. The upper platform of the powerhouse will be at an altitude of 139 meters and on it will be situated the transformer facilities for stepping up the generated voltage.

6. *Left-bank dam.* A concrete gravity dam, 250 meters long and with a volume of 1,100,000 m³ which will have blocked openings and connection for the construction of an intake intended for future expansion of the power station.

7. *Left-lateral dam.* A rock-fill dam with a crest at an altitude of 225 meters, 2,000 meters long and with a volume of 13,145,000 m³.

8. *Left-lateral dyke.* An earth dyke with a crest at an altitude of 225 meters, 3,000 meters long and with a volume of 3,115,000 m³.

9. *Hernandarias supplementary dyke.* A small earth dyke to be situated on the right bank approximately 4.5 kilometers west of the main dam, in the vicinity of the town of Hernandarias. The dyke will be designed to close off a depression into which the reservoir might overflow at maximum flood level.

10. *Sectioning substations.* Two sectioning substations to be situated one on each bank, approximately, 600 meters downstream from the powerhouse.

11. *Navigation works.* The project will include such works as may be necessary to meet the needs of river traffic, for example: land terminals and cognisance, locks, canals, elevators, and the like.

ANNEX C

FINANCIAL BASES OF ITAIPU AND CONDITIONS FOR THE PROVISION OF ITS ELECTRICAL SERVICES

I. DEFINITIONS

For the purposes of this annex:

I.1. "Entities" means ELETROBRÁS, ANDE or the Brazilian or Paraguayan enterprises or entities designated by them in accordance with article XIV of the Treaty signed by Brazil and Paraguay on 26 April 1973.

I.2. "Installed power" means the sum of the nominal plate power values expressed in kilowatts, of the alternators installed at the power station.

I.3. "Contracted power" means the power in kilowatts that ITAIPU will make available on a permanent basis to the purchasing entity for the periods and on the conditions specified in the respective contracts for the purchase and sale of electrical services.

I.4. "Finance charges" means all the interest, charges and commissions relating to the loans negotiated.

I.5. "Operating costs" means all costs chargeable to the provision of electrical services and covers direct operating and maintenance costs, including those for replacements necessitated by normal wear and tear, administrative costs and overhead costs, as well as insurance for the property and facilities of ITAIPU.

I.6. "Operating and billing period" means the calendar month.

I.7. "Operating account" means the annual balance of income and cost of services.

II. CONDITIONS FOR THE PROVISION OF ELECTRICAL SERVICES

II.1. The division of energy into equal parts as provided in article XIII of the Treaty, shall be effected by dividing the installed power at the power station.

II.2. In exercise of its right to utilize the installed power, each entity shall conclude contracts covering periods of 20 years each with ITAIPU for fractions of the installed power of the power station according to a time-schedule of utilization which will cover the contract period and will indicate the power to be used each year.

II.3. Each entity shall deliver the above-mentioned time-schedule to ITAIPU two years prior to the date on which the first generating unit of the power station is to begin commercial operation and two years prior to the expiry of the first and succeeding 20-year contracts.

II.4. Each entity shall be entitled to utilize the energy that can be produced by the power it has contracted for, up to the limit to be established, for each operating period, by ITAIPU. It is understood that each entity may utilize the said power contracted for by it for as long as it wishes during each operating period, provided that the energy it utilizes during the entire period does not exceed the aforesaid limit.

II.5. Where an entity decides not to utilize part of the contracted power or part of the energy corresponding to that power within the fixed limit, it may authorize ITAIPU to cede to the other entities that part of the power or energy which becomes available during the period referred to in II.4., on the conditions laid down in IV.3. below.

II.6. The energy produced by ITAIPU shall be delivered to the entities through the bar system in the power station, on the conditions laid down in the contracts of purchase and sale.

III. COST OF THE ELECTRICAL SERVICE

The cost of the electrical service shall consist of the following annual components:

III.1. The amount needed to pay to the parties constituting ITAIPU profits in the amount of 12 per cent per annum on their participation in the constituted capital, in accordance with article III, paragraph 1, of the Treaty and with article 6 of the Statute (annex A).

III.2. The amount needed to pay the finance charges on the loans obtained.

III.3. The amount needed to pay the amortization of the loans obtained.

III.4. The amount needed to pay royalties to the High Contracting parties, calculated at the equivalent of dollars US 650 gigawatt-hours, generated and measured at the power station. The said amount may not be less than dollars US 18 million per annum, at the rate of one half for each High Contracting Party. Royalties shall be paid monthly in the currency available to ITAIPU.

III.5. The amount needed to pay to ELETROBRÁS and ANDE in equal parts, reimbursement for administrative and supervisory expenses relating to ITAIPU, calculated at the equivalent of dollars US 50 per gigawatt-hour generated and measured at the power station.

III.6. The amount needed to cover operating expenses.

III.7. The amount of the balance, whether positive or negative, of the operating account for the preceding financial year.

III.8. The amount needed to pay compensation to one of the High Contracting Parties at a rate equivalent to dollars US 300 per gigawatt-hour ceded to the other High Contracting Party. Such compensation shall be paid monthly in the currency available to ITAIPU.

IV. INCOME

IV.1. Annual income from the contracts for the provision of electrical services shall be equal, each year, to the cost of the service established in this annex.

IV.2. The said cost shall be broken down in proportion to the power values contracted for by the entities receiving the services.

IV.3. Where the situation provided for in II.5. above occurs, the contracting entities shall be billed according to the power actually utilized.

IV.4. Where the situation provided for in II.5. does not occur and with due regard being given to the provisions of article XIII of the Treaty and of IV.2. above, the entity which contracted the purchase shall be responsible for the amount corresponding to the entire contracted power.

V. OTHER PROVISIONS

V.I. The Governing Council, acting after consultation with ELETROBRÁS and ANDE, shall regulate the norms of this annex, with a view to increasing ITAIPU's efficiency.

V.2. The value of the profits, the royalties, the reimbursement for expenses and the compensation referred to respectively in III.1., III.4., III.5., and III.8. above shall be kept constant in accordance with the provisions of article XV, paragraph 4 of the Treaty.

VI. REVIEW

The provisions of this annex shall be reviewed after 50 years have elapsed from the entry into force of the Treaty, due regard being given, *inter alia,* to the degree of amortization of the debts contracted by ITAIPU for the construction of the utilization scheme and the relation between the power values contracted for by the entities of the two countries.

From *Treaty Series: Treaties and International Agreements Registered or Filed and Recorded with the Secretariat of the United Nations,* vol. 923 (New York: United Nations, 1981), 92–110.

38
The Founding of Abuja

Nigeria

DID YOU KNOW . . . ?

➤ Construction of Abuja began in 1976, and it became the official capital of Nigeria in 1991.

➤ It covers an area of 2,824 square miles (7,314 square kilometers).

➤ Nigeria is the most populous nation in Africa, and among the 10 most populous countries in the world.

➤ The Nigerian currency is the *naira*.

➤ Abuja is 50 percent Muslim and 40 percent Christian.

➤ Abuja was named after the Zarian emir Abu Ja.

➤ The area of the new capital was not new to some indigenous peoples, who had been there for 40,000 years.

The energy crisis of the 1970s may have been in part responsible for the new capital of Abuja: Nigeria had ample funds as a result of the high price of petroleum. In the United States and Japan, new automobiles designed to use less energy eventually resulted in the hybrid vehicles of the early twenty-first century. But in the 1970s, high prices on the world petroleum market set the stage for an initiative that had long been contemplated in Nigeria: the building of a new capital.

HISTORY

In 1960, following independence, Nigeria began to discuss the idea of moving the capital from Lagos to another site. Lagos was congested and suffered from

Construction underway in Abuja. © Gilbert Liz / Corbis Sygma.

environmental pollution and inadequate infrastructure. Its climate was hot and humid, and its coastal location exposed it to possible attack. The idea of building a new, well-planned capital at a different location began to attract interest; other examples were cited such as Canberra (Australia), Brasília (Brazil), Islamabad (Pakistan), St. Petersburg (Russia), and Washington, DC (United States).

A Federal Capital Territory, in the geographic center of Nigeria, was created in 1976. At the time, the country's oil industry was enjoying a revenue boom, and Nigeria aspired to the role of a world power and leader of Africa. The government opted to build the infrastructure for a modern city of 1.6 million people, a figure expected to grow to 3 million. Boosters called it the biggest construction project in the world.

The document reproduced here as Document III is the 1977 agreement between the military government then in power in Nigeria and International Planning Associates, a U.S.-based consortium of firms that won the competition for the planning contract.

The new capital city was assigned a site in the northeast corner of the Federal Capital Territory, an area of 2,824 square miles (7,314 square kilometers). The site enjoys the advantages of ample space for expansion and, since it lies amid rolling hills at an elevation of 1,180 feet (360 meters), has a climate cooler and less humid than that of Lagos. Its location in an area not closely associated with any of the country's major ethnic groups was also important, since the capital was intended to be a symbol of unity. The new capital was named *Abuja* after a nearby town and emirate founded in 1828 by the Zarian ruler Abu Ja. The old town of Abuja was renamed Suleja.

By late 1991, with the basic infrastructure in place, the city was officially declared the country's capital. Many bureaucrats were initially reluctant to leave the big-city attractions of Lagos for what they thought of as dull Abuja. But by the end of the decade, all the federal ministries were established in the new capital, which had an international airport and highway links to other major Nigerian cities and was the site of the headquarters of the Economic Community of West African States. Democratic rule had been restored and a new legislature elected, adding to the sense of a fresh spirit in Abuja.

CULTURAL CONTEXT

The population of Nigeria was 55.6 million at the time of the 1963 census but by 1986 had grown to an estimated 106 million. With the population of cities growing at a rate five times greater than that of the countryside, the number of people living in Lagos was doubling every 10 years. A campaign to shift the capital from Lagos attracted more and more supporters.

A new capital would have many advantages. While Lagos was situated on the coast and was therefore somewhat vulnerable from a security point of view, a new inland capital would be less vulnerable. Lagos was dominated by entrenched power centers and ethnic biases; if a new capital were chosen in a more neutral area, it might draw off some of the population from overcrowded Lagos, especially with the enticement of new jobs and new housing in a new capital.

According to Andrew C. Lemer (1982), a leading member of the team that planned and designed Abuja, the push for a new capital came to the forefront after the Biafran civil war; there was a need to bring the country together. Building a new capital gained impetus for two reasons: it would ease the exploding population of Lagos, and it would inspire Nigerians to see a new vision of their country and its potential leadership role in Africa. "Nigeria's Federal Capital will be the largest free-standing new town project ever undertaken" (Lemer, 1982: 86).

There were concerns about the vulnerability of Lagos as a coastal city, similar to the uneasiness that prompted Brazil to move its capital from Rio de Janeiro to Brasília. Another concern involved the various ethnic groups: if a site in the center of Nigeria could be found that was not under the influence of a single dominant tribe, would it be possible to achieve a new vision of Nigerian unity? With the funds flowing in from Nigeria's oil resources, the expensive project seemed within realistic reach.

PLANNING

A suitable location had to be found. When government surveyors looked for a site with good climate, plenty of water and land, not too many people, and a central location, they settled on an area inhabited by several ethnic groups: the Gbagyi, Koro, Gade, and Gwandara, none of which was dominant. The Gbagyi had been in the area for 40,000 years. The trade routes that came through the region at the confluence of the Niger and Benue Rivers had given rise to many small states and independent groups. In 1967, the Federal Military Government created three states: Northwestern, Benue-Plateau, and Kwara. But everything changed in 1975.

When the government of General Yakubu Gowon ended on July 29, 1975, a different direction was pursued by the new head of state, General Murtala R. Mohammed. A few days later, he appointed a panel to evaluate the idea of moving the capital from Lagos. Should Lagos continue to be both a federal and a state center of government? The panel reported back that a new location for the federal capital would be in the best interests of Nigeria's growth and development. Site visits were launched to explore the existing 12 Nigerian states and their capitals, as well as overseas capitals.

Nigerians particularly liked Washington, and the city of Paris, as redesigned by Haussmann, was also much admired. It may be that the Nigerians found these two cities particularly attractive because of their wide boulevards; in the case of Paris, Haussmann had recommended wide streets to let air circulate freely, encouraging a healthful climate. Given the crowded environment of Lagos, the Nigerians may have been attentive to the importance of such aspects of design.

The decision was made to create a Federal Capital Territory (FCT). In a broadcast proclaiming Abuja as the federal capital on February 3, 1976 (reproduced here as Document I), General Mohammed stated, "The area is not within the control of any of the major ethnic groups in the country. We believe that the new capital created on such virgin lands, as suggested, will be for all Nigerians a symbol of their oneness and unity. The Federal Territory will belong to all Nigerians." General Mohammed decreed that the new capital could instill in every citizen a sense of national unity; he thanked everyone and sounded the call for "justice, peace, and stability." Just seven days later he was assassinated, and he never witnessed the realization of his vision. But changes in government did not deter the plan to move, although it was slowed.

General Mohammed's speech sounded many of the calls for unity and concern for stability voiced by General George Washington when the latter became president of the United States and sought to establish a neutral capital. At a dinner party hosted by Thomas Jefferson at which the location of the new capital was determined, Washington was notably absent, opting to remain at home near the Potomac River to quell land speculation that might arise following the announcement. So too General Mohammed had expressed concern about land speculation: "In order to avoid land speculation in the area, a decree is being promulgated immediately to vest all land in the Federal Territory in the Federal Government." In his speech, General Mohammed estimated 10–15 years would be needed to complete the transition.

A competition was held to identify a designer for the new capital; many international firms submitted plans. International Planning Associates (IPA), a consortium of firms, won the competition, as indicated in the agreement presented in this chapter. The planning for Abuja was extensive. Both a regional plan and a city plan were developed, and implementation information was also supplied. IPA submitted its master plan for Abuja in early 1979. Taking advantage of the terrain's natural contours, the design envisioned streets and neighborhoods with a less rectilinear and more individualized character than in, for example, Islamabad or Brasília. Two zones were stipulated: a central zone with government buildings and cultural institutions on broad avenues, and a residential and shopping zone.

When IPA designed the city, every aspect was detailed. The site was the first item for consideration; all parties wanted a location whose climate was not too hot and not too cold. Another consideration was water, and the chosen site was at the confluence of two rivers.

With the location settled, IPA could begin designing a new city. Most important was the seat of government and administration—the major reason for pushing forward the concept of a new capital. But along with government facilities came the creation of an urban environment: water supply, airport, schools, health care, and public transport. For transportation, a high-speed system that could quickly bring in workers from outlying areas was planned. But the key element of the design centered on monumental public buildings that would inspire patriotism and national unity. In fact, the slogan created for the new capital was Centre of Unity.

The city's growth, however, did not conform to the original plan, which foresaw an orderly, phased pattern of development through the gradual addition of new mini-cities that had their own business and residential areas. Squatter settlements began to appear as lower-income people encountered difficulty finding housing in high-priced Abuja.

BUILDING

Although the IPA design team had taken inspiration from foreign capitals like Paris, Nigerian culture remained the focus of design. The team paid particular attention to African social structure, visiting Nigerian settlements around the country and observing the fabric of community life. Because Nigeria has always had an urban tradition, it was possible to study the benefits and challenges of existing Nigerian cities (Taylor).

Building a new city is expensive, and aspects of the IPA plan had to be realized piece by piece. A logical sequence for construction was developed, and most of the construction contracts were let to local firms. First would come roads and an airport to facilitate construction of the federal buildings. Next would be government offices, then housing for civil servants and government employees.

Construction began in 1980, but slowed to a standstill between 1983 and 1985, as falling oil prices led to a collapse in Nigeria's oil revenues and subsequent austerity measures. There were other problems as well: civil unrest, a military coup in 1983, and pervasive corruption and mismanagement—all contributed to sluggish progress. Still, Abuja slowly took shape. By 1987, water and telephone systems were prepared for more than a million people, although the city's population started out at a mere 15,000. A university was founded in

1988. A new service industry was planned for development in the new central government location.

The actual transfer from Lagos to Abuja got underway in 1986 when the Federal Ministries of Internal Affairs, Trade and Industries, Agriculture, Water Resources, and Rural Development made the move. But it was not until 1991 when the real transfer of power occurred with the arrival in Abuja of General Ibrahim Babangida. When General Babangida set foot in the new capital, the Federal Capital Territory Decree No. 51 dated December 12, 1991, signaled the beginning of a new Nigerian vision, and a new center of power for the Federal Republic of Nigeria.

IPA planned a careful buildup of population, with a goal of 200,000, but people streamed into Abuja faster than housing could be built, quickly reaching 1.2 million residents. Abuja was pushed to keep up with housing demand and related systems: education, health care, sanitation, power, and transport. The need for an adequate water supply, which was ample initially, forced the building of dams to supply additional drinking water as well as more electricity.

While building houses for government employees was an immediate goal, housing for other residents was much slower. The number of housing units built for civil servants was 32,000. Cooperative agreements with the private sector were both prudent and essential to combat the sudden influx of people arriving in the FCT, most with no place to live, who then resorted to the kinds of informal housing arrangements that can lead to widespread health issues. According to government plans, private developers were invited to build low-cost housing in the Mbora District.

Along with housing came the priority to educate children as part of a national program launched by President Obasanjo called Universal Basic Education. By the early twenty-first century, there were more than 285 public schools, 80 private primary schools, and 65 secondary schools, of which 44 were government-run. Higher education is equally essential. George Washington hoped that institutions of advanced learning would be part of America's new capital in Washington, DC, and Abuja University represents a similar vision. Abuja's College of Education helps develop future teachers who can provide education for the growing capital's new families.

Sewage control was a pressing problem. As was the case in ancient Rome when the satirist Juvenal wrote of the problem, combating the habits of a rural lifestyle among people who now live in a city was a primary concern, and sewage systems became a high priority in later phases of developing the new capital as a true urban center.

More should be said about the displacement of the native people who resided in the former old town of Abuja before it became the FCT. Committees were empowered to look into relocation, and in Decree No. 6 of 1979, the federal government assumed the expense of displacement, with an accompanying rate of compensation for churches, mosques, and households. The total program cost one million *naira* for the Niger, Plateau, and Kwara states in a deal sealed on October 10, 1977. However, people did not move easily, and by 1980 the decision to move people involuntarily was considered.

From 1999 to 2002, several hospitals were built. At the same time, more than 300 doctors moved to Abuja. However, the country's health-care situation needed further attention, for not everyone could travel to Abuja for medical attention. So smaller hospitals in Abaji, Kubwa, Karki Bwari, and Nyana were readied for operation. At the same time, public health programs were active with initiatives for the control and eradication of leprosy, epidemic management, immunization campaigns, family planning, and health education, and a program for the health of mothers and children.

The lines demarcating the new capital went through several existing settlements. The residence dormitories for one high school fell within the FCT, but the village from which the students came remained outside the boundary. In another case, the line put one part of the village in the Gawu district and the rest of the community in the Niger state. In Zuba, the village was inside but their chief's house was outside. Numerous accommodations were required to resolve these unexpected quirks.

IMPORTANCE IN HISTORY

What will be Abuja's place in history? It is too early to say. It can take at least 100 years to determine the effects of moving a capital and to measure the ensuing impacts.

Can the precedents set by Abuja provide helpful examples for other countries considering a new capital? As South Korea wrestles with the choice of location for its new centralized government and national administration, four sites are under review. A budget of $45 billion has been allocated. South Korea seeks to decentralize economic power by creating a new capital that will invite 500,000 people to move there after construction is begun in 2007 and completed in 2030. Will a study of Abuja; Washington, DC; Brasília; and other capitals created as symbols of national unity yield valuable insight?

FOR FURTHER REFERENCE

Books and Articles

Adewale, Toyin, ed. *25 New Nigerian Poets*. Berkeley, CA: Ishmael Reed Publishing, 2000.
Chinweizu. *Decolonising the African Mind*. Lagos: Pero Press, 1987.
————. *The West and the Rest of Us*. New York: NOK Publishers, International, 1978.
Ellis, Joseph, J. *His Excellency: George Washington*. New York: Alfred A. Knopf, 2004.
Kirk-Greene, Anthony, and Douglas Rummer. *Nigeria Since 1970: A Political and Economic Outline*. New York: Africana Publishing, 1981.
Lemer, Andrew C. "Foreseeing the Problems of Developing Nigeria's New Federal Capital." In *Macro-engineering and the Future: A Management Perspective*,

edited by Frank P. Davidson and C. Lawrence Meador. Boulder, CO: Westview Press, 1982.

———. "Old Cities and New Towns for Tomorrow's Infrastructure." In *Macro-engineering: MIT Brunel Lectures on Global Infrastructure*, edited by Frank P. Davidson, Ernst G. Frankel, and C. Lawrence Meador. Horwood Series in Engineering Science. Chichester, England: Horwood Publishing, 1997.

Internet

For a study of Nigeria's cities, including the new city of Abuja, see an article by Robert W. Taylor, Department of Environmental, Urban, and Geographic Studies, Montclair State University, Upper Montclair, NJ, entitled "Urban Development Policies in Nigeria: Planning, Housing, and Land Policy," September 2000, http://alpha.Montclair.edu/~lebelp/CERAFRM002Taylor1998.pdf.

For information on Nigeria and the city of Abuja, see http://www.nigeria.gov.ng/ and http://www.abujacity.com.

For Nigeria's socio-political issues, see: http://www.dawodu.com.

If you are interested in popular culture in Nigeria, see http://www.mothernigeria.com/music.html.

For more on South Korea's proposed new capital, see http://news.bbc.co.uk/1/hi/world/asia-pacific/3554296.stm.

Documents of Authorization—I

EXCERPTS FROM A SPEECH BY GENERAL MURTALA MOHAMMED PROCLAIMING ABUJA AS THE FEDERAL CAPITAL OF NIGERIA, FEBRUARY 3, 1976

FELLOW Nigerians,

A joint meeting of the Supreme Military Council and the National council of States has just concluded sitting.

The meeting has declared on reports submitted by the following panels:

a) assets investigation of some former public officers;
b) abandoned properties in the three Eastern States which comprised the former Eastern Region;
c) location of the Federal Capital; and
d) creation of more states.

Those panels, as you will no doubt recall, were set-up by this administration at its inception. They were all given adequate terms of reference and sufficient time in which to deliberate and submit their reports and recommendation to the FMG.

I will like to seize this opportunity to thank each and every one of them for the excellent work they have done. They deserve the nation's gratitude.

In deliberation on these reports, I will like to emphasise that the joint meeting was guided solely by national interest and consideration for justice, peace and stability.

THE PANEL ON THE LOCATION OF FEDERAL CAPITAL

The panel on the location of the Federal Capital has recommended that the nation's capital should move out of Lagos to a federal territory of about 8,000 square kilometres in the central part of the country. The Supreme Military Council has accepted this recommendation. The site recommended satisfied the panels' criteria of centrality, good and tolerable climate, land availability and use, adequate water supply, low population density, physical planning convenience, security, and multi-access possibility. The area is not within the control of any of the major ethnic groups in the country. We believe that the new capital created on such virgin lands as suggested will be for all Nigerians a symbol of their oneness and unity. The Federal Territory will belong to all Nigerians.

The few local inhabitants in the area who need to be moved out of the territory for planning purposes will be resettled outside the area in places of their choice at government expense.

In order to avoid land speculation in the area, a decree is being promulgated immediately to vest all land in the Federal Territory in the Federal Government. A Federal Capital Development Authority is to be established to plan and administer the territory. An administrator for the Federal territory will soon be appointed to provide municipal services in the area.

The chairman of the Federal Government Authority of nine members will be of cabinet rank. The authority is expected to start work at once but the movement of the seat of the Federal Government out of Lagos is expected to take some ten to fifteen years. The present administration is firmly committed to ensuring that the necessary ground work is completed and construction work started within the next four years.

Lagos will, in the foreseeable future, remain the nation's commercial capital and one of its nerve centres. But in terms of servicing the present infrastructure alone the committed amount of money and effort required will be such that Lagos State will not be ready to cope. It will even be unfair to expect the state to bear this heavy burden on its own. It is therefore necessary for the Federal Government to continue to sustain the substantial investment in the area. The port facilities and other economic activities in the Lagos area have to be expanded. There is need in the circumstances for the Federal Government to maintain a special defence and security arrangement in Lagos which will henceforth be designated a special area. These arrangements will be carefully worked out and written into the new constitution.

From http://www.dawodu.com/murtala1.htm.

Documents of Authorization—II

The description of the land area that makes up the Federal Capital Territory of Nigeria starts from the village Izom on 7° E longitude and 9° 15' latitude, projected a straight line westwards to a point just North of Lefu on the Kemi River; then project a line along 6° 47¹ᐟ²' E southwards passing close to the villages called Semasu, Zui, and Bassa down to a place a little west of Ebagi, thence project a line along parallel 8° 27¹ᐟ²' North latitude to Ahinza village 7' E (on Kanama River); thence project a straight line to Bugu village on 8° 30' North latitude and 7° 20' E longitude; thence draw a line northwards joining the village of Odu, Karshi, and Karu. From Karu, the line should proceed along the boundary between the North-west and Benue-Plateau (Nassarawa) State as far as Karu; thence, the line should proceed along the boundary between North central (Kaduna) and North western (Niger) States up to the point just north of Bwari village; thence the line goes straight to Zuba village; and thence straight to Izom.

From Hon. Dr. Justice T. Akinola Aguda. *Report of the Committee on the Location of the Federal Capital of Nigeria,* December 20, 1975, p. 85fn. In: Baba, I. "Federal Capital Territory, Abuja: A Reflection into its past, Twenty Six Years After Creation." 2002.

Documents of Authorization—III

AGREEMENT FOR THE PRODUCTION OF THE MASTER PLAN FOR THE FEDERAL CAPITAL CITY AND ITS REGIONAL GRID

THIS AGREEMENT made the ____ day of ____ 1977
BETWEEN_____ for and on behalf of the Federal Capital Development Authority (hereinafter referred to as "the Employer" which expression shall, where the context so admits, include its successors and assigns) of the one part and

_____ for and on behalf of International Planning Associates (hereinafter referred to as "the Consultant" which expression shall, where the context so admits, include its successors and assigns) of the other part.

WHEREAS the Employer is desirous of having a Master Plan for the Federal Capital City and its regional grid,

AND WHEREAS the Consultant has agreed to prepare the said Master Plan for the Capital City, Now THIS AGREEMENT WITNESSETH as follows:

ARTICLE I—DEFINITIONS

1.01 In this Agreement, unless the context otherwise admits, the following terms have the meaning hereby assigned to them:

1.　"IPA" means International Planning Associates, a Consortium of Companies consisting of ARCHISYSTEMS, a division of SUMMA CORPORATION; PLANNING RESEARCH CORPORATION; and WALLACE, MCHARGE, ROBERTS AND TODD, INC.

2. "Government" means the Federal Military Government of the Federal Republic of Nigeria and any nominee or a combination of them as may be varied from time to time at the discretion of the said Federal Military Government.
3. "FCDA" means Federal Capital Development Authority.
4. "Survivor" means a successor or an assignee.

ARTICLE 2—ANNEXES

2.01 The Annexes A, B, C, D, and E shall be interpreted as forming an integral part of this Agreement.

2.02 The Annexes consist of the following:

Annex A—Terms of Reference.

Annex B—Schedule of the Scope of Services of the Consultant.

Annex C—Schedule of the Programme of Work.

Annex D—Schedule of Payment of Consultant's Fees.

Annex E—Schedule of Reimbursable Items and Fixed Costs.

ARTICLE 3—SCOPE OF WORK

3.01 The overall scope of work is as set forth in Annex B of this Agreement and includes the following items:

1. Environmental Assessment.
2. Preparation of Planning Programme, Development Criteria and Community Structure Analysis.
3. Urban Forms Hypotheses.
4. Alternative Capital City Site Evaluation.
5. Draft Regional Plan.
6. Draft Capital City Plan.
7. Draft Comprehensive Design and Development Manual.
8. Implementation Programme.
9. Draft Master Plans Report/Three-dimensional Models.
10. Advice and Assistance to FCDA.

ARTICLE 4—COMMENCEMENT

4.01 This Agreement shall take effect from the date of execution thereof, and commencement of the works on the Project shall take place not later than one calendar month after such date. During this month, the Consultant shall request and the Employer shall supply all initial pertinent data required by the Consultant. Such requisition shall not limit requests by the Consultant for additional data during the process of the Project provided that such data can reasonably be supplied by the Employer.

ARTICLE 5—PROGRAMME

5.01 The Project shall be carried out in accordance with the time and work schedule set out in Annex C—Schedule of the Programme of Work. The Consultant will prepare and submit the interim reports specified in Annex C to allow the Employer to examine the progress of Consultant's work.

5.02 The Consultant shall prepare and submit to the Employer within Month One as set out in Annex C, a revised schedule of work that may be used by the Employer as a guide to pinpoint major events as they may relate to overall completion and progress. The schedule of work may be revised by both parties so long as events during the course of the Project warrant a revision.

5.03 The Consultant shall also submit as soon as possible all working documents that may be requested by the Employer or its representative.

ARTICLE 6—ALTERATIONS

6.01 Should circumstances arise which call for alterations in the scope of the work and therefore modification, amendments, or additions to the Agreement, they shall be made by mutual written consent. Proposals in this respect from one party shall be given due consideration by the other party and a response will be given within thirty (30) days.

ARTICLE 7—COMPLETION

7.01 The Consultant shall proceed to execute the scope of work under this Agreement with all reasonable speed and dispatch in accordance with Annexes A, B and C provided that there are no delays caused through no fault of the Consultant.

ARTICLE 8—OBLIGATIONS OF THE EMPLOYER

8.01 It is recognised by the parties hereto that a substantial part of Consultant's work under this Agreement will be performed in its offices in the U.S.A. To facilitate the execution of the works in Nigeria, the Consultant shall procure the services, facilities, equipment, and personnel necessary for this work provided that the Employer assists the Consultant to procure any available accommodation, where needed, for use during the duration of the services to be performed under this Agreement. Additionally, the Employer agrees to provide at no cost to the Consultant lodging, subsistence, and travel while and when the Project staff are on visits to the Capital District.

8.02 The Employer shall give the Consultant such assistance as lies within its powers to facilitate the work of the Consultant and in connection with the work of the Consultant under the Agreement that may require the cooperation or assistance of other Governmental Agencies; the Employer shall provide the necessary liaison. The Employer shall also ensure that the Consultant has access to any information that may be required in connection with the services to be performed under this Agreement, insofar as such information is in the possession of the Employer or procurable with its assistance and is not protected by any privilege preventing its disclosure or procuration.

8.03 The Employer shall give such assistance as is legitimately possible to assist the Consultant to obtain from the appropriate authorities residence permits, work permits, and entry visas for expatriate staff in accordance with Immigration Regulations, and permits, where necessary, for temporary importation and subsequent exportation of instruments, tools, and equipment necessary for the proper execution of the Project.

8.04 The Employer shall introduce the Consultant to the appropriate bodies and communities, and shall give such information and assistance reasonably required by the Consultant for the performance of their duties under this Agreement, and shall afford adequate protection to the Consultant's field staff when required.

8.05 The Employer shall make available to the Consultant information from any other Consultants on any work related to this Agreement.

8.06 The Consultant shall have the right, subject to the Employer's approval, to publish descriptive articles and illustrations in professional journals subject to observing the usual professional codes of conduct.

ARTICLE 9—OBLIGATIONS OF THE CONSULTANT

9.01 The Consultant shall exercise all reasonable skill, care and diligence in the discharge of the services agreed to be performed by them. The Consultant shall respect local customs and traditions and desist from any conduct, whether on its part or that of its employees or agents, which could arouse or offend public feelings, and where in doubt shall seek advice from the Employer.

9.02 The Consultant shall provide adequate specialized staff to carry out their obligations under the terms of this Agreement. The Consultant shall, to the greatest extent practicable, involve Nigerian citizens in the performance of the program of work. The Consultant agrees to accept under his supervision and control Nigerian citizens selected by the Employer for use on the Project. It is agreed that the Employer will offer or furnish these Nigerian citizens at no cost to the Consultant.

ARTICLE 10—REMUNERATION OF CONSULTANT

10.01 The Employer shall pay the Consultant for the services to be performed under this Agreement at the times and in the manner set out in the "Schedule of Payment of Consultant's Fees" hereunder in Annex D and as provided in Article 13.01.

10.02 Payments under the Schedule of Payment shall become due in accordance with Annex D and upon presentation of an invoice by the Consultant to the Employer. However, the approval of the final report shall be obtained from the Employer before the last payment is due.

10.03 Should the Consultant be requested to undertake additional service, such services shall be undertaken on such terms and conditions as may be agreed in writing between the parties hereto.

10.04 It shall be the duty of the Consultant to maintain the progress of work as specified in Annex C of this Agreement. However, where delays are caused by Consultant's failure to obtain adequate reply, approval, or decision within thirty (30) days after a request has been made by the Consultant due to the following reasons:

(a)   the Employer's omission to reply, approve or decide within the said period of thirty (30) days; or

(b)   the failure of a third party, who has been assigned by the Employer to comply with the request to act or to perform within the specified period, the Employer shall adequately compensate the Consultant for any costs incurred as a result of the delay.

ARTICLE 11—REIMBURSABLE EXPENSES

11.01 In addition to the remuneration to be paid under Article 10 above, the Consultant shall be reimbursed by the Employer for reimbursable expenses at cost plus 2 1/2%, all as detailed in Annex E. The reimbursable expenses, including the handling charge, as shown in Annex E, cannot be exceeded unless approved by the Employer.

11.02 The Consultant shall submit to the Employer every calendar month, claims for all reimbursable items in Naira and/or applicable foreign currency, which set out the amounts due to the Consultant from the Employer as reimbursable expenses incurred during the preceding calendar month. Such accounts are payable within thirty (30) days of receipt of the invoices.

ARTICLE 12—DISPUTED ITEMS OF PAYMENT

12.01 In the event that any item of cost included in any notice of the Consultant is questioned or not approved by the Employer, the Employer shall promptly provide a statement to the Consultant identifying such item and setting forth the reason for questioning or not approving it. Payment of such items may be withheld by the Employer until such time as the Consultant provides satisfactory justification or explanation of any such item. All remaining items included in the invoice shall be approved and paid by the Employer in accordance with Article 11.02 above and shall not be subject to delay because of items that may be questioned or not approved. When the question has been resolved, the next payment shall reflect the credit or debit as the case may be.

12.02 Payments due to the Consultant under this Clause shall be made in Naira and/or applicable foreign currency by deposit in the Consultant Bank

Account set out under Article 13.01 below in the amount invoiced, subject only to adjustments in accordance with Article 12.01 above.

12.03 Together with every installment listed in Annex D—Schedule of Payment of Consultant's Fees, and the provisions under Article 11.02 above, the Consultant shall be entitled to charge interest for the time between the date of the expiry of this period and the date on which payment is received by the Consultant at the current Nigerian bank rate if the Employer fails to pay the Consultant within the periods specified under this Agreement.

ARTICLE 13—PAYMENT CONVERSION

13.01 It is the decision of the Government that the Consultant be permitted to remit in U.S. currency to its designated U.S. Bank Account eighty percent (80%) of its fixed fee as in Annex D and fifty percent (50%) of the reimbursable as in Annex E.

13.02 The Employer shall assist the Consultant in obtaining the necessary foreign exchange permits and approvals for the transfer of the sums to be repatriated under this Agreement.

ARTICLE 14—TAXATION

14.01 The Consultant shall not be subject to the payment of any company income tax, not being a Nigerian Company, and where such tax or taxes are deductible and/or have been deducted, same shall constitute an item or items of reimbursement. No taxes of any kind shall be paid on the salaries, income or allowances of expatriate personnel by the Consultant provided always that when the Consultant or any of its said personnel have, in fact, paid such taxes, the Employer shall reimburse the Consultant for the actual sum of money so paid.

ARTICLE 15—INSURANCE

15.01 During the performance of the work covered by this Agreement, the Consultant shall maintain at all times suitable insurance coverage for statutory workmen's compensation; and the comprehensive general liability insurance, including insurance protecting it from claims or damages because of bodily injury, including personal injury, sickness, disease, or death of any of the Consultant's staff allocated and/or employed for the Project; and professional indemnity. The Consultant shall submit certificates of insurance to the Employer for scrutiny, evidencing that claims brought under such policies are enforceable in Nigeria.

ARTICLE 16—DEFAULT OF THE CONSULTANT

16.01 If the Consultant shall, before the completion of the duties under this Agreement, pass a resolution or an order shall be made for the winding up of the Consultant, or a receiver shall be appointed for any part of the assets of the Consultant, and the Consultant shall therefore otherwise become unable to perform the duties under this Agreement, then its survivor shall, if requested to do so by the Employer, hand over all surveys, drawings, maps, statements and other documents whatsoever relating to the Consultant's duties herein and all such documents shall become the property of the Employer. PROVIDED that there shall be a lien on them until such equitable portion of any unpaid part of the remuneration hereinbefore specified be agreed upon and paid or determined by arbitration in the manner hitherto mentioned, and duly paid.

Provided that if the Employer and the survivor so agree, the survivor may continue to discharge the duties of the Consultant under this Agreement on terms no more onerous or less beneficial than those in this Agreement.

16.02 If the Consultant shall in any manner delay or neglect, refuse or be unable from any cause within their control at any time to proceed with or finish the programme of work or an identified part thereof within the time prescribed for the purpose under the aforementioned Article 5.01 and 5.02, it shall be lawful for the Employer by notice in writing to require the Consultant to proceed with the Project or an identified part thereof and complete the same within a reasonable period of time specified in such notice.

If the Consultant fails to comply with the notice, the Employer may employ any other Consultant to continue and complete the Programme of work or an identified part thereof upon any terms and conditions agreed between the Employer and any such other Consultant and the former Consultant shall in such case be liable to pay consequential loss resulting from their breach.

ARTICLE 17—POSTPONEMENT OR ABANDONMENT OF THE PROJECT

17.01 In the event of the whole or any part of the Project being postponed, canceled, abandoned or materially altered at the instance of the Employer, then the Employer shall pay the Consultant all installments due on the date of such postponement, cancellation or abandonment together with all reimbursable expenses incurred or reasonably committed to be incurred on the date of such postponement, cancellation or abandonment, including pro rata, the earned portion of the installment which would have become due on the next scheduled payment as specified in Annex D and, in addition thereto, the Consultant shall be adequately compensated for any costs incurred as a result of such postponement, cancellation or abandonment.

ARTICLE 18—*FORCE MAJEURE*

18.01 If, during the period of the execution of this Agreement, war or civil commotion, riot, rebellion, revolution, insurrection, strikes, epidemics, and any other causes similar to the kind enumerated shall materially affect the execution of the responsibilities of either party to this Agreement, then provided such party affected gives notice and files particulars of such acts of *Force Majeure* by writing to the other party as soon as possible after the occurrence of the cause related, the obligation of such affected party, as far as they are affected by *Force Majeure,* shall be suspended during the continuance of any disability caused by *Force Majeure*. Should the period of disabling due to *Force Majeure* materially alter the Consultant's or the Employer's ability to perform or impair the value of this Agreement, in such event, either party at its election, may terminate this Agreement.

ARTICLE 19—COMMUNICATIONS

19.01 In all matters relating to this Project the Consultant shall communicate with the Executive Secretary.

19.02 All notices, requests, and authorizations provided for herein shall be in writing and shall have been delivered or sent by prepaid registered mail, cable, or telex to addresses as follows:

For the Employer:	Federal Capital Development Authority
	P.M.B. 12534, LAGOS, NIGERIA.
	Cable Address: FECADA
For the Consultant:	International Planning Associates (IPA)
	7798 Old Springhouse Road
	McLean, Virginia 22101, U.S.A.
	Telex: 899105
Alternative address:	7, Gbajumo Street
Mail, Cables	(P.O. Box 4506)
and Telex:	Lagos, Nigeria

or addressed to either party at such other addresses as such party shall hereafter furnish to the other party in writing. Each such notice, request, or authorization shall be deemed to have been duly given or made on proof that it was duly posted or delivered by hand, mail, or cable to the party to which it is required to be given or made at such party's address specified as aforesaid.

ARTICLE 20—NON-ASSIGNMENT

20.01 The Consultant shall not have the right to assign or transfer the benefit or obligations of this Agreement or any part thereof, without the written consent of the Employer.

ARTICLE 21—RECORDS

21.01 The Master Plan Report, including presentation materials, models and supporting statistical data and technical materials, shall be the property of the Employer.

ARTICLE 22—SETTLEMENT OF DISPUTES

22.01 If any dispute or difference of any kind whatsoever shall arise out of this Agreement which cannot be amicably settled by the parties hereto, the same shall be referred to an arbitrator mutually agreed upon by the parties for determination in accordance with the provisions of the Arbitration Act of the Federal Republic of Nigeria.

22.02 If the parties cannot mutually agree on an arbitrator or arbitrators, each of the parties shall appoint one arbitrator and the two arbitrators shall appoint the third who shall be the Chairman. If within thirty (30) days after the receipt of a request for arbitration either party has not appointed an arbitrator, or if within thirty (30) days of the appointment of the two arbitrators the third arbitrator has not been appointed, either party may request the High Court of any of the Judicial Divisions of the Federal Republic of Nigeria to appoint an arbitrator. The arbitration award shall be accepted by the parties as the final binding adjudication of the dispute. In the event of arbitration, the venue of arbitration shall be Nigeria.

ARTICLE 23—TERMINATION OF AGREEMENT

23.01 Without prejudice to accrued rights or obligations of both parties, either party shall be at liberty to terminate this Agreement by serving a notice in writing on the other one calendar month prior to the effective date of such termination.

23.02 In the event of the termination of the Agreement as provided in this clause, the party seeking to terminate the Agreement shall pay or reimburse to the other within two calendar months a proportion of the agreed fee under Article 10 and Annex D as shall adequately compensate the other for the termination of the Agreement.

23.03 Upon the due termination of this Agreement as provided in this clause, the Consultant shall deliver to the Employer all documents, preparations, drawings, reports, and other information obtained or prepared by the Consultant in connection with the Project.

ARTICLE 24—APPLICABLE LAW

24.01 This Agreement shall be governed by the Laws of Nigeria. The Parties shall comply with these Laws.

ARTICLE 25—LANGUAGE

25.01 English shall be the official language of this Agreement.

IN WITNESS whereof the parties hereto have hereunto set their hands and seals the day and year first above written.

SIGNED, SEALED AND DELIVERED by within named

_____

for and on behalf of the Federal Capital Development Authority.
In the presence of
Name:
Address:
Occupation:
SIGNED, SEALED AND DELIVERED by the within named

_____

for an on behalf of International Planning Associates.
In the presence of
Name:
Address:
Occupation:

ANNEX A

TERMS OF REFERENCE

PREAMBLE

1. The Government of the Federal Republic of Nigeria has decided to develop a new city to serve as the nation's Capital in place of the current Capital City of Lagos. The area designated for this development is made up of approximately 8,000 square kilometers of territory about 200 airline miles from Lagos, the present Capital. The Territory is, by comparison with other relatively more urbanized areas of Nigeria, virgin territory located at about the geographical centre of the country. The first phase of development is expected to hold a population of approximately 500,000 and to be completed by 1997.

2. As the national Capital, the Federal Capital City shall have the following functions:

    (i) The Federal Capital City shall be the nation's political and administrative centre with normal supporting secondary and tertiary services. In that capacity it shall also serve as a symbol of the nation's aspiration for national unity and integration, development, and progress.

(ii)   In support of (i) above, the city shall be the focus of other national activities which may encourage and support its status as the seat of the Federal Government. Such national activities will include but not be limited to:

    (a)   the promotion of intellectual and cultural activities at the highest possible levels nationally and internationally through the following institutions: a university with strong emphasis on post-graduate and post-doctoral research; a national library of reference; a national museum of art and culture; a national theatre for the performing arts; a national archives; a national conference centre;

    (b)   the promotion of international and national amity through sports.

(iii)   The city shall serve as a major communications centre and will be expected to develop commercial and financial networks of national and international significance.

(iv)   In order to provide for the needs set out above and be a living community for its inhabitants, the plan of the city and its regional grid will provide an economic base for a self-sustaining development of the Federal Capital Territory as a whole and provide within the Federal Capital City itself (as distinct from the rest of the Territory), pollution-free industry.

3.   Bearing these functions in mind, the Consultant shall proceed to prepare the Master Plan for the Federal Capital City and its regional grid on the basis of the more specific terms of reference set out hereunder:

(i)   The Consultant shall collect and review all data necessary, both factual and graphic, for producing a City and Regional Plan for the Federal Capital Territory (8,000 square kilometers).

(ii)   Select alternative Capital City sites with corresponding urban forms within 6 months of commissioning for consideration by the Federal Capital Development Authority.

(iii)   With (ii) prepare alternative City and Regional Concept Plans for the selected Capital City site showing in schematic form:
—location of Government areas
—position of central Area
—district centres
—residential areas
—circulation system, public transport systems, etc.
—basic utility systems
—green belts, open spaces, and recreation areas; and
—any other plan features

(iv)   Prepare Draft Master Plan for the Capital City (1979–2000) at a scale of 1:10,000 assuming an estimated initial population of

500,000–1,000,000 with a cut-off population of 3 million after which new towns have to be planned within the Federal Capital Territory area to absorb further growth. (Best estimates and further details will be provided after the on-going demographic survey has been completed.) The Draft Master Plan shall define physical components such as:

(a)　Land use (Federal government facilities, residential areas, central business).

(b)　Transportation and circulation—strong emphasis on public transport (bus routes and rapid transit system), residential roads, arterial roads, highways, etc.

(c)　Utilities and services such as water supply, sewerage and drainage network, solid waste disposal, the generation and distribution of power, telecommunication system.

(d)　Recreation and other open spaces—parks, green belts, sports centres, leisure facilities, etc.

(e)　Community facilities—City Hall, libraries, civic centres, religious institutions, neighbourhood and district centres, etc.

(f)　Public services and administration—City administration, educational, and health institutions.

(g)　Industry—service, light manufacturing, and construction industries.

(h)　National facilities—National Library, conference centre, stadium, archives, theatre, etc.

(i)　Foreign representational facilities.

Also, to be defined shall be other programmes such as:

(j)　Health

(k)　Educational and vocational training

(l)　Safety and security

(m)　Economic base

(n)　Planning, administrative, and implementational framework—building codes, zonal codes, and bylaws.

(v)　The Consultant shall prepare a Regional Plan of the entire Federal Capital Territory at a scale of 1:50,000 (for the period 1979–2000 and beyond). The Regional Plan shall define the physical components associated with the plan such as:

(a)　Land use—industry, agriculture, etc.

(b)　Transportation and circulation—roads, railway facilities, airport. A multi-modal transportation study of links with the rest of the country shall be an essential element of this plan.

(c)　Utilities essential for regional development.

(d)　Open spaces, conservation areas.

(e)　Regional settlement structure—the Consultant shall give consideration in the Master Plan to the question of the desirability of limiting the ultimate size of the initial metropolitan area of the Capital City.

In the Regional Master Plan, the Consultant shall put forward suggestions for the growth or development of urban centres within the Federal Capital Territory.

The Plan shall also define non-physical components such as:

- (f) Health and education
- (g) Safety
- (h) Vocational training
- (i) Economic development
- (j) Administration and implementation
- (k) Environmental management including "peripheral control"
- (l) Development phasing and programmes of implementation with particular reference to agricultural business and industry.

(vi) The Draft Master Plans for the Capital City and region shall be accompanied by:

- (a) A comprehensive design and development Standards Manual.
- (b) An implementation and phasing plan—for the period 1979–2000 in 5-year medium-term plans. The phasing plan shall include a physical plan, financial programme, and logistics of implementation.
- (c) Models of alternative Concept Plans.
- (d) Two, three-dimensional scale models of the Capital City and Regional Plans.
- (e) A Landscape Plan of the city with recommendations on the location and development of plant nurseries within the region.

(vii) Throughout the studies and investigations, the Consultant shall advise and assist the Federal Capital Development Authority on:

- (a) The most suitable training programmes for its staff who will be involved in implementation and Phase II development.
- (b) The management of all data collected during the planning process.
- (c) The co-ordination of the work of others employed to detail specific sections of the Master Plan during the first phase of development.
- (d) The employment of the services of qualified Nigerian professionals in consultation with the Federal Capital Development Authority in carrying out all essential surveys, investigations, and studies.
- (e) Accept on its studies and design teams Federal Capital Development Authority staff as may from time to time be assigned to them.
- (f) Early actions relative to the development of the Capital.

(viii) The Consultant shall, if required by the Federal Capital Development Authority:

   (a) Make recommendations on an orientation system which would include street naming and house numbering in the Capital City;

   (b) Provide detail plans of such areas as are specified by the Federal Capital Development Authority as a test of the functionality of the Capital City Master Plan and, in addition, produce three-dimensional models of such specific areas; and

   (c) Prepare a documentary historical film of the design and physical development process.

ANNEX B

SCHEDULE OF THE SCOPE OF SERVICES OF THE CONSULTANT

The Scope of Work to produce a Draft Regional Plan of the Federal Capital Territory, and a Draft Capital City Plan for an urbanized portion thereof consists of ten major integrally related multi-discipline tasks, described herewith and on the attached "Work Flow Diagram" as follows:

TASK 1. ENVIRONMENTAL ASSESSMENT

This task shall consist of Consultant's environmental assessment work leading to the location and function of alternative sites for the new Federal Capital city within the designated Federal Territory area, and shall include the following considerations:

a. Field trips to the territory for direct reconnaissance of the characteristics, problems, and potential for alternative site location selection.

b. Assembly of existing available data on the environmental character and condition of the Federal Capital Territory.

c. Determination by inspection, research, and interviews of data gaps in the available environmental information.

d. Recommended scopes of work to produce the needed environmental data, together with generalized cost and production time estimates.

e. Analysis, interpretation, and preparation of maps and studies illustrating the interrelationships for planning purposes of the environmental data (land, water, air, flora, vistas, ecology, archaeology, etc.).

f. Synthesis of the environmental analysis by illustration, determination of future land use suitabilities, and evaluation ratings to determine optimum locations for alternative site designations.

TASK 2. PREPARATION OF PLANNING PROGRAM, DEVELOPMENT CRITERIA, AND COMMUNITY STRUCTURE ANALYSIS

This task shall include an analysis of the demographic, social, economic, cultural, administrative, etc., considerations bearing upon and relating to physical planning and development standards for the federal territory and the urbanized area as follows:

a. Assembly and analysis of existing available demographic, social, economic, cultural, administrative, etc., data from inspection, surveys, research, and interviews with key resource individuals and agencies.

b. Determination of data gaps and preparation of scopes of work for producing the needed data, together with generalized cost and production schedules.

c. Determination of evaluation criteria for selection of the types and concentration of land uses, regional and urban functions and activities, social and institutional requirements, and economic base studies leading to forecasts of land requirements for all major community structure needs in the federal territory and urbanized portion therein.

Included in this study will be forecasts of future population and employment levels: residential, commercial, and industrial densities; generalized economic linkages of the Federal Capital Territory with its trade areas.

d. Development of physical planning standards from the aforementioned investigations as a basis for testing alternative sites for optimum locations and functional land use potentials, to include the following features:

     i.   urban land use and density standards
     ii.   regional and local multi-modal transportation standards
     iii.   environmental quality standards
     iv.   residential, commercial, industrial land development standards
     v.   regional and urban open space standards
     vi.   generalized utilities/communications requirements
     vii.   administrative, educational, cultural, health service standards
     viii.   community facility standards
     ix.   safety and security standards designated by the military government
     x.   federal governmental space standards
     xi.   foreign representation space standards.

TASK 3. URBAN FORM HYPOTHESES

This task essentially consists of testing the capacity of selected candidate sites for the new capital city to accommodate urban development. By hypothesizing a series of alternative urban forms for each site, the Consultant can bring functional design and community structure considerations to bear upon environmental quality and planning standards as a basis for final site selection evaluation. This range of coordinated tests will extend FCDA's choice and options for its recommendations to the Government of Nigeria on the optimum site.

Generalized urban form configurations of the candidate sites will include, in schematic form, the following features:

a. Location of federal, state, municipal government administration areas.
b. Town centre (day-time and night-time use considerations)
c. Residential, commercial area/employment centres
d. Multi-modal transportation features (functional separation of air, highway, water, rail transit, vehicular, pedestrian ways).
e. Environmental/aesthetic considerations.
f. Major utility transmission systems.
g. Parks, recreation, and open spaces.
h. Energy and resource conservation considerations.
i. Regional settlement structure and expansion areas for growth beyond the year 2000.

TASK 6. DRAFT CAPITAL CITY PLAN

Proceeding simultaneously with the Draft Regional Plan study, but extending beyond that effort in greater detail, the preparation of the Draft Capital City Plan will dominate the latter half of the year-long planning program. The Draft Capital City Plan will be prepared for the site selected by the Employer after review and evaluation of the choice presented by the Consultant in Task 4.

The Draft Capital City Plan will include all of the features described in Task 3, Urban Form Hypotheses, for alternative sites, as adapted to the finally selected site. The Draft Capital City Plan will be supported by a full range of implementation recommendations aimed at achieving the nation's goals and objectives for planning, the time schedule for population settlement, and the inauguration date scheduled for 1986.

TASK 7. DRAFT COMPREHENSIVE DESIGN AND DEVELOPMENT MANUAL

The Comprehensive Design and Development Manual will be FCDA's single most important technical device for controlling the functions and physical development quality of both regional and urban planning activities in the new Federal Capital Territory. The preliminary urban design standards in the manual will establish the guidelines for architectural and landscaping studies to come after the first year of site selection and preliminary general planning work are completed.

The Design and Development Manual will provide applicable ranges of development standards evolved in Task 2 of this work scope, with selected illustrations of residential density arrangements, urban and community open spaces, building height and coverage standards, protection of conservation areas, and vistas, separation of vehicular and pedestrian ways, etc.

The Design and Development Manual would be prepared to an easy-to-use form suitable for continual updating throughout the planning, design, and construction phases of the new capital city.

TASK 8. IMPLEMENTATION PROGRAM

The implementation program for the Draft Regional Plan and Draft Capital City Plan will cover a wide range of recommendations for achieving the goals, objectives, and deadlines for construction of the new city, and will include the following features:

a. Generalized cost estimates for major physical improvements (utilities infrastructure, transportation facilities, residential communities, land improvements, etc.).

b. Five-year staged development schedules of the year 2000 for various sectors of the plans.

c. Alternative governmental arrangements for administering the phased growth of the new city.

d. Selected design and development standards (summarized from the Design and Development Manual produced in Task 7).

e. Alternative methods for financing the construction of the new city.

f. Suggested methods of land use control, preservation of open space, and joint public/private uses of land.

TASK 9. DRAFT MASTER PLANS REPORT/THREE-DIMEN-SIONAL MODELS

The final task in the first year of planning work will be the preparation and publication of a report on the Draft Regional Plan and the Draft Capital City Plan.

This report will be accompanied by three-dimensional models of both plans in a size and scale suitable for policy discussions at the highest levels of Nigerian government.

The report will provide a comprehensive account of the planning process, and a record of the data, program criteria, environmental assessment, findings, goals, and assumptions for planning (from Tasks 1 and 2); alternative site evaluation and urban form hypotheses studies (from Tasks 3 and 4); a full description of the Draft Regional Plan and Draft Capital City Plan (from Tasks 5 and 6); design and development standards, and implementation recommendations (from Tasks 7 and 8).

The report will be both a working and an historic document which will provide a foundation of great ideas and development objectives for many decades to come.

TASK 10. ADVICE AND ASSISTANCE TO FCDA

Throughout the studies and investigations, the Consultant shall advise and assist the Federal Capital Development Authority on:

(a)   The most suitable training programmes for its staff who will be involved in implementation and Phase II development.

(b)   The management of all data collected during the planning process.

(c)   The co-ordination of the work of other consultants employed to detail specific sections of the Master Plan during the first phase of development.

(d)   The employment of the services of qualified Nigerian professionals in consultation with the Federal Capital Development Authority in carrying out all essential surveys, investigations, and studies.

(e)   Accept on its studies and design teams Federal Capital Development Authority staff as may from time to time be assigned to them.

(f)   Early actions relative to the development of the Capital.

ANNEX C

SCHEDULE OF THE PROGRAMME OF WORK

The Schedule of the Programme of Work is demonstrated in two ways.

1.   The following bar chart, Work Schedule, demonstrates the time frame in which the work described in Annex B will be performed and reviewed by FCDA within the one-year period of the FCDA-IPA Agreement.

2.   The accompanying list of work products shows the monthly output of the work effort, and a suggested bimonthly review schedule between the Employer and the Consultant. Review sessions can be held in any of the IPA offices in the United States, or in any location in Nigeria, subject to mutually agreed-upon time and cost considerations by the parties to this Agreement.

WORK PRODUCTS/REPORTS/MODELS NIGERIA'S FEDERAL CAPITAL PLANNING PROGRAM

MONTH a   IPA Team mobilization
(After receipt of FCDA's approval to proceed).

MONTH 1   Report on information gaps; scopes of programs for new data needs, together with costs and time schedules.

MONTH 2   Preliminary draft of Planning Program, Development Criteria and Community Structure Analysis. (Includes statements on goals, objectives, assumptions, and forecasts for planning.)

FCDA REVIEW

MONTH 3   Environmental assessment Map Studies. Final draft of Planning Program, Development Criteria and Community Structure Analysis.

MONTH 4   Preliminary draft of Environmental Assessment (synthesis) findings; criteria for site identification and selection; preliminary urban form concepts and hypotheses.

FCDA REVIEW

MONTH 5   Preliminary draft of sites evaluation report; land suitability studies synthesis.

MONTH 6   Report on alternative sites evaluation and recommendations; regional development and urban form concepts; study models.

FCDA REVIEW AND SITE SELECTION

MONTH 7   Outline draft of Design and Development Manual; refine planning program and development criteria for the Draft Regional Plan and the Draft Capital City Plan.

MONTH 8   Preliminary Report on the Draft Regional Plan; outline draft of the Final Report on total effort.

FCDA REVIEW

MONTH 9   Preliminary Report on the Capital City Plan, staged development and cost consideration; and the implementation program, to be included in the first draft of the Final Report on the entire first year planning effort.

MONTH 10   Final draft of the Design and Development Manual; final draft of the Final Report; demonstrate Three-dimensional Study Models.

FCDA REVIEW AND APPROVAL TO PUBLISH

MONTH 11   Complete Implementation Program. Complete all text and graphics of the Final Report for publication.

MONTH 12   Publish Final Report.
Complete Three-dimensional Models.

PRESENTATION TO FCDA OF ALL WORK

From "Agreement for the Production of the Master Plan for the Federal Capital City and Its Regional Grid." International Planning Associates: McLean, Virginia, 1977.

39
The Channel Tunnel

England and France

DID YOU KNOW . . . ?

➤ Napoleon is said to have discussed a channel tunnel—with a midchannel artificial island "to breathe the horses"—when English statesman Charles James Fox came to Paris in 1802.

➤ The tunnel is 31.35 miles (50.45 kilometers) long, connecting Folkestone, England, and Sangatte, France.

➤ 24 miles (38 kilometers) of the tunnel lie underwater, with an average depth below the seabed of 150 feet (46 meters).

➤ The tunneling machines were so large they spanned more than two football fields and weighed 1,200 tons each.

➤ The project employed approximately 15,000 people.

➤ Estimated in 1959 to cost $100 million, the actual cost of building—using the same basic design—was $15 billion upon completion in 1994.

➤ The Chunnel (the colloquial name for "Channel Tunnel") carries 10 million tons of freight and millions of passengers annually.

➤ The travel time between Paris and London has been reduced to 2 hours, 20 minutes.

➤ The Chunnel was 700 times more expensive than San Francisco's Golden Gate Bridge.

Eurotunnel, the company that operates the Channel Tunnel, by providing a reliable, all-weather trip time of 2 hours and 20 minutes from London to Paris, may foreshadow a paradigmatic change in global transportation. Air travel has

A Eurostar train enters the Channel Tunnel in Pas de Calais, northern France, 1995. Courtesy of AP / Wide World Photos.

become increasingly problematic with commuting delays and extra pollution. Japan's Seikan Tunnel, completed for rail traffic in 1988, confirms the convenience of rapid rail service; it largely replaced ferryboats across the stormy Tsuguru Strait separating Honshu and Hokkaido. As train technology reaches for supersonic speeds—the next step beyond the French TGV (*Train de Grande Vitesse*) and the Japanese Shinkansen—we shall witness the kind of competition experienced by our forebears when trains challenged the primacy of stagecoaches and other horse-drawn vehicles for long-distance journeys.

HISTORY

The English Channel, or *La Manche*, as it is known in French (*Le Canal de France*, on earlier French maps), is slightly more than 21 miles (33 kilometers) wide at its narrowest point—in the Strait of Dover, or *Pas de Calais*. The idea of tunneling under the English Channel was discussed as early as 1802 when Napoleon became first consul. He is said to have reviewed a scheme for two tunnels that would meet in the middle of the channel at a man-made island to rest the horses. Various proposals were presented for providing a reliable physical connection between continental Europe and the island of Great Britain. In 1803, for instance, the prescient Hector Horeau designed an "immersed tube," a pipeline that could be laid in a dredged trench. Thomé de Gamond, a brilliant polytechnician, envisaged a tunnel, bridge, and even dams and dikes in a half-century of ardent and able innovation and promotion.

The subsequent flowering of railway technology gave further impetus to such a tunnel; British and French ventures were formed to carry out the project. The British and French parliaments authorized preliminary work in 1875, as shown in the Draft Anglo-French Treaty of 1876 (reproduced here as Document I). In the ensuing years, shafts were sunk on both sides of the Channel. The project was halted as the result of objections from the British War Office.

By 1924, the issue of a tunnel had come up again, and British premier Ramsay MacDonald brought all four former British prime ministers—Balfour, Asquith, Lloyd George, and Baldwin—to debate the issue. It took just 40 minutes for them to concur in their disagreement and flatly reject the proposal. Churchill later commented, "There is no doubt about their promptitude. The question is: Was their decision right or wrong? I do not hesitate to say that it was wrong" (Abel, 34).

The 1950s saw renewed interest in a permanent channel link. In 1955, Prime Minister Harold Macmillan was asked, in the House of Commons, whether there remained any military objections to a channel tunnel. "Scarcely at all," was his terse reply (Abel, 40). In 1957 the Submarine Tunnel Study Group was formed. It consisted of a five participants: the Channel Tunnel Company, Ltd.; the *Société Concessionnaire du Chemin de Fer Sous-Marin entre la France et l'Angleterre*; the International Road Federation (Paris office); the Suez Canal Company; and a New York company, Technical Studies, Inc. (its protocol of agreement and list of participants are included here as Document II). Meanwhile, Cyril C. Means Jr., a professor of international law, rendered immense service by presenting the project in Europe to the head of what became the Suez Financial Company (after the nationalization of the Suez Canal). On August 29, 1958, the United States became the first nation to approve support for the project at the highest level: a presidential determination was signed by the International Cooperation Administration authorizing W. O. Smith of the U.S. Geological Survey to join British and French colleagues in observation and analysis of the channel seabed.

A treaty providing for construction by two British and French private companies, as well as establishing a joint British-French operating authority, was signed in November 1973 by Great Britain and France; thereafter, the British and French parliaments passed bills authorizing start-up work, and construction commenced in 1974. About a mile of tunnel from the English coast—the segment constitutes part of the system now in daily use—was completed.

Then construction was halted, and history seemed to repeat itself. In 1975, Britain—as it had nearly a century before—suddenly backed away from the project, just as the treaty was about to be ratified by Parliament. The country was in the grips of a severe economic crisis and had undergone a change of government the previous year. A few years later, discussion of a cross-channel link was revived, with the proviso that it be entirely privately financed. This time conditions were right for the project to be completed.

A variety of ways to connect the two countries was considered, among them road and rail options with a tunnel, or a bridge, or a bridge-and-tunnel combination. The final choice was the rail tunnel, as basically designed in 1957–59 by a team headed by Charles Dunn, president of International Engineering Company, a division of Morrison-Knudsen, which was assigned this task by a consortium of companies selected by the Channel Tunnel Study Group. The rail tunnel would consist of three bored, interconnected, concrete-lined tubes: two outer tubes, each with a single-track rail line, and a narrower middle tube for use as a service tunnel. Motor vehicles would not drive through but would be carried on railway cars.

In February 1986, British prime minister Margaret Thatcher and French president François Mitterand signed the Treaty of Canterbury (reproduced here as Document III), authorizing "the construction and operation by private concessionaires of a channel fixed link"; enabling legislation was passed by the two countries' parliaments the following year.

CULTURAL CONTEXT

The English attitude toward the channel tunnel, so lamented by Churchill, can be summed up in an 1858 comment by Lord Palmerston, who served as foreign minister and subsequently as prime minister: "What! You pretend to ask us to contribute to a work the object of which is to shorten a distance which we find already too short!" (Whiteside, 24). He was not alone in that belief. Lord Randolph Churchill, in an 1889 speech before the British House of Commons, proclaimed, "The reputation of England has hitherto depended upon her being, as it were, *virgo intacta*" (Whiteside, 13).

When Prime Minister William Gladstone permitted Sir Edward Watkin to start building a tunnel across the channel in the 1880s, Lieutenant General Sir Garnet Wolseley, adjutant general of the British Army, had a nightmare about a French invasion that made use of such a tunnel. He awoke, put on his full dress uniform, called on Gladstone, and declared that he could not be responsible for the defense of the island if a tunnel were built. Wolseley was then driven to the offices of the *London Times* and said the same thing. Prior to this incident, the *Times* had backed construction of such a tunnel, but it now withdrew support. Wolseley's stand on the issue was not to be taken lightly, as he had the full backing of the Duke of Cambridge, commander of the British Army at that time—and brother of the king. Sir Edward Watkin, promoter of the tunnel, was forced to quit, and the work stopped in 1881.

In World War I, both Marshal Foch and Sir John French, head of the British General Staff, said that if there had been a channel tunnel, they could have resupplied the British army far more easily and thereby shortened the war. Winston Churchill quoted them in his 1936 *Daily Mail* article, which strongly urged construction of the Channel Tunnel.

In the prologue to his 1962 book *Adventure Underground*, Joseph Gies reported on "Lunch at Luchow's," a seemingly incidental event that would have major ramifications on the channel tunnel project. It was in November

1956 that Cyril C. Means Jr., then arbitration director of the New York Stock Exchange and later a professor of international law, came for lunch with a college friend (a coeditor of this book). In the course of lunch, the subject of the channel tunnel came up. Because both young men were heavily committed to other work, they decided to employ a research specialist they knew, Joan Reiter, to prepare a concise history of efforts to date to build a channel tunnel. When completed, this document (reproduced here as Related Cultural Document I) was handed to Dean Jay at the 23 Wall Street offices of the Morgan Guaranty Trust Company of New York. Jay introduced the document, and its bearer, to Thomas S. Lamont, vice chairman and a leading shareholder of the bank. When Lamont took a personal and constructive interest in the project, what had hitherto been regarded as a nearly impossible dream suddenly became a matter of intense curiosity and interest for key members of the investment world. Morgan Stanley and Dillon Read began attending meetings in New York. Morgan Grenfell, a leading London merchant bank joined the team, along with Erlanger's, a bank headed by Leo d'Erlanger, whose family had helped found a channel tunnel company in the nineteenth century! In France, De Rothschild Frères came on board, due to their participation in the SNCF, the French national railways.

Eurotunnel terminal in France. Courtesy of Shutterstock.

PLANNING

World War II, with its constant and increasing use of air power, diminished the usefulness of the English Channel as a moat defensive. In the aftermath of World War I, both Marshal Foch and Sir John French lamented the absence of a channel tunnel, which, had it existed, would have facilitated the supply of the allied armies.

With a project so obviously dependent on the approval of two national legislative bodies and on the drafting and ratification of a treaty between two major sovereign powers, it would be difficult to exaggerate the role of public opinion. In this sense, the emergence of a consensus in favor of a scheme that had failed to win approval for more than a century deserves an explanation. It would be wrong to conclude that defense considerations alone led to antitunnel sentiment among the British public. The problem was, rather, sheer boredom with a topic that had been debated too often and too long. What finally stirred up positive public interest in the scheme was the result of a series of accidental events.

In 1962, British Rail decided to build a three-dimensional working model of the English terminal for the channel tunnel, exhibited to enthusiastic crowds at Charing Cross tube station in London. The BBC decided to produce a four-minute film clip of the model itself. Thereupon, towns from all over the United Kingdom expressed interest in seeing the model. At this time British public opinion began to veer into a supportive position for the project. The sense of *ennui* and frustration was dispelled by images that allowed the average individual to imagine himself or herself entering a channel tunnel train in person, or in a vehicle that could be driven up a simple ramp onto a specially designed railway car.

Of course, there had to be competent technical planning. This occurred when Thomas Lamont, perhaps the most trusted and admired banker in the United States, took an interest in having the project reviewed by three senior engineering firms: Bechtel Corporation of San Francisco, California; Brown & Root, of Houston, Texas; and Morrison Knudsen Company, of Boise, Idaho. The heads of these firms had lunch with Thomas Lamont at the Harvard Club of New York City. When asked to provide a cost estimate, the heads of the engineering firms advised that the scheme could not be estimated until it was designed. At the request of the recently formed Channel Tunnel Study Group, and of Mr. Lamont, they agreed to design it.

Technical Studies, Inc., the New York company that initiated the Submarine Tunnel Study Group (popularly known as the International Channel Tunnel Study Group), had as its founding chairman Arnaud F. de Vitry, an honors graduate of the *Ecole Polytechnique* and the Harvard Business School. His successor as Technical Studies chairman was Georges F. Doriot, a professor at the Harvard Business School who, during World War II, was appointed Brigadier-General and Deputy Director for Research and Development of the War Department General Staff. The company's international legal counsel was

Alfred E. Davidson, a principal draughtsman of the Lend-Lease Act, who had been General Counsel of the Foreign Economic Administration in Washington, DC and, for several years, director of the European headquarters of the United Nations Children's Emergency Fund (UNICEF).

Promotion of the post–World War II scheme came to a halt in 1974 when the British government, embroiled in an economic crisis, decided not to complete ratification of the Anglo-French Tunnel Treaty already signed with France. There ensued an 11-year hiatus. The single mile of tunnel already built by companies established through the initiative of the Channel Tunnel Study Group was ultimately included in the tunnel system now in daily use.

In 1981, when Prime Minister Thatcher met for a decisive summit with President Mitterand of France, a joint commission was named to revive and study the cross-channel link; ultimately, the decision was made to implement the design first propounded in the 1959 report of the Channel Tunnel Study Group, a report endorsed and amplified by the group's eminent engineering chiefs, Monsieur René Malcor and Sir Harold Harding.

Perhaps the one individual deserving the principal credit for the design of the project is Charles Dunn, president of International Engineering Company of San Francisco, who led the team that produced the 1959 report, and who was later consulted by the group's engineers when very difficult issues had to be resolved.

BUILDING

The Channel Tunnel is 31.35 miles (50.45 kilometers) long, connecting Folkestone, England, and Sangatte, France. About 24 miles (38 kilometers) of its length lie under water, with an average depth below the seabed of 150 feet (46 meters). It took three years for gigantic tunnel-boring machines to cut through the earth, racing in competition with each other from both ends toward the middle (the British won). The tunneling machines were so large they spanned more than two football fields and were capable of churning through 250 feet of earth per day. Altogether, 11 tunnel-boring machines, each weighing as much as 1,200 tons, were used to cut through the soft chalk marl under the Strait of Dover. At its peak, the $15 billion job employed roughly 15,000 people.

The rail tubes have two crossover points where trains can switch from one track to the other. The service tube provides an escape route in the event of an emergency, another innovation designed by Charles Dunn. The immense complex is kept dry by five pumping stations: three under the sea and one on each coast. To keep the heat of fast trains from building up, the tunnels are cooled by chilled water.

The Channel Tunnel was officially opened in May 1994. Trains started transporting passengers in September, and vehicles in March of the following year. In the first five years of operation the tunnel carried 28 million people and 12 million tons of freight. Those who bring cars can stay in their vehicles and are offered a train especially designed to carry automobiles. Truckers whose heavy vehicles are transported via a different train disembark from their cabs and congregate in a Club Car lounge where they are served a meal and can relax. The flagship line, however, is the Eurostar, a luxurious and fast passenger train for business and social travelers who are whisked at 186 miles (300 kilometers) per hour between London's Waterloo station and the Paris Gare du Nord or Brussels Midi station. Some commute for meetings and return the same day.

If construction of the Channel Tunnel had commenced in 1959 when Charles Dunn completed his seminal report—the same design that was finally implemented in 1993 after years of political delays—the project would have cost $100 million. When one of the original supporters of the tunnel looked at the final $15 billion cost, and compared that figure to the $100 million originally estimated by the three firms that had prepared the 1959 report for the Channel Tunnel Study Group, he remarked, "Time is the ultimate magnifier" (Litwin, 123).

However, in 1985 the governments invited bids from a binational consortium, and in January 1986, when the winning design and bid were announced, the costs of the project had dramatically escalated. Five French and five British construction companies, supported by three French and two British banks, formed the team. The banks owned 40 percent of the founder shareholders' shares, and the 10 contractors owned 60 percent of the shares (Litwin, 115). Sir Alistair Morton and Monsieur André Bénard, joint chairmen representing England and France, played important roles in the financial and construction management.

The cost of building had continued to rise: by 1989 it had risen 40 percent, and the banks required a full contract amendment. By the time the banks had been satisfied, there were over 200 banks included in the Eurotunnel syndication. According to Bénard, many of the construction decisions had to be approved by 90 percent of the syndicate, and in a number of cases some decisions had to be approved by an almost unachievable 100 percent (Litwin, 118).

Even in 2005, after 11 years of successful operation, Eurotunnel still struggled with a substantial debt. The company mounted an effort to renegotiate its debt before 2006 when Eurotunnel's contract required the payment of all interest on the debt in cash. Measures were taken to reduce costs by a workforce reduction of 3,000, and a new, all-French management team was organized after angry shareholders stormed a 2004 April board meeting in Paris.

More than 10 years after completion of the Channel Tunnel, the approaches on the British side have been improved by a Channel Tunnel Rail Link. That 46-mile (74-kilometer) southern addition reduced travel time from London to Paris to two hours and 15 minutes.

IMPORTANCE IN HISTORY

The Channel Tunnel was voted chief among the wonders of the modern world by the American Society of Civil Engineers. More countries may follow the example of the Channel Tunnel. A study group jointly led by the kings of Spain and Morocco was established several years ago to investigate a Gibraltar tunnel. A Bering Strait tunnel has been discussed, but if built, it would require trains that run at supersonic speeds over the vast unpopulated territories to be traversed. Cheng Du University and other research centers in China are said to be interested in the development of prototypes for a supersonic train that would run in a tube. In 1999, Daryl Oster was granted a U.S. patent for a supersonic train system, expanding on ideas originally put forward before World War I by Robert Goddard, the space pioneer.

Several factors influenced the Channel Tunnel's success. First, Professor Harold Edgerton of MIT and other specialists help determine precise geological conditions below the seabed. Second, the design of the tunnel included a third tube specifically as a safety escape route. Third, diplomacy was critical to the process. Political considerations had a major impact on the time it took to move from idea to construction. Once the mental shift had been accomplished, the tunnel project began to move ahead. The Channel Tunnel Study Group secured the leadership of Sir Ivone Augustine Kirkpatrick of England and Ambassador René Massigli of France. As co-presidents of the Channel Tunnel Study Group, these two diplomats guided the project forward and in the process helped attract the support of leaders from both sides.

To some degree, the Channel Tunnel Study Group exemplified the philosophy of the late Harold D. Lasswell, University of Chicago professor of psychology and later a member of the faculty of the Yale Law School, who proposed what came to be known as the *decision seminar*. Lasswell believed that many academic conferences are apt to be a waste of time, because those with the power to implement the ideas under discussion are not present at the conference. Lasswell suggested that many conferences should include both experts with functional knowledge as well as individuals with power to authorize implementation. The Channel Tunnel Study Group was one of the first examples of such a decision seminar.

The Channel Tunnel is an example of how construction can reduce pollution. Ships and airplanes cause enormous amounts of pollution; the Channel Tunnel has reduced the number of ships and aircraft plying the channel. Across the earth's great oceans, ships pollute in many ways, not all of them when oil spills occur. The sound of a ship's engine vibrations disturbs the environment of whales and dolphins, which are sensitive to sound, and an undersea tunnel is one way to combat this noise pollution. It is generally assumed that engineering and construction cause environmental pollution, but the Channel Tunnel actually reduced pollution and improved the environment.

FOR FURTHER REFERENCE

Books and Articles

Abel, Deryck. *Channel Underground: A New Survey of the Channel Tunnel Question*. London: Pall Mall Press, 1961.

Bechtel Corporation, Brown & Root, Inc., and Morrison-Knudsen Company, Inc. *The Channel Tunnel: Design and Construction of a Channel Tunnel Recommended by Three Engineer-Constructors*. November 1959. A copy of this is available in the Channel Tunnel archives at the Historical Collections, Baker Library, School of Business Administration, Harvard University, Cambridge, Massachusetts.

Bénard, André. "Financial Engineering of Eurotunnel." In *Macro-engineering: MIT Brunel Lectures on Global Infrastructure*, edited by Frank P. Davidson, Ernst G. Frankel, and C. Lawrence Meador. Horwood Series in Engineering Science. Chichester, England: Horwood Publishing, 1997.

Bonnaud, Laurent. *Le Tunnel sous la Manche: Deux Siècles de Passions*. New York: Hachette, 1994.

Davidson, Frank P. "An Express of the (Near) Future." *Air and Space*, December 1995/January 1996, 22–24. Further information on the 1910 invention by Robert Goddard.

Fetherston, Drew. *Chunnel: The Amazing Story of the Undersea Crossing of the English Channel*. New York: Random House, 1997.

Gies, Joseph. *Adventure Underground: The Story of the World's Great Tunnels*. New York: Doubleday, 1962.

Harding, Sir Harold. *Tunnelling History and My Own Involvement*. Toronto: Golder Associates, 1981.

Hunt, Donald. *The Tunnel: The Story of the Channel Tunnel, 1802–1994*. Malvern, England: Images Publishing, 1994.

Lemoine, Bertrand. *Le Tunnel sous La Manche*. Paris: Le Moniteur, 1994.

Litwin, George H., John J. Bray, and Kathleen Lusk Brooke. *Mobilizing the Organization: Bringing Strategy to Life*. London: Prentice Hall, 1996.

Macaulay, David. *Building Big*. New York: Houghton-Mifflin, 2000.

Slater, Humphrey, and Correlli Barnett, in collaboration with R.H. Géneau. *The Channel Tunnel*. London: Allan Wingate, 1957.

Whiteside, Thomas. *The Tunnel under the Channel*. London: Rupert Hart-Davis, 1962.

Internet

The Eurotunnel Web site provides information about the company's history, management, and board and tunnel traffic; see http://www.eurotunnel.com/.

For a summary of the Channel Tunnel, see http://www.pbs.org/wgbh/buildingbig/wonder/structure/channel.html.

For records related to planning done in the 1950s and 1960s, see Technical Studies, Inc. Records from 1957–1994, Historical Collections, Baker Library, School of Business Administration, Harvard University: http://www.library.hbs.edu/hc/additions/tsi.shtml.

For the ASCE list of the modern wonders of the world, including the Channel Tunnel, see http://www.asce.org/history/7_wonders.cfm.

Film and Television

Building Big. Videocassette box set. Hosted by David Macaulay. PBS television, WG965. Boston: WGBH, http://www.pbs.org/wgbh/buildingbig/shop/video.html. The video that deals specifically with tunnels is WG994.

Eurotunnel—Two Years On. Videocassette. Pasadena, CA: Association of American Railroads, 1993. http://www.pentrex.com/ chunnel.html. This short video focuses on the first two years of building the Channel Tunnel.

Documents of Authorization—I

BILL NO. C.157B, SESSION NO. 1876 ANGLO-FRENCH TREATY OF 1876

FOR A CHANNEL TUNNEL AND SUBMARINE RAILWAY

1. The boundary between England and France in the Tunnel shall be halfway between low-water mark (above the tunnel) on the coast of England, and low-water mark (above the tunnel) on the coast of France. The said boundary shall be ascertained and marked out under the direction of the International Commission to be appointed, as mentioned in Article 4, before the Submarine Railway is opened for public traffic. The definition of boundary provided for by this Article shall have reference to the Tunnel and Submarine Railway only, and shall not in any way affect any question of the nationality of, or any rights of navigation, fishing, anchoring, or other rights in, the sea above the Tunnel, or elsewhere than in the Tunnel itself.

2. The French section of the Submarine Railway shall be constructed, maintained, and worked in conformity with the French laws, and with that of the 2nd of August, 1875, in particular, subject to the provisions of the Treaty to be concluded between the two Governments. The English section of the Submarine Railway shall, subject to the provisions of the Treaty to be concluded between the two Governments, be constructed, maintained, and worked in accordance with such conditions as Her Majesty may by Order in Council hereafter impose in connection with the undertaking of the said Company (as specified in the Channel Tunnel Company Limited Act, 1875), with such, if any, modifications as may be hereafter made by Act of Parliament.

3. Within five years from the 2nd of August, 1875, the French Company shall be bound to conclude an agreement in writing with an English Company, and reciprocally, the English Company shall be bound to conclude an agreement in writing with a French Company, with a view to the construction, maintenance and working of the Submarine Railway. This term, 'Submarine Railway,' applies throughout the present Protocol to the Tunnel, to the Railway, and to all the works connected therewith, such railway being bounded in France by its junction with the railway from Boulogne to Calais, and in England by its junctions with the South Eastern and London, Chatham and Dover Railways.

This term does not include the works mentioned hereafter in Article 16.

4.   There shall be constituted an International Commission to consist of six members, three of whom shall be nominated by the British Government and three by the French Government.

    The International Commission shall advise the two Governments on all questions relating to the construction, the maintenance, and the working of the Submarine Railway, and shall have power, on giving notice to the respective Companies, to make such inspections as they consider necessary, and the Companies shall be bound in every way to facilitate such inspections, and to cause their delegates to be present. Each Company shall render annually to its Government an account of its receipts and expenses in such form as the Governments shall approve, after hearing the International Commission, and shall, if required, afford to its Government the necessary facilities for comparing such accounts with the books of the Company. If at any time any difference shall arise between the two Companies as regards the construction, maintenance, or working of the Submarine Railway, such difference shall be settled by the two Governments after having taken the opinion of the International Commission, subject to such legal actions as the Companies may bring in conformity with the Conventions concluded between them and with the legislation of the two countries.

    The Commission shall meet at all times when it shall consider it convenient to do so, and at least twice in each year. It shall also meet at any time at the request of either Government. But no meeting shall be valid unless there be present at least two Members appointed by each Government. If, at any meeting of the International Commission, the Members present of the one nationality shall differ in opinion from the Members present of the other nationality, reference shall be made to the respective Governments.

    The International Commission shall report every year to the respective Governments, both upon its own proceedings and upon questions connected with the Submarine Railway. It shall, moreover, submit to the two Governments its proposals for Supplementary Conventions with respect:—

(a) To the apprehension and trial of alleged criminals for offences committed in the Tunnel or in trains which have passed through it, and the summoning of witnesses.

(b) To Customs, police, and postal arrangements, and other matters which it may be found convenient to deal with.

5.   On the completion of the Submarine Railway, the International Commission shall cause it to be inspected as they may see fit on behalf of

the two Governments, and after such inspection, and on receiving from the International Commission their recommendation in writing, but not before, the Submarine Railway shall be opened for traffic.

6. One set of regulations shall be applicable to the Submarine Railway as a whole; the regulations to be subject to the approval of the two Governments on the recommendation of the International Commission; the tariff of maximum charges shall be fixed in accordance with the Tariff hereto annexed.

7. Each Company shall be responsible for keeping in good and substantial repair the portion of the Submarine Railway situated within its own country; and in case of default, the two Governments, on the recommendation of the International Commission, shall have power, each in its own country, to execute, as may seem right, all necessary works and repairs. The two Governments shall also have power, each in its own country, to receive all moneys payable to the Companies, until the expenses of such works and repairs are covered. These moneys shall be collected in each country in accordance with the existing laws.

8. The concession granted by each Government shall be for a term of ninety-nine years from the opening of the Submarine Railway. At the date fixed for the termination of the concession, or at an earlier period, in the event of the forfeiture of the concession, pronounced in the manner laid down in Article 10 below, each Government shall become possessed of all the rights of the Company, established on its territory, in an over the Submarine Railway in such country, and shall enter immediately into enjoyment of all the revenues of the Company.

The Company, in each country, shall be bound to hand over to the Government in a good state of repair the portion of the Submarine Railway in such country.

During the five years preceding the date fixed for the end of the concession, the Government of each country shall have the right to receive the revenues of the Company established in its own country, in order to apply them to the maintenance of the said portion, unless the Company takes steps to carry out this engagement fully and entirely.

With regard to the rolling stock, movables, and stores of all kinds, the furniture and tools of workshops and stations, each Government shall be bound, at the request of the Company established in its own country, to take all the above-mentioned objects at a valuation, which shall be made in such manner as may be provided by the laws of the country; and reciprocally, if the Government requires it, the Company shall be bound to give up, under the same conditions, the rolling stock and other things above-mentioned.

The Government, however, will only be bound to take over the stores necessary for working the railway for six months from the end of the Concession.

9. The works of exploration shall be commenced within one year from the 1st July, 1876.

   If within five years from the 2nd of August, 1875, the concessionaires have not been able to conclude the agreement referred to in Article 3, or if, in consequence of the result of the borings and other preparatory works, they recognize the impossibility of carrying out the undertaking, the Companies shall have the right of abandoning the concessions.

   Within five years from the 2nd of August, 1887, each Company is to declare to its own Government whether such Company proposes to retain the concession. This period of five years can, however, on the application of the Company, be extended in either country by the Government, at its discretion, for three further years, that is to say, for eight years from the 2nd of August, 1875.

   In default of such declaration having been made by either Company within the above periods, and also if either Company should declare its intention of abandoning the undertaking, the concession to the Company making such default or declaration shall be considered as null and void; and action shall be taken in accordance with the provisions of Article 10. If one of the two Companies abandons its concession, the two Governments shall consult as to the measures to be adopted, without the other Company being entitled to raise any objection or to lay claim to any indemnity.

   Twenty years, to date from the day on which the Company shall declare its intention to retain the concession, shall be allowed for the completion of the Submarine Railway and the opening of the said Railway for public traffic.

10. At the expiration of each of the periods mentioned in the preceding Article, the Companies shall cease to have the right to commence or to execute the works which should have been commenced or executed within the period which has so expired, and if at any time after the works have been commenced the Companies shall for a period of one year, without such cause as the respective Governments, after hearing the International Commission, may consider reasonable cease to carry on the works, and if the Submarine Railway be not opened for public traffic before the expiration of the period of twenty years mentioned in the preceding Article, or if at any time the Companies, without such cause as the respective Governments, after hearing the International Commission, may consider reasonable, cease for a period of six months to work the Submarine Railway, in conformity with the rules laid down by their Governments, then, and in any of such cases, the concessions granted to the Company in fault shall be liable to forfeiture, which forfeiture shall be enforced according to the laws for the time being of each country respectively.

    The forfeiture can only be pronounced by a Government against a Company after the necessity of that forfeiture has been recognised by the joint agreement of the two Governments on the recommendation of the International Commission.

11. Each Company may, at any time during the construction of the works, abandon its concession, on proving to the satisfaction of its Government the impossibility of continuing the said works.

In such case, forfeiture shall be declared and enforced according to the provisions of the Law granting the concession in France or of the Act of Parliament in Great Britain.

12. At any time after the end of thirty years from the opening of the Submarine Railway, each Government shall have the right to purchase the undertaking of the Company established on its territory. This right shall not however be exercised excepting after a joint agreement between the two Governments, and after six calendar months' notice in writing has been given to the Companies. In the event of purchase, the rights of each Government in and over the soil, works, and undertaking, shall be limited to its own territory, as defined in Article 1.

13. The amount of the purchase-money in each country shall be determined as follows, under the supervision of the International Commission. The net receipts of the Company during the seven years immediately preceding the year in which the purchase is effected shall be ascertained; the two years of minimum receipts shall be excluded, and the mean of the annual net receipts during the other five years shall be taken. That mean net receipt will form the amount of an annuity, to be payable to the Company for the unexpired term of the concession, or, at the option of the British Government, for the purchase of the English concession, the basis of the calculation of a capital sum representing the value of the annuity at the time of purchase. In any case the amount of the annuity to be so payable, or which is to form the basis of such calculation as aforesaid, is not to be less than the amount of the net receipts during the year immediately preceding the year of purchase.

Each Government is to provide and pay the annuity or capital sum which will be due to the Company established on its territory.

The Company shall receive, in addition, the payments to which they may be entitled at the date fixed for the expiration of the concession, in accordance with paragraph 4 of Article 8.

14. The working and maintenance of the Submarine Railway after either the purchase, or the termination, or the forfeiture of the concession in either country, shall be provided for by a Supplementary Convention then to be made between the two Governments.

15. Each Government shall have the right to suspend the working of the Submarine Railway and the passage through the Tunnel whenever such Government shall, in the interest of its own country, think necessary to do so. And each Government shall have power, to be exercised if and when such Government may deem it necessary, to damage or destroy the works of the

Tunnel or Submarine Railway, or any part of them, in the territory of such Government, and also to flood the Tunnel with water. If any of the powers of this Article are exercised by either of the Governments, then and in every such case neither the other Government nor either of the Companies shall have any claim to any other indemnity or compensation than the following. If any such power is exercised during the term and currency of the concession to either Company, the period of concession to that Company is to be extended for a term equal to that during which the working of the Submarine Railway has been suspended in consequence of the exercise of any of the powers mentioned in this Article. If any such power is exercised before the expiration of the period during which the French Government has engaged not to grant any rival concession, the term of this period shall be extended in like manner as that of the concession.

Each Government, however, reserves to itself the right, if it should think fit, to grant the Company established in its own country, but not to the Company established in the other country, such compensation for damage actually done by its order to the works of each Company as such Government may in its discretion think proper.

From British Official Publications Collaborative Reader Information Service: http://www.bopcris.ac.uk/bopall/ref5090.html.

Documents of Authorization—II

Although only a selection of the document is presented below, it still offers a valuable historical prespective.

DOCUMENTS ESTABLISHING THE SUBMARINE TUNNEL STUDY GROUP AND MINUTES OF THE MEETING OF THE BOARD OF CONTROL OF JULY 26, 1957 IN PARIS SUBMARINE TUNNEL STUDY GROUP PROTOCOL OF AGREEMENT

Among the undersigned:
I
The Channel Tunnel Company, Ltd., represented by Mr. Leo d'Erlanger, hereinafter called: the British Group";
II
(1) The *Société Concessionnaire du Chemin de Fer Sous-Marin entre la France et l'Angleterre*, represented by MM. Louis Armand and Jacques Getten,
(2) The International Road Federation (Paris office), represented by M. Charles de Wouters,
constituting, for the purpose hereof, the "French Group";
III
The Universal Suez Canal Company, represented by M. Jacques Georges-Picot;
IV
Technical Studies, Incorporated, represented by Mr. Frank P. Davidson;

(the above-mentioned organizations being hereinafter called "the partici-pants"); it has been agreed and decided as follows:

ARTICLE I

The participants have decided to study jointly the conditions under which the construction and operation of a submarine railway and/or roadway tunnel, link-ing British territory with Continental Europe, could be brought about.

For the carrying out of this contract, a Board of Control will bring together the representatives of the participants. The studies will be entrusted to a Man-ager of the Group, who will pursue them under the direction of the Board of Control, with the assistance, on occasion, of a Technical Committee.

ARTICLE II

The participants bind themselves to commit an initial lump sum of £100,000 to the studies jointly undertaken, the share of each of them being fixed under the following conditions:

British Group:

30%,

it being here specified that The Channel Tunnel Company may associate with itself one or several third parties as co-participant(s), but with the approval of a majority of the Board of Control;

French Group:

30%,

divided, 25% to the *Société Concessionaire du Chemin de Fer Sous-Marin entre la France et l'Angleterre,*
and 5% to the International Road Federation (Paris office);

Suez Company:

30%,

Technical Studies, Inc.:

10%,

By decision of the Board of Control, a complementary share not exceed-ing 15% may be offered to the other North American participants. This share would be created by deducting, at that time, equal fractions from the shares of the British Group, the French Group, and the Suez Company.

The funds will be deposited in accounts of the Manager of the Group, opened in France, in Great Britain and in the United States, as chosen by the participants, at times and in the manner decided upon by the Board of Control in accordance with the amounts needed in the successive stages of the studies.

If experience reveals the necessity for additional funds over and above the £100,000 hereinabove provided for, the Board of Control may propose a call for them. Such decision must be taken by the participants unanimously.

ARTICLE III

Each participant binds itself:

(a) To place at the disposal of the Manager of the Group, without delay or restriction, all elements in his possession likely to render the studies easier and more effective, and particularly all information, documents and previous projects.

(b) Without compromising its rights of ownership or user, to permit the Manager of the Group, subject to the necessary administrative permits, to make use of such of its lands, works, and equipment as may be of use for the studies, as well as the rights and powers which it holds.

ARTICLE IV

The administration of the joint undertaking is provided by a Board of Control, composed of eleven members, distributed as follows:

One Chairman,
three representatives of the British Group,
three representatives of the French Group,
three representatives of the Suez Company,
one representative of Technical Studies, Inc.

In the event that the American participation should be increased as provided for in Article II, *supra,* its representation may be secured by the addition of one or two members to the Board of Control, approved by the other members of the said Board.

The Chairman will be chosen by agreement amongst the participants, but none of the above-mentioned representatives shall be eligible. Should the Chairman be absent from one of the meetings of the Board, it will elect a Chairman of the meeting.

Notices of meetings of the Board of Control shall be directed to the representatives by the Chairman.

Decisions shall be reached by majority vote; in case of equal division the Chairman, or Chairman of the meeting, having a casting vote.

Each representative shall have the right to represent any absent representative or representatives of his group.

ARTICLE V

The Board of Control will designate, for the duration of the contract, a Manager of the Group and, if it be deemed necessary, a Technical Committee composed of three or four engineers.

The Manager and the members of the Technical Committee may be drawn from outside the Board of Control.

ARTICLE VI

The Manager of the Group will be responsible for the organization and conduct of the studies and for the executing of all instruments necessary to carry them out.

He will act in conformity with instructions given him by the Board of Control and under its supervision; accordingly, he will carry out the decisions reached by the Board within the scope of his managerial and representative functions and the means, in particular the financial means, placed at his disposal. He will submit to the Board of Control, for its prior approval, every transaction involving an expenditure of more than £2,000.

With regard to third persons, he will always act either in his own name or in that of the participant which will sign with him his employment contract. In that case, such participant will itself act on behalf and for the account of all the others.

The participants bind themselves to act as guarantors of the Manager of the group and his apparent principal whenever they may be put in jeopardy of personal inability by third parties by reason of something done within the scope of the Manager's employment.

ARTICLE VII

The Technical Committee will proceed with the studies wherewith, on occasion, the Board of control entrusts it. In particular, the Committee will give the Board its advice concerning the programs proposed to the Board by the Manager of the group, and concerning the progress and the findings of the studies whose development it will follow at regular intervals.

ARTICLE VIII

The present contract has been entered into for a period of two years; this period will be extended by one year unless the Board of Control decides to the contrary.

Every subsequent extension will require an express decision of the Board.

Each participant may renounce the benefit of the present contract at the end of each of the periods hereinabove provided for, on condition that it informs each of its co-participants three months before the expiration date of the aforesaid period. In such case, its contribution in the form of money will be returned to it, after deduction of a portion, proportional to its contribution, of the expenses already effected or contracted to be made; on the other hand, the contributions already effected in the fulfillment of Article III, *supra,* will remain at the disposal of the Manger of the Group until the close of the Group's operations.

ARTICLE IX

At the expiration of this contract, the studies accomplished will remain the joint property of the participants, each of which undertakes not to use them for purposes of its own without the prior agreement of the others.

If the funds placed at the disposal of the Manager of the group as provided for in Article II shall not have been wholly used up, the remainder will be divided amongst the participants in proportion to their respective contributions of money.

ARTICLE X

Disputes arising from the carrying out of the present contract will be finally settled by arbitrators deciding as *amiables compositeurs,* without any possibility of appeal or other recourse.

Each party to a dispute shall designate an arbitrator of his choice; in the absence of such designation within fifteen days after notice, such arbitrator shall be named by the Presiding Judge of the Commercial Tribunal of the Seine.

The board of arbitrators must deliver its award unanimously; if it be unable to do so, a third arbitrator, designated by agreement of the first two or, in the absence thereof, by the Chairman of the International Bank for Reconstruction and Development, will decide the questions remaining in dispute.

Done, in the City of Paris, in two copies, the twenty-sixth of July, A.D. 1957.

CHANNEL TUNNEL COMPANY, Limited
(signed) Leo d'Erlanger
Société Concessionnaire du Chemin de Fer Sous-Marin Entre La France et l'Angleterre
(signé) L. Armand.
(signé) J. Getten.
COMPAGNIE UNIVERSELLE DU CANAL MARITIME DE SUEZ
(signé) J. Georges-Picot.
FEDERATION ROUTIERE INTERNATIONALE (BUREAU DE PARIS)
(signé) (C. de) Wouters.
TECHNICAL STUDIES, INCORPORATED
(signed) Frank P. Davidson

From Technical Studies, Inc. Records from 1957–1994. Cambridge, MA: Harvard University: Historical Collections, Baker Library, School of Business Administration.

Documents of Authorization—III

THE TREATY OF CANTERBURY

No. 25792

FRANCE
and

UNITED KINGDOM OF GREAT BRITAIN
AND NORTHERN IRELAND

Treaty concerning the construction and operation by private concession-aires of a channel fixed link. Signed at Canterbury on 12 February 1986
Authentic texts: French and English.
Registered by France on 16 March 1988.

FRANCE
et

ROYAUME-UNI DE GRANDE-BRETAGNE
ET D'IRLANDE DU NORD

Traité concernant la construction et l'exploitation par des
sociétés privées concessionnaires d'une liaison fixe
transmanche. Signé á Cantorbéry le 12 février 1986
Textes authentiques : français et anglais.
Enregistré par la France le 16 mars 1988.

TREATY[1] BETWEEN THE UNITED KINGDOM OF GREAT BRITAIN AND NORTHERN IRELAND AND THE FRENCH REPUBLIC CONCERNING THE CONSTRUCTION AND OPERATION BY PRIVATE CONCESSIONAIRES OF A CHANNEL FIXED LINK

Her Majesty The Queen of the United Kingdom of Great Britain and Northern Ireland and of Her other Realms and Territories, Head of the Commonwealth and the President of the French Republic,

---

1. Came into force on 29 July 1987 by the exchange of the instruments of ratifications, which took place at Paris, in accordance with article 20.

781

Confident that a Channel fixed link will greatly improve communications between the United Kingdom and France and give fresh impetus to relations between the two countries,

Desiring to contribute to the development of relations and of exchanges between the Member States of the European Communities and more generally between European States,

Desiring also to permit the construction and operation of a Channel fixed link by private enterprise in accordance with the criteria laid down by the Government of the United Kingdom and the French Government,

Have decided to conclude a Treaty and to this end have appointed as their Plenipotentiaries:

Her Majesty The Queen of the United Kingdom of Great Britain and Northern Ireland and of Her other Realms and Territories, Head of the Commonwealth: The Right Honourable Sir Geoffrey Howe, OC, MP, Her Majesty's Principal Secretary of State for Foreign and Commonwealth Affairs:

The President of the French Republic: His Excellency Monsieur Roland Dumas, Minister for External Relations;

who, having presented their full powers, found in good and due form, have agreed as follows:

ARTICLE 1. OBJECT AND DEFINITIONS

(1)   The High Contracting Parties undertake to permit the construction and operation by private concessionaires (hereinafter referred to as "the Concessionaires") of a Channel fixed link in accordance with the provisions of this Treaty, of its supplementary Protocols and arrangements and of a concession between the two Governments and the Concessionaires (hereinafter referred to as "the Concession"). The Channel fixed link shall be financed without recourse to government funds or to government guarantees of a financial or commercial nature.

(2)   The Channel fixed link (hereinafter referred to as "the Fixed Link"), which shall be more particularly described in the Concession, means a twin bored tunnel rail link, with associated service tunnel, under the English Channel between Cheriton in Kent and Fréthun in the Pas-de-Calais, together with the terminal areas for control of access to, and egress from, the tunnels, and shall include any freight or other facility, and any road link between the United Kingdom and France, which may here-after be agreed between the High Contracting parties to form part of the Fixed Link.

ARTICLE 2. INTERNATIONAL, LEGISLATIVE AND REGULATORY MEASURES

(1)   The High Contracting Parties shall take the measures which are necessary to ensure that the construction and operation of the Fixed Link shall be

consistent with their international obligations. They shall co-operate in making any necessary approaches to the relevant international organizations.

(2) The High Contracting Parties shall adopt such legislative and regulatory measures, and take such steps, as are necessary for the construction and operation of the Fixed Link by the Concessionaires in accordance with the Concession.

ARTICLE 3. FRONTIER AND JURISDICTION

(1) As regards any matter relating to the Fixed Link, the frontier between the United Kingdom and France shall be the vertical projection of the line defined in the Agreement signed at London on 24 June 1982 relating to the delimitation of the Continental Shelf in the area east of 30 minutes West of the Greenwich meridian,[2] and the respective States shall exercise jurisdiction accordingly, subject to the provisions of paragraph 3 of this Article and any Protocol or particular arrangements made pursuant to Articles 4, 5, 7 and 8 below.

(2) The frontier in the Fixed Link shall be marked by a Joint Commission, composed of representatives of the two States, as soon as possible after the completion of the relevant section of the Fixed Link and in any event before the Fixed Link comes into operation.

(3) If in the construction of the Fixed Link any works carried out from one of the two States extend beyond the line of the frontier, the law that applies in that part which so extends shall, in relation to matters occurring before that part is effectively connected with works which project from the other State, be the law of the first mentioned State.

(4) Rights to any natural resources discovered in the course of construction of the Fixed Link shall be governed by the law of the State in the territory, or in the continental shelf, of which the resources lie.

ARTICLE 4. POLICE AND FRONTIER CONTROLS

(1) The frontier controls shall be organized in a way which will reconcile, as far as possible, the rapid flow of traffic with the efficiency of the controls.

(2) Provisions for the exercise of police, immigration, customs and health controls, including animal and plant health controls, and of other controls which might appear necessary, will be the subject of a supplementary Protocol or other arrangements.

(3) Such a Protocol or arrangements will make provision to enable public authorities to exercise their functions in an area in the territory of the other State where controls are juxtaposed. They will also include provisions for the free circulation throughout the Fixed Link of public officials and other persons, so far as is necessary for the exercise of their functions in relation to

---

2. United Nations, *Treaty Series*, vol. 1316, p.119.

the construction and operation of the Fixed Link, and for the protection and assistance to be accorded to them.

(4) The construction and maintenance of the buildings and installations necessary for frontier controls will be at the charge of the Concessionaires on terms prescribed in the Concession.

(5) Each Government shall be responsible for the payment or recovery of the costs of its own controls.

ARTICLE 5. DEFENCE AND SECURITY

(1) Defence and security matters relating to the Fixed Link and the implementation of this Treaty shall be the subject of special arrangements between the two Governments. Such arrangements shall include provisions for the free circulation for the exercise of their functions in relation to the defence and security of the Fixed Link, and for the protection and assistance to be accorded to them.

(2) Such arrangements will provide for the designation by each Government of the authorities empowered to take any decision necessitated by the defence and security of the Fixed Link. The authorities so designated by the two Governments, or their agents, will so far as possible co-ordinate their activities within the framework of such arrangements.

(3) The Concessionaires shall submit to the two Governments for their approval any proposed designs, plans or arrangements affecting the defence and security of the Fixed Link and the two Governments shall agree a joint response to any such proposals.

(4) The Concessionaires shall, if required by the two Governments, take measures necessary for the defence and security of the Fixed Link. Save in exceptional circumstances of the kind envisaged in Article 6, the two Governments shall consult each other before requiring the Concessionaires to take such measures, and shall act jointly.

ARTICLE 6. EXCEPTIONAL CIRCUMSTANCES

(1) In the event of any exceptional circumstances, such as natural disasters, acts of terrorism or armed conflict, or the threat thereof, each Government, after consultation with the other if circumstances permit, may take measures derogating from its obligations under this Treaty, its supplementary Protocols and arrangements, or the Concession.

(2) Such measures may include closure of the Fixed Link, but shall be limited to the extent required by the exigencies of the situation and shall be notified immediately to the other Government and, as appropriate, to the Concessionaires.

ARTICLE 7. SOCIAL SECURITY, SAFETY AND LABOUR LAWS

The two Governments may make provision, by supplementary Protocol or other arrangements, relating to the application of laws on social security, employment and health and safety at work to the construction or operation of the Fixed Link.

ARTICLE 8. MUTUAL ASSISTANCE IN LEGAL MATTERS

The two Governments may make provision, by supplementary Protocol or other arrangements, for mutual assistance in the enforcement of civil, commercial, criminal, and administrative law and for the law applicable to the construction and operation of the Fixed Link.

ARTICLE 9. FISCAL, CUSTOMS AND MONETARY REGIME

(1)   The taxation by the two States of profits and gains derived from the construction or operation of the Fixed Link shall be in accordance with the laws of the two States, including any Convention for the avoidance of double taxation and the prevention of fiscal evasion with respect to taxes on income[3] that is in force for the time being and any Protocol thereto.

(2)   The two States shall observe the principle of non-discrimination in relation to taxes on charges made to users of transport which is in direct competition for cross-channel traffic.

(3)   The transfers of funds and financial settlements necessitated by the construction or operation of the Fixed Link, whether between the two States or from or to third countries, shall be permitted subject to the procedures, if any, prescribed by national laws made consistent with Community law. Conversions shall be made at the market rate applicable to similar transactions. The two States shall not levy any tax on such transfers of funds or financial settlements other than generally applicable taxes on the payments which they represent.

(4)   Both Governments intend, so far as may be consistent with their international obligations, to allow to travellers through the Fixed Link from the mainland of one State to that of the other duty-free facilities which are comparable to those available to persons travelling from one State to the other by sea or air.

ARTICLE 10. INTERGOVERNMENTAL COMMISSION

(1)   An Intergovernmental Commission shall be established to supervise, in the name and on behalf of the two Governments, all matters concerning the construction and operation of the Fixed Link.

(2)   With regard to the Concessionaires, the two Governments shall exercise through the Intergovernmental Commission their rights and obligations under the Concession, other than those relating to the amendment, extension, suspension, termination or assignment of the latter.

(3)   The functions of the Intergovernmental Commission shall include:

---

3. United Nations, *Treaty Series*, vol. 725, p.3.

(a)   Monitoring the construction and operation of the Fixed Link;

(b)   Undertaking necessary consultations with the Concessionaires;

(c)   Taking decisions in the name of the two Governments for the implementation of the Concession;

(d)   Approving proposals made by the Safety Authority as provided by Article 11;

(e)   Drawing up, or participating in the preparation of, regulations applicable to the Fixed Link, including regulations relating to maritime matters and the environment, and monitoring their subsequent implementation;

(f)   Considering any matter referred to it by the Governments or the Safety Authority or any other matter which appears to it to be necessary to consider;

(g)   Giving advice and making recommendations to the two Governments or the Concessionaires.

(4)   Each Government shall appoint half the members of the Intergovernmental Commission, which shall comprise at most 16 members including at least two representatives of the Safety Authority. The Chairmanship of the Commission shall be held for a period of one year by the head of each delegation alternately.

(5)   The decisions of the Intergovernmental Commission shall be taken by agreement between the heads of the British and French delegations. In the event of disagreement between them, the procedure for consultation between Governments provided for in Article 18 shall apply.

(6)   The Intergovernmental Commission shall draw up its own rules of procedure and submit them for the approval of the two Governments.

(7)   For the purpose of carrying out its functions the Intergovernmental Commission may invoke the assistance of the authorities of each Government or any body or expert of its choice.

(8)   The Governments shall take all necessary measures to ensure that regulations applicable to the Fixed Link have the necessary force and effect within their national laws and shall grant to the Intergovernmental Commission such powers of investigation, inspection and direction as are necessary for the performance of its functions.

(9)   The expenses of the Intergovernmental Commission shall be met by the Concessionaires as provided in the Concession.

ARTICLE 11. SAFETY AUTHORITY

(1)   A Safety Authority shall be established to:

(a)   Advise and assist the Intergovernmental Commission on all matters concerning safety in the construction and operation of the Fixed Link.

For this purpose, the Safety Authority shall:

(i)   Give advice or make proposals to the Intergovernmental Commission, at the request of the Intergovernmental Commission or on its own initiative;

(ii)   Participate in the drawing-up of any regulations applicable to safety of the Fixed Link and present them to the Intergovernmental Commission;

(iii)   Discharge, within the scope of its own powers, any function delegated to it by the Intergovernmental Commission;

(b)   Ensure that the safety measures and practices applicable to the Fixed Link comply with the national or international laws in force, enforce such laws, monitor their implementation, and report thereon to the Intergovernmental Commission; and

(c)   Examine reports concerning any incident affecting safety within the Fixed Link, make such investigations as are necessary, and report thereon to the Intergovernmental Commission.

(2)   The Safety Authority shall undertake necessary consultations with the Concessionaires.

(3)   In an emergency, the Chairman of the Safety Authority or his agent shall take the measures necessary for the safety of persons and property within the Fixed Link. He shall report any measures taken to the two Governments and to the Intergovernmental Commission.

(4)   The composition of the Safety Authority shall be determined by the two Governments by agreement. Each Government shall appoint half its members. The Chairmanship of the Safety Authority shall be held for a period of one year by the head of each delegation alternately.

(5)   The Safety Authority shall draw up its own rules of procedure and shall submit them through the Intergovernmental Commission for the approval of the two Governments.

(6)   For the purpose of carrying out its functions, the Safety Authority may invoke the assistance of the authorities of each Government or any body or expert of its choice.

(7)   The Safety Authority may, where it considers it necessary to do so, make a report to the two Governments at the same time as it reports to the Intergovernmental Commission.

(8)   The two Governments shall grant to the Safety Authority and its members and agents such powers of investigation, inspection and direction as are necessary for the performance of its functions.

(9)   The expenses of the Safety Authority shall be met by the Concessionaires as provided in the Concession. The budget of the Safety Authority shall be determined by the Intergovernment Commission after consultation with the Authority.

ARTICLE 12. FREEDOM OF MANAGEMENT OF THE CONCESSIONAIRES

(1)   The two Governments shall ensure that the Concessionaires are free, within the framework of national and Community laws, to determine their commercial policy, their tariffs and the type of service to be offered, during the term of the Concession.

(2) In particular laws relating to the control of prices and tariffs shall not apply to the Fixed Link during the term of the Concession.

(3) These provisions shall not, however, exclude the application of national or Community rules concerning competition or abuse of a dominant position.

ARTICLE 13. OBLIGATIONS OF THE CONCESSIONAIRES

The Concession will include provisions which give effect to the following principles:

(1) In their dealings with the Governments, the Concessionaires shall act jointly and shall be represented by a single executive. They shall be responsible jointly and severally during the whole term of the Concession.

The Concessionaires shall appoint one or more independent project managers for the construction of the Fixed Link.

(2) The Concessionaires shall comply with the provisions of the Concession, with the laws and regulations in force in each of the two States, and with the Community rules applicable to the construction and operation of the Fixed Link; they shall comply with those provisions of this Treaty and of the supplementary Protocols and arrangements which are applicable to them. They shall not take any action which would result in either State being in breach of its international obligations.

(3) The Concessionaires shall ensure the continued flow of traffic in the Fixed Link under satisfactory safety conditions and subject to any decision which may be taken pursuant to Articles 4, 5, 6, 10 and 11 of this Treaty.

(4) In transactions concerned with the construction and operation of the Fixed Link, the British Concessionaires and the French Concessionaires shall apply the principle of equal division of costs and revenue between themselves taking into account, to the extent necessary, indirect taxation.

(5) The Concessionaires shall take such measures as may be necessary to prevent accidents. The Concessionaires shall be liable for damage caused to users and to third parties resulting from the construction, the existence or the operation of the Fixed Link in accordance with the law applicable on the part of the Fixed Link where the event giving rise to the damage takes place. The Concessionaires shall take out and maintain insurance, or other financial security, which is adequate and reasonably appropriate to the relevant risk.

(6) If the construction of the Fixed Link is not satisfactorily completed or if its operation has ceased for whatever reason, the Concessionaires shall, if required by the Governments as provided in the Concession, at their own expense ensure that any part of the Fixed Link which is abandoned or unserviceable is removed or made safe.

This provision shall not apply in cases where the Governments have taken action to terminate the Concession other than for reasons of national defence, or for a failure by the Concessionaires to satisfy or comply with the terms of the Concession, or under the powers conferred by Article 6.

The two Governments are not obliged to complete the construction or to operate the Fixed Link.

(7) The Concessionaires shall negotiate agreements with the British and French telecommunications operators relating to the use by them of the Fixed Link.

ARTICLE 14. MODIFICATION OF CONCESSION

No modification of the terms of the Concession shall be made without the prior approval of both Governments.

ARTICLE 15. COMPENSATION OF CONCESSIONAIRES

When the term of the Concession ends, no compensation of whatever kind shall be due to the Concessionaires except as expressly provided in the Concession.

The two States undertake not to interrupt or terminate the construction or operation of the Fixed Link by the Concessionaires throughout the term of the Concession save on the grounds of national defence, or in the case of a failure by the Concessionaires to satisfy or comply with the terms of, and as provided in, the Concession or under the powers referred to in Article 6. Any breach by a State of this obligation would give the Concessionaires a right to compensation in accordance with the provisions of the Concession and consistent with international law.

If a State interrupts or terminates the construction or operation of the Fixed Link by the Concessionaires on grounds of national defence, the Concessionaires shall be eligible for compensation as provided under the law of the State concerned. In those cases where both States are liable under this provision and where the Concessionaires make a claim for compensation against both States, they may not receive from each State more than half of the amount of compensation payable in accordance with the law of the State.

Each State shall bear the cost of the payment of the compensation to the Concessionaires in proportion to its responsibility, if any, in accordance with international law.

ARTICLE 16. COMPENSATION BETWEEN STATES

In the event of either State unilaterally interrupting or terminating the construction or operation of the Fixed Link by the Concessionaires during the term of the Concession, the other State shall be entitled to compensation. Such compensation shall be limited to the actual and direct loss suffered by that other State and shall exclude any indirect loss or damage; in particular it shall exclude any loss of taxation or other benefits derived from the establishment or operation of the Fixed Link. No compensation shall be payable in respect of an interruption or termination of the construction or operation of the Fixed Link on grounds of national defence where it serves the defence interests of both States.

ARTICLE 17. Rights Of Governments On Termination Of Concession

Where the Concession terminates, whether by effluxion of time or prematurely for whatever reason, the rights enjoyed by the Concessionaires in that part of the structure, land and fixed installations of the Fixed Link within the jurisdiction of each State will revert to that State. Other property relating to the Fixed Link should become the joint property of the two States under the conditions provided for in the Concession. If the two Governments decide to continue to operate the Fixed Link together, they will do so on the basis of equality of rights and obligations, including the upkeep of the structure and installations of the Fixed Link.

ARTICLE 18. Consultation Between Governments

The two Governments shall consult, at the request of either:

On any matter relating to the interpretation or the implementation of this Treaty or the Concession;

On the consequences of any measures announced or taken which could substantially affect the construction or the operation of the Fixed Link;

On any action proposed in relation to any rights or obligations of the States under the Treaty or the Concession,

Upon the termination of the Concession for any reason, on the future use of the Fixed Link, its continued development and its continued operation.

ARTICLE 19. Arbitration

An arbitral tribunal shall be constituted to settle:

Disputes between the two States relating to the interpretation or application of this Treaty which are not settled through consultations under Article 18 within three months;

Disputes between the Governments and the Concessionaires relating to the Concession;

Disputes between the Concessionaires relating to the interpretation or application of this Treaty.

The arbitral tribunal shall be constituted for each case in the following manner:

Within two months of the receipt of the request for arbitration each Government shall appoint one arbitrator.

The two arbitrators shall, within a period of two months of the appointment of the second, appoint, by mutual agreement, a national of a third State as third arbitrator, who shall act as chairman of the tribunal.

If within the time limits specified above any appointment has not been made, a party may, in the absence of any other agreement, request the President of the Court of Justice of the European Communities to make any necessary appointment.

If the President of the Court is a national of either State, or if he is otherwise unable to act, the Presidents of the Chambers of the Court in order of seniority shall be requested to make the appointment.

If the latter are nationals of one of the States or are likewise unable to act, the member of the Court next in seniority who is not a national of either State or otherwise unable to act shall be requested to make the appointment.

In any case to which the Concessionaires are parties they shall be entitled to appoint two additional arbitrators. The two arbitrators appointed by the Governments shall appoint the chairman of the tribunal by agreement with the two arbitrators appointed by the Concessionaires. In default of agreement within the time limit specified in sub-paragraph (*b*), the chairman shall be appointed in accordance with the procedure prescribed in sub-paragraphs (*c*), (*d*) and (*e*) of this paragraph. The arbitrators appointed by the Concessionaires shall not participate in that part of any decision relating to the interpretation or application of the Treaty.

The arbitral tribunal shall take decisions by a majority vote. No arbitrator may abstain. In the event of the votes being equally divided the chairman shall have a casting vote. The tribunal may, at the request of one of the parties, interpret its own decisions. Decisions of the tribunal shall be final and binding on the parties.

Each party shall bear the costs of the arbitrator appointed by it, or appointed on its behalf, and an equal share of the costs of the chairman; the other costs of the arbitration process shall be borne in a manner determined by the tribunal.

In order to resolve any disputes regarding the Treaty, the tribunal shall have regard to the Treaty and the relevant principles of international law.

In order to resolve any disputes regarding the Concession, the relevant provisions of the treaty and the Concession shall be applied. The rules of English law or the rules of French law may, as appropriate, be applied when recourse to these rules is necessary for the implementation of particular obligations under English law or French law. In general recourse may also be had to the relevant principles of international law, and if the parties in dispute agree, to principles of equity.

ARTICLE 20. RATIFICATION AND ENTRY INTO FORCE

This Treaty is subject to ratification. It shall enter into force on the date of the exchange of instruments of ratification, which shall take place at Paris.

EN FOI DE QUOI, les Plénipotentiaires respectifs ont signé le present Traité et y ont apposé leurs sceaux.

IN WITNESS WHEREOF, the respective Plenipotentiaries have signed this Treaty and have affixed thereto their seals.

FAIT à Cantorbéry, le 12 février 1986, en double exemplaire, chacun en langues française et anglaise, les deux texts faisant également foi.

DONE in duplicate at Canterbury, on 12 February 1986, in the French and English languages, both texts being equally authoritative.

Pour le Président de la   For the President of the
República française: French Republic:

[Signé—Signed][4]

Pour Sa Majesté britannique: For Her Britannic Majesty:

[Signé—Signed][5]

Related Cultural Documents—I

Joan Reiter report,

"THE CHANNEL TUNNEL"

December 1, 1956

THE CHANNEL TUNNEL

The Nineteenth Century

Napoleon was impressed

Perhaps the single best capsule history of the English Channel tunnel idea is contained in Willy Ley's recent (1954) book, *Engineers' Dreams*. A favorite "engineers' dream" is described in Chapter One, titled "Forbidden Tunnel."

"The original father of the idea," says Willy Ley, "was a French engineer named Mathieu-Favier. He wrote a report about his plan, which involved the creation of a small island near the center of the strait, where a shoal rises to within 100 feet of the surface. Tunnels were to go to this island from either side, each tunnel about a dozen miles long. With horse-drawn cars running on rails, each tunnel trip would require about an hour—or a little more than two hours would be required for the whole trip from France to England.

"Engineer Mathieu-Favier submitted his report to the First Consul of the French Republic, a gentleman by the name of Napoleon Bonaparte. The year was 1802.

---

4. Signé par Roland Dumas—Signed by Roland Dumas.
5. Signé par Geoffrey Howe—Signed by Geoffrey Howe.

"Napoleon was impressed. At the earliest opportunity, he discussed the plan with the English statesman Charles James Fox. 'Think,' he said, 'what great things we could do together.' Fox replied politely and diplomatically that 'a France-British union could rule the world.' Whether Fox actually liked the plan is not certain, but even if he had, there was little he could have done. His relations with the king of England were strained, and, moreover, the peace between England and France was in itself precarious. Only a few years later, Napoleon was looking across the Channel at "the enemy."

What Mathieu-Favier didn't know is that conditions for a tunnel from Cape Gris Nez (near Calais) to Folkestone (near Dover) are about as favorable as possible. The headlands on both sides rest on a solid layer of grayish chalk several hundred feet thick. Lower Gray Chalk, as it has been designated, is 1) soft enough to be worked without blasting, 2) impermeable to water (i.e., will stay dry except for moisture carried in by air), and 3) stretches all the way across with no apparent break or discontinuation.

Mathieu-Favier's plan now seems to us to have been unworkable, lacking power machinery as he did; ventilation, for example, requires plenty of power. But Mathieu-Favier's *idea* took hold immediately and has never been given up since.

Thomé de Gamond

Meanwhile other ideas for spanning the Channel had developed and had gathered supporters. One was a pre-fabricated tunnel to be laid, pipeline-wise, across the bottom of the Channel. A bridge was suggested. Somebody else proposed a dam with three wide gaps for ships to pass through, these gaps to be spanned by high bridges.

"Actually, the true tunnel, if you have the mechanical means for digging it, is the simplest and consequently the least expensive solution. . . . The problems," Willy Ley goes on, "are merely ventilation and removal of debris.

"The first man to realize this clearly and to say so was a Frenchman, Thomé de Gamond, who for the whole second half of the 19th century was the chief advocate of the Channel Tunnel idea. . . ." In 1856 de Gamond presented his old friend Louis Napoleon, now Emperor Napoleon III, with a plan for a Channel Tunnel. Napoleon III supported the plan and the political atmosphere was good. But this first proposal by de Gamond met strong resistance from seafaring people. The plan called for thirteen artificial islands in the Channel to be used as dumping places for excavated material and as bases for ventilating "chimneys." Maritime circles said that navigation in the Channel was bad enough without adding thirteen new obstacles.

Thomé de Gamond set about revising his plans. A major problem still was how to get fresh air in and debris out. The experience of English mining engineers proved helpful. They had developed the technique of digging their mining tunnels in pairs, with frequent cross connection. After ten years of work, de Gamond finished an improved version of his original scheme.

In 1868 the Anglo French Tunnel Committee was formed, including Thomé de Gamond; the Scottish engineer William Lowe; the English engineers Joseph Lock, Robert Stephenson, and Isambard Brunel, and—somewhat later—Sir John Hawkshaw. The political situation was even better than the first time around. Gladstone was Prime Minister; he favored the tunnel. So did Lord Lansdowne and Lord Salisbury. And Queen Victoria, who received de Gamond at Buckingham Palace. And of course Emperor Napoleon III.

But any project of such size requires a long organizational period. Before much could be done, the Franco-Prussian War of 1870–71 began. At war's end Napoleon was out and France was a republic once more. But this did not constitute a setback for the Channel Tunnel; the Republic was as receptive to the idea as the Emperor had been. A convention between Queen Victoria's government and the French Republic, to regulate Channel Tunnel problems which would arise, was signed in 1875. In both countries, companies for financing the work were formed.

The French had always felt that the Channel Tunnel was really theirs (most of the plans had originated in France) and were perfectly willing and able to finance the tunnel themselves. The French company (*Comité Francais du Tunnel*) was backed by the House of Rothschild and by the *Chemins de Fer du Nord*, aided and abetted by the Suez Canal Company. But the English did not wish to be left out of the financial end. Various corporate maneuvers and mergers resulted in the Channel Tunnel Company.[1] Its money—a quarter-million pounds—had come from London financiers and from the South Eastern Railway Co.

In 1877 the companies submitted to their respective governments the final plan as drawn up by the three chief engineers, Sir John Hawkshaw, Sir James Brunlees, and Thomé de Gamond. The latter had, in the meantime, died; but by now the project had attained sufficient momentum to go forward without him.

(Details of the 1877 plan are available in Ley's book, but are not included in this research because the 1930 plan—details of which will be given later—was felt to be of superior interest.)

The Tunnel is begun

Actual work began around 1880 with the sinking of two vertical shafts at the foot of Shakespeare Cliff and at Sangatte. These shafts were to house the

---

1. Corporate development on the English side of the Channel, as detailed in the *Encyclopedia Britannica*, seems to have taken approximately this course: in 1875 a Channel Tunnel Company was formed and obtained an act authorizing it to undertake certain work. This company failed to raise the necessary money and was bought up in 1886 by the Submarine Continental Railway Company which had been formed in 1881. In 1881 the South Eastern Railway Company obtained an act giving powers for experimental borings and carried out some preliminary works which were taken over in 1882 by the Submarine Continental Railway Company in return for cash and shares. In 1886, as stated earlier, the Submarine company absorbed the Channel Tunnel Company, and in 1887 it changed its name to the Channel Tunnel Company, Limited. The Channel Tunnel Company, Limited still exists today, apparently alive and functioning. Information on the company from *The Stock Exchange Official Yearbook* will be quoted later in this research.

pumping ducts later; at first they were to serve as bases for drilling test tunnels under the Channel bed. The French test tunnel was 4800 feet long; the English, built under the supervision of William Lowe, had a length of 6500 feet. Visitors to these test tunnels—on the British side they were very important personages like Queen Victoria, the Prince of Wales, Gladstone, Disraeli—were greatly impressed by the electric illumination and by the hand-drawn trolley cars.

The main work of digging the tunnel proper had not yet begun. Engineers and geologists were still checking and testing. But the public on both sides of the Channel followed the news avidly and expectantly. Engineers said the project would take between six and eight years; the tunnel should be finished in 1888, or at the latest in 1890.

There was one group in England which did not favor the project   the British War Office. In 1882 the War Office lowered the boom. First the *Times* published a strongly-worded editorial opposing completion of the tunnel. Lord Wolseley, British Chief of Staff, issued a blue book denouncing the tunnel as a military liability, a "permanent menace." A number of famous men were induced to add their voices to the anti-tunnel chorus—the Duke of Wellington, Herbert Spencer, Robert Browning, Thomas Huxley, the Duke of Marlborough. Then Lord Wolseley made a speech to the House of Lords; the matter was referred to a Parliamentary commission, which in due time rejected the tunnel by a vote of 6 to 4. The Channel Tunnel Company Limited was called upon to prove that an invasion of England could be prevented with absolute certainty; its suggestions were one and all vetoed by Lord Wolseley as "not foolproof."

In 1884 work at the British end was discontinued. The French kept on only a little longer.

The Twentieth Century

New attempts and old arguments

Two decades after the War Office had abruptly brought the scheme to an end, tunnel enthusiasts in both countries tried to revive the project. But in 1906, when actual discussions began, the Prime Minister, Sir Henry Campbell-Bannerman, rejected the idea, quoting Lord Wolseley's arguments and adding a few embellishments of his own. Lord Asquith was heard to remark, during the 1906 discussion, that a tunnel would be a nice thing to have in case of a sea blockade.

In 1913 the engineer Sir Francis Fox produced a revised plan for the Channel Tunnel, worked out with Albert Sartiaux of the French company. This proposal (like the 1906 one) *almost* received Parliamentary majority. Lord Asquith, now Prime Minister, had apparently changed his mind as well as his office; he rejected the proposal. The London *Times* took the same view, saying that the tunnel would end Britain's isolation—this despite the fact that four years earlier, on July 25, 1909, Louis Blériot had made his historic flight across the Channel.

World War I made many Englishmen feel that a railroad under the Channel, out of reach of Zeppelins or submarines, would have been decidedly useful.

Meanwhile additional safety suggestions, designed to prevent or delay surprise invasions, were being made by engineers. It seemed certain then that the tunnel would be built after the war.

In 1920 the British government rejected the plan.

In 1924 the British government rejected the plan. Ramsey MacDonald, the Prime Minister, called a conference of all former prime ministers still living. Lord Wolseley's ghost indubitably attended too. They met for forty minutes and decided on rejection, even though the Members of Parliament supported the tunnel with a majority of 2 to 1. Winston Churchill—not then in office—raised his voice. "There is no doubt about their promptitude. The question is: was their decision right or wrong? I do not hesitate to say that it was wrong."

A famous Frenchman had spoken his piece two years earlier. Said Marshal Foch in 1922: "If the Channel tunnel had been built it might have prevented the war, and in any event it would have shortened its duration by one half." But the prime ministers did not believe Foch.

L'affaire 1930

In 1928 and 1929 debate flared up again—always in England, of course, for the French tunnel company has the approval of its government and can go ahead any time it wants to, but the English company cannot proceed without an Act of Parliament. This "great debate" of the Channel tunnel idea in England during 1929 and 1930 aroused intense interest in the American press. One of the most complete and lucid reports appears in *Foreign Affairs,* July 1930:

"When it was first proposed to construct a tunnel connecting England and France, the military aspect was given most attention. In spite of the fact that provision was made in the plans for flooding five miles of the tunnel from either end at a second's notice, high military authorities in England strongly objected to the scheme. But the ease with which Germany conducted air raids over London and other cities during the war showed that England's position was no longer impregnable behind the ramparts of surrounding seas. Accordingly it was realized that the economic advantages—or disadvantages—of the tunnel should receive first consideration. In April 1929 Prime Minster Baldwin appointed a Committee 'to examine and report on the economic aspects of proposals for the construction of a Channel tunnel or other new form of cross-Channel communication.'

"In its report (published March 14, 1930, Cmd. 3513) the Committee condemns as unsatisfactory the various alternatives to a Channel railroad tunnel—a cross-Channel passenger and freight train ferry; a tunnel for both automobile and railroad traffic;[2] a bridge for both automobile and railroad traffic;[3] a rail-

2. "No vehicular traffic would be practicable, on account of limitation of ventilation." *Saturday Evening Post,* June 28, 1930, p. 28.
3. "The bridge scheme, which has numerous supporters, would cost $375,000,000 . . ." *Business Week,* June 11, 1930, p. 31.

road tube across the bed of the sea; and a cross-Channel jetty. Two schemes for a Channel tunnel were laid before the Committee. One of them provided for the construction of the tunnel and also of an entirely new railroad from London to the coast and another from Paris to the coast; the estimates of this company were: $292,500,000 for the English section of railroad, $500,000,000 for the French section, $154,000,000 for the tunnel proper. The Committee rejected this scheme because of its prohibitive costs.

"The proposals of the English Channel Tunnel Company are quite different. This company, formed in 1872, favors tunneling through the gray chalk from Sangatte (between Boulogne and Calais) to Shakespeare's Cliff (between Dover and Folkestone), the point of greatest depth being 95 feet below the channel bed, or 260 feet below sea level. Such a tunnel would be 36 miles long—24 miles under the Straits of Dover and 12 miles of approach tunnels. (The Simplon Tunnel, the longest in the world is 12 miles long.) The approach tunnel in England the English half of the submarine tunnel would be constructed by the English Channel Tunnel Company, the other half of the construction to be done by the French Channel Tunnel Company, which holds a concession from the French government enabling it to begin work at a moment's notice.

"The tunnel would consist of two traffic tubes, each of a diameter of 18-1/2 feet, and a 10-foot pilot tunnel for drainage and ventilation. Electricity to operate trains through the tunnel would be supplied from power stations at each end of the tunnel. Heretofore it has been assumed that the *Chemins de Fer du Nord* would insist upon the use of its larger loading gauge; but, since the broader and higher French cars are too big for the English railroad platforms, tunnels, and bridges, the latest English proposals are based on the assumption that the steel cars to be used in the tunnel will conform to the British loading gauge.

"There will be no insuperable geological difficulties in the construction of the tunnel if the chalk formation extends uninterruptedly across the Straits. The Committee is in favor of the construction of the pilot tunnel, which, if successful, will settle the question of the practicability of constructing the two traffic tunnels. Estimated cost of the preliminary or pilot tunnel is $28,000,000 and of the two traffic tunnels, about $125,000,000. The total cost would be divided equally between the English and French companies; the Committee favors construction by private enterprise, without state aid.[4] The pilot tunnel could be constructed in 2-1/2 years and the traffic tunnels in 4-1/2 years.

"The passage of a train through the tunnel would take about three quarters of an hour, as contrasted with an hour and a half for the present steamer crossing between Dover and Calais. Another three quarters of an hour would be saved if customs official examined baggage on the train . . . instead of at the points of

---

4. According to *Living Age*, the *Manchester Guardian* attacked the proposal to have the project privately financed. "Unforeseen obstacles may arise," said the *Guardian*, "and experience has proven, as in the case of the Suez Canal, that large-scale enterprises must eventually receive assistance from the state. (The proposed expenditure) represents a considerable speculation for any group of private individuals to undertake."

embarkation and disembarkation. This would shorten the trip from London to Paris—now nearly eight hours—by an hour and a half, and would therefore, it is believed, result in more passengers crossing the Channel to visit England or the Continent. The Committee estimates that the tunnel, if it is opened in 1938, will secure 2,357,000 passengers during the first year of its operation, and that the development of air services will not seriously interfere with the traffic prospects.

"The movement of freight across the Channel is at present relatively unimportant, being confined almost entirely to perishable foodstuffs and goods of high value. Freight imported, exported, and reexported via Dover, Folkestone, and New-Haven amounts to less than 3% of England's foreign trade. If this traffic remains the same in the future, construction of the proposed tunnel will bring but slight benefit to English trade. But the Committee argues that, as the development of communications during the past century resulted in great increases in the amount of traffic, so construction of the Channel tunnel will reap the benefit of increasing production and commerce in the future. The Committee believes that it could be built, maintained, and operated at a cost that would allow of freight and passenger rates not higher than those now in force in cross-Channel traffic.[5]

"The trades that might be adversely affected by the operation of a Channel tunnel are important, but they are not numerous and they represent but a small percentage of the trades and industries of the entire country. The agricultural producer in Southeastern England, who fears the increased competition from the Continent, can improve his position by developing better methods of marketing his produce; furthermore, he has an initial advantage over his competitor in his proximity to the London market. Concerns operating trans-Channel cargo boats are threatened with the loss, partial or absolute, of their business; accordingly, the Committee recommends that careful consideration be given to their plight, both for their own sake and also for the reason that, if they are driven out of business, freight rates might rise because of lack of competition. The railroads hold that, while they may lose their trans-Channel boat and train traffic, they will benefit from the general growth of traffic resulting from the operation of the tunnel.

"Construction and operation of the tunnel would have only slight effect on the labor market. About 250 men would be employed in the construction of the pilot tunnel and about 1,550 men in the construction of the traffic tunnels. In addition, it is estimated that work on the pilot tunnel would indirectly provide employment for 750 men and on the main tunnels for 4,500 men. The number of laborers thrown out of work when the tunnel is opened would be relatively small.

"The French public has always favored the construction of the tunnel. . . .

---

5. The *Saturday Evening Post* commented: "This (recommendation that the tunnel be built by private capital, without governmental financial assistance) will strike many Americans as strange. The reason given is that of being desirable on economic grounds. The capital cost is regarded as not in excess of $150,000,000, and this is believed to lie within the scope of private enterprise. It is suggested that the overhead, the maintenance, and the running cost would be relatively low; naturally the rates would be relatively high. The rates would be higher by tunnel than by boat, but lower than by airship."

"In England, however, opinion has always been divided. Though Gladstone, John Bright, and Lord Salisbury favored the project, Joseph Chamberlain and the War Office opposed it. No British government thus far has supported it. The report of the Channel Tunnel Committee favors the project, but the dissenting minute of one member of the Committee (Lord Ebbisham—*Living Age*) and also some hostile press comment are indications that the plan is still subject to controversy. Prime Minister MacDonald recently informed Parliament that the Committee of Imperial Defense would be asked for its opinion of the report and that, as soon as the government had reached a decision, it would be communicated to the House of Commons.

"In general," the *Foreign Affairs* article continues, "Conservative periodicals oppose the scheme while Liberal periodicals favor it. The Conservatives argue that, if the danger of invasion can be ruled out in a discussion of the tunnel, so can the necessity for the tunnel itself, for crossing the Channel today is a much easier and faster trip than it was when the project was first discussed. The only justification for constructing the tunnel, they continue, is unmistakable proof that it would add to England's comfort and prosperity, promote foreign trade, and relieve unemployment—and this is not found in the report." (The Tories glumly predicted, according to Willy Ley, that all the tourists would go one way—to France, and all the freight would go the other way—to England.)

It is apparent, from contemporary articles in *Popular Mechanics* and *Living Age*, that the "great debate" was not confined to *economic* aspects of the tunnel; quite a few people did not agree that "the danger of invasion can be ruled out." According to *Popular Mechanics*, "the military continued to object on the grounds that the "railroad connexion with the mainland would imperil England in time of war and require a huge standing army to guard against a surprise attack by train, in the event of war with France. In answer, the tunnel proponents pointed out that the plans provide for a sump near either shore, so that either England or France could flood a section of the tunnel at will. . . ." The *Manchester Guardian*, quoted in *Living Age*, tartly commented, "Practically there is no more in the argument that the tunnel would facilitate an invasion of England than there was in the Duke of Wellington's objection, on exactly the same ground, to the first railway between London and Portsmouth." *Popular Mechanics* quoted the argument that the tunnel was a potential military *asset*, "an uninterrupted means of transporting troops to the continent and getting food into England in event of another war, as neither surface ships nor submarines could interfere with the tunnel trains."

Rebuttals were quickly made to the Tories' economic arguments as well. "The Liberals," according to *Foreign Affairs*, "reply that [the Conservative] attitude is founded on prejudice and opposition to progress."[6] Furthermore, they suggest that, as imports from the Continent now exceed exports via the Channel, the railroads might offer especially attractive rates to English

---

6. "The traffic to be sought would be passengers, mail, express and fast package freight. It is not believed that bulky freight could be carried as cheaply as on the sea route ..." *Saturday Evening Post*, June 28, 1930, p. 28.

exporters in order to fill freight cars that would otherwise return empty to the Continent. The Committee's report may not be so enthusiastic as to persuade conservative people to invest money in the project, but, as one writer suggests, the romance and glamour of adventure in it might well attract a number of investors.

"So far we have considered only the material aspects: what about the spiritual results of constructing the tunnel? By putting an end forever to the tourist's terror of a rough Channel crossing, the tunnel should encourage visitors to England from foreign lands. Traveling Americans, it is agreed, most of whom visit Paris, will be more inclined to include England in their tour of Europe; and the French, who are not fond of crossing the sea, will become more familiar with their neighbor. Thus a better Anglo-French understanding, and probably a better understanding of Great Britain by foreigners generally, might result if the project ever materialized."

1930, The end of the affair

The course of the debate, and the fate of the Channel tunnel proposal, are succinctly outlined in the following entries in the 1930 *New York Times Index:*

British Com will recommend Dover-Calais project to Commons; cost estimated, March 12; editorial, March 13

British engnrng experts stress economic benefits; principal conclusions of Com, March 15

British militarists see danger of French invasion in plan, March 16

Brit Govt studying idea, March 21

Possibility discussed, March 23, III

Feature article; illus of Dover cliffs, April 6, IX

British Com of Imperial Defence vetoes plan, May 30

Premier MacDonald promises free vote on project if desired, June 3

MacDonald decides against tunnel; issues will be debated by Commons, June 6

Brit Govt White Paper explains rejection, June 7

French ports pleased with Brit decision, June 8, II; proposal discussed, June 15, IX

House of Commons rejects plan after Premier MacDonald gives reasons for opposition; Channel Tunnel Co still hopes, says d'Erlanger, pres., July 1

World War II

Willy Ley relates that the Allied Supreme War Council, meeting in the fall of 1939, immediately began to consider the possibility of a tunnel under the English Channel. They called in experts: was there still time to build the tunnel? And how much would it cost? The engineers declined to answer the first question; in December 1939 they said that the cost of the tunnel would be 60 million pounds sterling. The defense budget of England and France together was 6 million pounds per day. There was never a chance to discover if a completed tunnel might shorten the war by at least 10 days; the Germans, turning their offensive to the west, answered that enigmatic first question once for all.

At Dunkirk, Ley points out, the British lost relatively few men but most of their equipment. It is his opinion that had the Channel tunnel existed, the British would probably have escaped with fewer losses and gotten away with most of their equipment.

During World War II British airmen, returning from flights over enemy territory, reported building activity near Calais and Sangatte. An alarming thought suddenly came to the British authorities: it would be quite possible for the Germans, if they so desired, to dig that tunnel right under the Channel to England. Listening devices were installed in the old British pilot tunnel so that German drilling, if it did begin, could be detected instantly. But the Germans were not constructing tunnels, they were building launching sites for V-1's and V-2's. Nevertheless, they were not too preoccupied to neglect installing listening devices in *their* pilot tunnel on the French side of the Channel!

Recent developments

The "engineers' dream" is not yet dead, according to the following story in the March, 1953, *United Nations World*:

"High authorities of SHAPE headquarters in Paris are now seriously examining a new version of the age-old plan to construct a tunnel between England and France. The project was submitted to SHAPE in November, 1951, by British M.P. Ernest Thurtle and French architect André Basdevant, who has been a pioneer of the tunnel idea for over 20 years. Gen. Alfred M. Gruenther, SHAPE Chief of Staff, thanked Mr. Thurtle for the documentation in a personal letter Dec. 13, 1951, and promised to circulate it among his officers for extensive study. This study took the better part of 1952, but a few weeks ago M. Basdevant came in for a talk. He asked a number of specific questions about the Frenchman's project and indicated that SHAPE would help promote it in the course of 1953.

"A leak brought this information to the knowledge of the *'Société Concessionaire du Chemins de Fer Sous-Marin entre la France et l'Angleterre,'* which was founded on August 2, 1875, with a capital of 20 million gold francs (and has not distributed any dividends since). Aroused by the news, the company's Board of Directors held an extraordinary session in Paris on November 27, 1952, under the chairmanship of M. René Claudon. The directors reminded interested parties that their firm was still the only French entity legally entitled to build the tunnel and that it 'had kept abreast of the latest technical developments to carry out the job if and when requested.'

"What these gentlemen meant was that they had in their files a copy of architect Basdevant's new scheme. This draft provides for a 35-mile-long tunnel in the form of an extended 'W' that would connect Bazinghen, near Calais, with Folkestone in Kent. The tunnel would avoid permeable layers and consist of two stories—one for car traffic, the other for a double-track railroad.

"According to competent sources, SHAPE is interested in such a link for numerous logistic reasons. In case of conflict, the tunnel could serve as a subway for troops and supplies that would be exposed to heavy air attacks if moved up by surface transport. The experts believe it would be difficult to bomb the planned underwater road. On the other hand, it could easily be rendered unusable if an enemy force approached it.

"Another argument in favor of the project is that geologists have now located abundant stores of oil and uranium ore under the Channel. 'The only way these treasures can be reached is by digging a tunnel,' M. Basdevant points out. 'That's one of the main reasons why we ought to start work as early as possible.'

"There remains one major handicap: the ambiguous attitude of the British. They, too, founded the Channel Tunnel Co. in 1875, but its concession was withdrawn seven years later when General Wolseley, then British Chief of Staff, violently protested against the whole idea. 'How can we be expected to reduce a distance that's already much too short for our taste!' the General exclaimed.

"This viewpoint is still being shared by many insular-minded Britons, but others feel that it is no longer justified in the age of A-bombs and guided missiles. More than 260 M.P.'s have created a Tunnel Committee, headed by Mr. Thurtle, who urges that a new building license be granted without further delay.

"One of the advocates of the scheme is Prime Minister Winston Churchill himself. He has repeatedly championed the idea in public speeches. That's why SHAPE and M. Basdevant hope to persuade the British government in 1953 that a tunnel under the Channel would form a precious strategic asset for Western European defense."

"The two pilot tunnels," Willey Ley reports, "are now being watched by one man each, one in England and one in France. They are as dry as they were when Queen Victoria was shown around."

This appears to be the most recent word on the subject—but history would indicate that it will not be the last.

BIBLIOGRAPHY

*Business Week, June 11, 1930, pl. 31.

Christian Science Monitor Magazine, January 13, 1940, pl. 6. "That chunnel bobs up again." Argus.

Contemporary Review, July, 1949, pp. 51–4. "Channel tunnel." G. Glasgow.

*Encyclopedia Britannica.

Foreign Affairs, July, 1930, pp. 65–8.

*Ley, Willy, Engineers' Dreams (1954). Chapter 1: "Forbidden Tunnel," p. 15–43.

*Literary Digest, February 9, 1929, p. 19.

Literary Digest, November 6, 1937, p. 30. "Tunnel under the sea." A. Black.

*Living Age, April 15, 1930, pp. 195–6.

Living Age, September, 1939, pp. 71–4. "Dryshod under the Channel." D'Erlanger.

*New York Times Index. 1930.

*Popular Mechanics, May, 1929, pp. 767–8.

*Saturday Evening Post, June 28, 1930, p. 28.

*Saturday Evening Post, May 13, 1950, pp. 28–9 et al. "Forgotten tunnel from England to France." J.P. O'Donnell. Photostatted for inclusion in this report.

*Saturday Review, June 5, 1954, p. 19.

Science Digest, September, 1944, pp. 61–5. "Channel tunnel may change history." W. Ley.

*Stock Exchange Official Yearbook, 1955. (London) Volume 2, p. 1980.

Time, January 3, 1944, pl. 30. "Death of a dreamer."

*United Nations World, March, 1953, p. 7.

Asterisks indicate sources used for this report.

The card catalogue of the New York Public Library lists 59 individual books, pamphlets, articles in antique learned journals, etc., on the subject of a tunnel under the Channel. The majority of the titles are English, but many are French, a few are German, and at least one is Italian.

The following are samples of what is available:

Albert, Prince de Monaco, *À propos du tunnel sous la Manche,* 1991.

Anglo-French Sub-Marine Railway Company, *Tunnel sous-marin entre la France et l'Angleterre . . . sur le systeme des deux galleries jumelles de M. William Low . . . ,* 1870.

Great Britain, Channel Tunnel Committee, *Report . . .* London, 1930.

Harrington, J., . . . *The Channel tunnel and ferry . . . ,* 1949.

Ryves, Reginald Arthur, *The Channel tunnel project; a brief history,* London, 1929.

Economics librarians of the New York Public Library and librarians of the British Information Service were consulted in an effort to find statistics on traffic (passenger and freight) across the English Channel. Nothing was turned up. "Trade doesn't go that way, you know," said the man at B.I.S.

However, one clue was found while perusing the annual BRITISH TRANSPORT COMMISSION REPORT & ACCOUNTS: the group which appears to control at least some Channel shipping is the Railway Executive, an arm of the aforesaid British Transport Commission. Unfortunately, for our purposes the Railway Executive's REPORT & ACCOUNTS has two drawbacks: 1) it does not distinguish Channel traffic from Irish, Channel Island, and coastal traffic, all are lumped together, and 2) there is no way of telling how much of the total Channel traffic is under the aegis of the British Transport Commission's Railway Executive.

From Technical Studies, Inc. Records from 1957–1994. Cambridge, MA: Harvard University: Historical Collections, Baker Library, School of Business Administration.

Related Cultural Documents—II

PROJECT ESTIMATE

Our estimate of the cost of the project described in this report is £82,695,000 sterling. It includes all engineering costs and provision for the construction contractor's fee or profit. It excludes taxes of all types, the cost of land, interest during construction, other project financing costs, and the owners' administrative and corporate costs.

The estimate is broken down into major items on the table below. Costs as anticipated in 1960 are used as the basis. Provision for escalation during construction is estimated based on the proposed time schedule.

A number of currencies will be involved in carrying on the work. For convenience, they are all converted to sterling at the official rate of exchange.

This estimate includes provision for all construction difficulties that might reasonably be expected and is considered to be adequate to complete the project as planned, provided it is administered as described in the chapter on Project Time Schedule.

Estimated Costs for Channel Tunnel Project (Pounds Sterling)

	Est. cost 1960 basis	Provision for escalation during construction	Est. completed cost
Two 6.7-meter inside diameter concrete-lined tunnels and one 4.0-meter inside diameter concrete-lined service tunnel including all cross-overs, shafts, and other structural elements	£75,405,000	£5,597,000	£81,002,000
Two service buildings, one at each service shaft	99,000	7,000	106,000
Tunnel ventilation system	114,000	9,000	123,000
Tunnel drainage system	769,000	57,000	826,000
Electrical system	571,000	43,000	614,000
Telephone system	23,000	1,000	24,000
Total	**£76,981,000**	**£5,714,000**	**£82,695,000**

From Bechtel Corporation, Brown & Root, Inc., and Morrison-Knudsen Company, Inc., *The Channel Tunnel: Design and Construction of a Channel Tunnel as Recommended by Three Engineer-Constructors* (November 1959). A copy of this is available in the Channel Tunnel archives at the Historical Collections, Baker Library, School of Business Administration, Harvard University, Cambridge, Massachusetts.

40
SEMATECH

DID YOU KNOW . . . ?

➤ SEMATECH is derived from *semiconductor manufacturing technology.*
➤ The consortium was formed in 1987 by 14 original founding semiconductor companies, and was later joined by 5 international members later.
➤ Congress authorized $100 million to fund SEMATECH.
➤ By 1992, U.S. semiconductor manufacturers, whose world market share had fallen to 40 percent, held a 53 percent share.
➤ The microchip is an American invention.

Desperation, military considerations, and a fear of competition inspired one of the most cooperative associations in history, turning failure into success. SEMATECH, a consortium of 14 companies that opted to bind their fates together, was a grand vision born of dark times.

HISTORY

Although the microchip—a tiny piece of semiconductor material such as silicon bearing an entire integrated circuit—was originally an American invention, by the 1980s the Japanese had moved into a dominant position in semiconductor production, beginning with the memory-chip market. Semiconductor sales were a big business that was just getting off the ground in 1975. In 1980 sales were barely at $1 billion, but quickly climbed worldwide.

Sematech chief executive Robert Noyce holds up a semiconductor wafer at Sematech in Texas, 1989. An enlarged section of a semiconductor circuit is seen on the monitor in background. Courtesy of AP / Wide World Photos.

The U.S. Congress sought to foster corporate collaborations via the National Cooperative Research Act of 1984, which freed research consortia from several antitrust restrictions. Among such consortia, the most important was SEMATECH (derived from *semiconductor manufacturing technology*), formed in 1987 by 14 major semiconductor manufacturers. As a nonprofit organization, SEMATECH was intended to pursue advances in semiconductor manufacturing at the precompetitive stage—that is, to raise the level of manufacturing competence rather than develop specific products. Congress worried that a waning domestic U.S. semiconductor industry could undermine the nation's military capability as well as its economic vigor, so it acted to protect the vitality of this strategic sector. It authorized federal assistance to SEMATECH, reflecting a major policy decision to support commercially focused private research in technology.

Reproduced here is the legislation authorizing federal grants to SEMATECH and establishing the terms under which the consortium was to operate. The U.S. Congress was concerned with "improving manufacturing productivity of

United States semiconductor firms" and with meeting the country's "national security and commercial needs" for new "high-quality, high-yield semiconductor manufacturing techniques." Congress envisioned SEMATECH's purpose as twofold: research on advanced manufacturing techniques and the development of techniques for product manufacture. Federal funds from the Defense Department were limited to $100 million for 1988. In general, government funding, whether federal, state, or local, was required not to exceed half of SEMATECH's research and development costs for a given fiscal year.

U.S. suppliers of equipment for the semiconductor industry were in particularly dire straits, and SEMATECH, based in Austin, Texas, quickly targeted its efforts to this area with positive results. Its success stories included more uniform specifications for equipment, improvements in manufacturing systems, and better communications between equipment users and suppliers. By 1992, U.S. semiconductor-equipment manufacturers, whose share of the world market had fallen from 75 percent to 40 percent in the previous decade, held a 53 percent share, compared to 38 percent for Japan. SEMATECH was widely hailed as a major factor contributing to this turnabout.

In the mid-1990s, with the U.S. semiconductor industry restored to good health, federal assistance to SEMATECH came to an end, and the consortium was funded by its member companies—a shift that gave it greater control over its research program and assuaged objections to non-U.S. companies joining the consortium. A division called International SEMATECH was formed in 1998 to permit foreign participation, and at the beginning of 2000 the consortium officially changed its name to International SEMATECH. Today, International SEMATECH's members include firms from Asia and Europe, all with full access to the consortium's R&D programs for improving chip-manufacturing techniques. In 2000, the consortium had 13 members: Agere, Advanced Micro Devices, Conexant, Hewlett-Packard, Hynix, IBM, Infineon, Intel, Motorola, Philips, STMicroelectronics, Taiwan Semiconductor, and Texas Instruments.

CULTURAL CONTEXT

The introduction of the microchip was like discovering a new continent. It opened up a world of technological innovation and immense commercial opportunities; the ensuing computer revolution made microchips ubiquitous. Chips and related products became objects of fierce competition in the marketplace.

Japan had already gained a major share of the automobile market, at the expense of U.S. carmakers, by focusing first on manufacturing technology and then on design. Fears were voiced in the United States that the same scenario was unfolding with semiconductors, with the Japanese initially taking control of the high-volume, process-intensive memory-chip sector before moving on to more design-intensive microprocessors. *Competitive* emerged as the watchword for many commentators on the American economy, and considerable attention was directed to the role played by the Japanese government, notably the Ministry of Trade and International Development, which helped Japanese firms work together to master new technologies.

PLANNING

The spark behind the success of SEMATECH was the insight that competitors could still cooperate if they shared only some information. All the companies that were interested in joining SEMATECH were fierce competitors in the field of inventing microchips and patenting related technologies. But what they all had in common was manufacturing and production. It was decided that manufacturing processes could be an area of shared information, but inventions would remain proprietary and could be pursued for competitive advantage.

This breakthrough about what to share and what to keep secret was a key idea. It was a common platform that created strength, for in most industries there is an awkward period of growth where competing standards vie for dominance. A classic example of competing standards was Betamax/VHS in the videocassette-recorder industry. Another example of standards agreement that enabled an entire industry to move forward was the adoption of ASCII (American Standard Code for Information Interchange) as the universal computer language, thereby allowing file transfers and compatibility across computer systems. SEMATECH quickly recognized the importance of common standards and made that platform one of its goals.

SEMATECH's financial sharing formula was creative. Member firms had only to contribute one percent of their semiconductor sales revenues. The minimum contribution of $1 million proved too expensive for some smaller firms that would have liked to join, but it was no problem for larger industry giants who were likely to affect the world market. For such giants, the maximum contribution of $15 million ensured against the upside possibilities of rampant success. The consortium received $100 million in matching government funds until 1997 (Gillespie). In any case, the value of SEMATECH membership outweighed the costs because while the government laid out funds from 1987 to 1996, when the federal monies were no longer needed, member companies continued to fund the entire effort themselves.

There was an insider benefit as well: member companies were able to buy newly designed manufacturing equipment as soon as it became ready for the market. That meant SEMATECH companies had the latest equipment six months before the rest of the industry.

Even more significant than financial sharing may have been the personnel exchange. SEMATECH members are required to send personnel to SEMATECH, which then dispatches them to assignments in various facilities. Those people stay in the assignment two years before returning to their home employer. In some cases, one-third of all personnel in a given facility are on loan from SEMATECH members. According to Lieutenant Colonel Jack Dempsey, of the Industrial College of the Armed Forces, "This arrangement greatly aids in the dissemination of information throughout the semiconductor industry" (Dempsey and Lyons, 32).

BUILDING

By 1995, sales were almost at $200 billion and were forecast to reach $1.1 trillion in 2010 (Powell). SEMATECH had succeeded because of organizational and cooperative competencies, not just technical know-how.

According to Browning and Shetler, the turnaround in U.S. leadership of the chip industry by 1994 was due to two practices: (1) SEMATECH identified manufacturing problems, then matched those problems with member companies who had demonstrated operating success in that area; and (2) once best

Table 40.1   Founding Members, International Members of SEMATECH

Company Name	Country	Date of Membership
Advanced Micro Devices (AMD)	U.S.	1987–present
AT&T (Lucent/Agere) (1)	U.S.	1987–present
Hewlett-Packard	U.S.	1987–present
Intel Corp.	U.S.	1987–present
IBM	U.S.	1987–present
Motorola	U.S.	1987–present
Rockwell (Conexant)	U.S.	1987–present
Texas Instruments	U.S.	1987–present
Digital Equipment Co. (DEC) (2)	U.S.	1987–2000
National Semiconductor	U.S.	1987–1998
Harris Corporation	U.S.	1987–1993
LSI Logic	U.S.	1987–1992
Micron Technology	U.S.	1987–1992
NCR (acquired by AT&T, 1991)	U.S.	1987–1991
International Members since 1998 (3)		
Hyundai Electronics (Hynix)	Korea	1998–present
Siemens (Infineon)	Germany	1998–present
Philips	Netherlands	1998–present
STMicroelectronics	Italy/France	1998–present
Taiwan Semiconductor (TSMC)	Taiwan	1998–present

1. AT&T later sold Symbios Logic, NCR's semiconductor division, to Hyundai in 1994.

2. Digital was acquired by Compaq Computers in 1997; Compaq assumed Digital's membership until 2000.

3. The five non-U.S. firms that joined International SEMATECH participated in SEMATECH's I3001 subsidiary during 1996–98. LG Semicon and Samsung Electronics participated in I3001, but did not join International SEMATECH at the time.

Silverman, Brian S., Ziedonis, Arvids A., and Ziedonis, Rosemarie Ham. "Diffusion of Research from SEMATECH: Evidence from Patent Citations."

practice had been identified, SEMATECH spread the solution widely throughout the industry via its members.

According to a report from the National Defense University, another advantage achieved by SEMATECH was protection against antitrust complaints and lawsuits. Dempsey and Lyons point out that when the Department of Defense is a working partner, no one wants to bring a lawsuit or complaint about information sharing (34).

The founding members of SEMATECH numbered 14. In 1998, five international members joined. Before they joined as full members, the international firms had been part of SEMATECH's I3001 subsidiary from 1996 to 1998. In addition to the original five, LG Semicon and Samsung Electronics were part of the I3001 group but elected not to join because of financial issues. Table 1 lists the members as of 2004.

Did SEMATECH succeed? Evidence weighs in the affirmative. The government made a relatively modest investment of $100 million per year and rebuilt a failing industry. Member companies derived enough benefit that they continued to fund the program on their own after the government stopped its financial support in 1997. With the addition of worldwide members in 1998, SEMATECH grew beyond its original national boundaries to become an international cooperative. Finally, SEMATECH companies were able to spend less on R&D because of benefits derived from their cooperative relationships with member companies.

IMPORTANCE IN HISTORY

Two professors at the University of Chicago suggested that when member companies cut their R&D it might have had a negative impact on the industry. According to Douglas A. Irwin and Peter J. Klenow, the rebound in the U.S. share of the semiconductor industry could just as well have been traced to the declining U.S. dollar, or to trade deals with Japan, or to changes in the technology of computer memory worldwide. Irwin and Klenow also observed that SEMATECH members did not appear to have an advantage in research productivity. The professors compared member and nonmember companies in the semiconductor industry and found that SEMATECH partners spent less on R&D. The professors interpreted modest R&D spending as a sign of failure. Others disagree, arguing that the ability to share R&D meant that less money was spent individually. In any case, the fact that SEMATECH remained vigorous even after government funding ended and that SEMATECH expanded to include international members are both indicative of ongoing interest and success.

Besides wealth for its member companies, what other benefits have derived from SEMATECH? Certainly some innovations were developed that are now industry landmarks. For example, SEMATECH developed a program that enables a company to determine the cost of retooling. The company runs the simulation application to determine what equipment they need in order to manufacture specific quantities of semiconductors using the most economical methods. While some companies see no need for this kind of testing, it is critical for small production runs that are common in "boutique" firms that specialize in custom applications.

Another SEMATECH innovation that has affected the entire industry is the open architecture used for integrated computer manufacturing systems. It has allowed more automated material production processes and instilled better control, according to studies at the Industrial College of the Armed Forces of the National Defense University in Washington, DC.

It was cooperation among several sectors—government, private companies, and educational institutions—that formed the kernel of SEMATECH's success. The role of universities was critical. As SEMATECH developed relationships with the large laboratories that conduct much of the world's research, those laboratories recruited colleges and universities to join the effort.

Eventually, the cooperation between universities, laboratories, and companies resulted in the most recent innovation: Centers of Excellence. The State University of New York at Albany is one such center. Albany's Center for Environmental Sciences and Technology Management (CESTM) expanded to become SEMATECH North. The university attracted the partnership through its Center of Excellence in Nanoelectronics, a developing technology for computer chips that may transform the future of computing. More than 250 scientists and technical experts relocated their offices to facilitate working together in a new building that is referred to as a "technology accelerator."

Can the paradigm of SEMATECH inspire initiatives in other industries? With a modicum of leadership and incentive, governments can persuade companies to work together. SEMATECH is an example of how a government can step in to rescue and revive a failing industry. Does this strategy work only for high-tech industries? A similar strategy could be applied to mature low-tech industries such as railroads. If there were a consortium for the railroad industry—perhaps combined with a college or institute for railway technology and management—the rail industry might be helped to identify ways to escape the deteriorating situation it faces today in the United States. In other countries, such as Japan and France, competitive success has already been achieved—and their problem will be to avoid complacency and therefore miss out on the next generation of supersonic rail technology. How would such a cooperative begin? One factor to be addressed is reinsurance. Insurance companies are needed to guarantee prototypes and applications against natural or man-made disasters.

SEMATECH serves as a clear example of what it takes to launch, grow, and continue to develop a truly multipartner, public-private, international industry on a macro scale. It is not just technology but organization that is critical to success. As the president and chief executive officer of International SEMATECH, C. Robert Helms, explained: "The technical challenges that face the semiconductor industry are daunting and growing in complexity, and are too great for any single organization, nation, or region to solve alone" (http://www.albany.edu/pr/ualbanymagfall02/sematech.htm). Can institutes like the Center for Macro Projects and Diplomacy at Roger Williams University in Bristol, Rhode

Island, develop training and conferences to help industries work together in the new paradigm suggested by International SEMATECH?

SEMATECH is an example of the role diplomacy can play in macro cooperatives. According to Mark Melliar-Smith, president of SEMATECH from 1990 to 1996, "The job is tougher on the diplomatic front, not on the technical front." When Melliar-Smith was asked about his job as he prepared to hand over executive direction to incoming president Helms, the former president commented that members from various countries "all have slightly different cultures and different perspectives on life; coalescing to get synergy is a tough job. . . . I've got tremendous respect for Kofi Annan at the United Nations" (*Micro*, June, 2001)

FOR FURTHER REFERENCE

Books and Articles

Browning, Larry D., and Judy C. Shetler. *SEMATECH: Saving the U.S. Semiconductor Industry*. Austin: Texas A&M University Press, 2000.

Dempsey, Jack, and Robert E. Lyons. "Semiconductors and SEMATECH: Rebirth of a Strategic Industry?" Monograph. Washington, DC: National Defense University, Industrial College of the Armed Forces, 1993, 40pp. To read this report online, including footnotes referencing relevant articles, see http://www.ndu.edu/library/ic6/93S14.pdf.

Irwin, Douglas A., and Peter J. Klenow. "SEMATECH: Purpose and Performance." *Proceedings of the National Academy of Science* 93 (November 1996): 12739–42. http://www.pnas.org/cgi/content/full/93/23/12739/. This study is referred to in many discussions of SEMATECH, and provides references.

Silverman, Brian S., Ziedonis, Arvids A., and Ziedonis, Rosemarie Ham. "Diffusion of Research from SEMATECH: Evidence from Patent Citations." Working paper, July 2001, completed January 2002. Paper presented at the Second Annual Wharton Technology Mini-Conference, the Emerging Technologies Management Research Program, The Wharton School, University of Pennsylvania. April 12–13, 2002. Available at http://emertech.wharton.upenn.edu/TechMiniConf2d/html. From this website go to Tech Mini-Conference Reading Room and choose title of paper.

Ziedonis, A.A., R.H. Ziedonis, and B.S. Silverman. "Research Consortia and the Dissemination of Technological Knowledge: Insights from the SEMATECH." Working paper, Stephen M. Ross School of Business, University of Michigan, and the Joseph L. Rotman School of Management, University of Toronto, 2005.

Internet

If you want to know the latest developments at SEMATECH, see http://www.sematech org/.

For a graph developed by Beverly Powell, TPIC chair, for a presentation on February 7, 2000, depicting worldwide semiconductor sales growth from 1975 to 2010, see http://www.tpic.org/Conference00/04-SEMATECH.pdf.

To learn more about International SEMATECH North, see http://www.albany.edu/pr/ualbanymagfall02/sematech.htm.

For the *Micro* magazine article containing an interview with former SEMATECH CEO Mark Melliar-Smith and CEO Robert Helms, see http://www.micromagazine. com/archive/01/06/leadnews.html.

For photos see http://www.national.com/company/pressroom/gallery/historical.html.

Documents of Authorization

SEC. 271. FINDINGS, PURPOSES, AND DEFINITIONS.

(a) FINDINGS.—The Congress finds that it is in the national economic and security interests of the United States for the Department of Defense to provide financial assistance to the industry consortium known as SEMATECH for research and development activities in the field of semiconductor manufacturing technology.

(b) PURPOSES.—The purposes of this part are—

(1) to encourage the semiconductor industry in the United States—

(A) to conduct research on advanced semiconductor manufacturing techniques; and

(B) to develop techniques to use manufacturing expertise for the manufacture of a variety of semiconductor products; and

(2) in order to achieve the purpose set out in paragraph (1), to provide a grant program for the financial support of semiconductor research activities conducted by SEMATECH.

(c) DEFINITIONS.—In this part:

(1) The terms "Advisory Council on Federal Participation in SEMATECH" and "Council" mean the advisory council established by section 273.

(2) The term "SEMATECH" means a consortium of firms in the United States semiconductor industry established for the purposes of (A) conducting research concerning advanced semiconductor manufacturing techniques, and (B) developing techniques to adapt manufacturing expertise to a variety of semiconductor products.

SEC. 272. GRANTS TO SEMATECH.

(a) AUTHORITY TO MAKE GRANTS.—The Secretary of Defense shall make grants, in accordance with section 6304 of title 31, United States Code, to SEMATECH in order to defray expenses incurred by SEMATECH in conducting research on and development of semiconductor manufacturing technology. The grants shall be made in accordance with a memorandum of understanding entered into under subsection (b).

(b) MEMORANDUM OF UNDERSTANDING.—The Secretary of Defense shall enter into a memorandum of understanding with SEMATECH for the purposes of this part. The memorandum of understanding shall require the following:

(1) That SEMATECH have—

(A) a charter agreed to by all representatives of the semiconductor industry that are participating members of SEMATECH; and

(B) an annual operating plan that is developed in consultation with the Secretary of Defense and the Advisory Council on Federal Participation in SEMATECH.

(2) That the total amount of funds made available to SEMATECH by Federal, State, and local government agencies for any fiscal year for the support of research and development activities of SEMATECH under this section may not exceed 50 percent of the total cost of such activities.

(3) That SEMATECH, in conducting research and development activities pursuant to the memorandum of understanding, cooperate with and draw on the expertise of the national laboratories of the Department of Energy and of colleges and universities in the United States in the field of semiconductor manufacturing technology.

(4) That an independent, commercial auditor be retained (A) to determine the extent to which the funds made available to SEMATECH by the United States for the research and development activities of SEMATECH have been expended in a manner that is consistent with the purposes of this part, the charter of SEMATECH, and the annual operating plan of SEMATECH, and (B) to submit to the Secretary of Defense, SEMATECH, and the Comptroller General of the United States an annual report containing the findings and determinations of such auditor.

(5) That (A) the Secretary of Defense be permitted to use intellectual property, trade secrets, and technical data owned and developed by SEMATECH in the same manner as a participant in SEMATECH and to transfer such intellectual property, trade secrets, and technical data to Department of Defense contractors for use in connection with the Department of Defense requirements, and (B) the Secretary not be permitted to transfer such property to any person for commercial use.

(6) That SEMATECH take all steps necessary to maximize the expeditious and timely transfer of technology developed and owned by SEMATECH to the participants in SEMATECH in accordance with the agreement between SEMATECH and those participants and for the purpose of improving manufacturing productivity of United States semiconductor firms.

(c) CONSTRUCTION OF MEMORANDUM OF UNDERSTANDING.—The memorandum of understanding entered into under subsection (b) shall not be considered to be a contract for the purpose of any law or regulation relating to the formation, content, and administration of contracts awarded by the Federal Government and subcontracts under such contracts, including section 2306a of title 10, United States Code, section 719 of the Defense Production Act of 1950 (50 U.S.C. App. 2168), and the Federal Acquisition Regulations, and such provisions of law and regulation shall not apply with respect to the memorandum of understanding.

(d) FUNDING FOR FY88.—Of the amounts appropriated to the Defense Agencies for fiscal year 1988 for research, development, test, and evaluation, $100,000,000 may be obligated only to make grants under this section.

SEC. 273. ADVISORY COUNCIL.

(a) ESTABLISHMENT.—There is established the Advisory Council on Federal Participation in SEMATECH.

(b) FUNCTIONS.—(1) The Council shall advise SEMATECH and the Secretary of Defense on appropriate technology goals for the research and development activities of SEMATECH and a plan to achieve those goals. The plan shall provide for the development of high-quality, high-yield semiconductor manufacturing technologies that meet the national security and commercial needs of the United States.

(2) The Council shall—

(A) conduct an annual review of the activities of SEMATECH for the purpose of determining the extent of the progress made by SEMATECH in carrying out the plan referred to in paragraph (1); and

(B) on the basis of its determinations under subparagraph (A), submit to SEMATECH any recommendations for modification of the plan or the technological goals in the plan considered appropriate by the Council.

(3) The Council shall review the research activities of SEMATECH and shall submit to the Secretary of Defense and the Committees on Armed Services of the Senate and the House of Representatives an annual report containing a description of the extent to which SEMATECH is achieving its research and development goals.

(c) MEMBERSHIP.—The Council shall be composed of 12 members as follows:

(1) The Under Secretary of Defense for Acquisition, who shall be Chairman of the Council.

(2) The Director of Energy Research of the Department of Energy.

(3) The Director of the National Science Foundation.

(4) The Under Secretary of Commerce for Economic Affairs.

(5) The Chairman of the Federal Laboratory Consortium for Technology Transfer.

(6) Seven members appointed by the President as follows:

(A) Four members who are eminent individuals in the semiconductor industry and related industries.

(B) Two members who are eminent individuals in the fields of technology and defense.

(C) One member who represents small businesses.

(d) TERMS OF MEMBERSHIPS.—Each member of the Council appointed under subsection (c)(6) shall be appointed for a term of three years, except that of the members first appointed, two shall be appointed for a term of one year, two shall be appointed for a term of two years, and three shall be appointed for a term of three years, as designated by the President at the time of appointment. A member of the Council may serve after the expiration of the member's term until a successor has taken office.

(e) VACANCIES.—A vacancy in the Council shall not affect its powers but, in the case of a member appointed under subsection (c)(6), shall be filled in the same manner as the original appointment was made. Any member appointed to fill a vacancy for an unexpired term shall be appointed for the remainder of such term.

(f) QUORUM.—Seven members of the Council shall constitute a quorum.

(g) MEETINGS.—The Council shall meet at the call of the Chairman or a majority of its members.

(h) COMPENSATION.—(1) Each member of the Council shall serve without compensation.

(2) While away from their homes or regular places of business in the performance of duties for the Council, members of the Council shall be allowed travel expenses, including per diem in lieu of subsistence, at rates authorized for employees of agencies under sections 5702 and 5703 of title 5, United States Code.

(i) FEDERAL ADVISORY COMMITTEE ACT.—Section 14 of the Federal Advisory Committee Act (5 U.S.C. App. 2) shall not apply to the Council.

SEC. 274. RESPONSIBILITIES OF THE COMPTROLLER GENERAL.

The Comptroller General of the United States shall—

(1) review the annual reports of the auditor submitted to the Comptroller General in accordance with section 272(b)(4); and

(2) transmit to the Committees on Armed Services of the Senate and the House of Representatives his comments on the accuracy and completeness of the reports and any additional comments on the report that the Comptroller General considers appropriate.

SEC. 275. EXPORT OF SEMICONDUCTOR MANUFACTURING.

Any export of materials, equipment, and technology developed by SEMATECH in whole or in part with financial assistance provided under section 272(a) shall be subject to the Export Administration Act of 1979 (50 U.S.C. App. 2401 et seq.) and shall not be subject to the Arms Export Control Act.

SEC. 276. PROTECTION OF INFORMATION.

(a) FREEDOM OF INFORMATION ACT.—Section 552 of title 5, United States Code, shall not apply to information obtained by the Federal Government on a confidential basis under section 272(b)(5).

(b) INTELLECTUAL PROPERTY.—Notwithstanding any other provision of law, intellectual property, trade secrets, and technical data owned and developed by SEMATECH or any of the participants in SEMATECH may not be disclosed by any officer or employee of the Department of Defense except as provided in the provision included in the memorandum of understanding pursuant to section 272(b)(5).

From P.L. 100–180, National Defense Authorization Act for FY88 and FY89.

41
The Central Artery/
Tunnel Project ("Big Dig")

United States

DID YOU KNOW . . . ?

➤ The seven-building ventilation system is the largest highway-tunnel ventilation system in the world.

➤ The tunnel contains enough reinforcing steel for a one-inch-thick steel ribbon to wrap around circumference of the earth.

➤ 16 million cubic yards of earth have been excavated—enough to fill Boston's professional football stadium 15 times over.

➤ Enough concrete has been used to build a three-foot (1-meter)-wide sidewalk with a four-inch (10-centimeter)-thick sidewall from Boston to San Francisco—three times over.

➤ During peak construction years, the Big Dig spent nearly $110 million a month.

➤ City officials warned that construction could dislodge millions of rats from deep under the city.

➤ The Big Dig gives Boston over 200 acres of new parks near the waterfront.

➤ The city of Boston originally occupied only 478 acres.

Popularly known as the "Big Dig," the Central Artery/Tunnel Project has been called the largest and most complex road project in U.S. history. The project's real claim to fame was the extraordinary difficulty of its assigned objectives: (1) replace the Central Artery, the major but obsolete elevated highway running through the heart of Boston, with an underground, larger-capacity,

Construction on Boston's Interstate 93 surrounds the Fleet Center (now known as TD Banknorth Garden), site of the 2004 Democratic National Convention. © William B. Plowman / Getty Images.

modern road system; (2) complete the task with minimal disturbance to surrounding communities, including the Italian North End and Chinatown, while preserving the labyrinth of existing infrastructure and numerous subway lines; (3) devise new technologies for building and installing connecting links under Boston Harbor to the city's airport; (4) demolish and rebuild connectors to the major highways exiting Boston to the west and south; and (5) accomplish all the foregoing while this main artery through the city remained in constant use.

HISTORY

Like many towns in New England whose names derive from English towns, Boston takes its name from a town in Lincolnshire, England, also built on fens (marshy swampland). Boston originally occupied a mere 478 acres of land, virtually surrounded by water and extensive marshes, and connected to the mainland by a narrow isthmus. Soon there was a need for more space for the town's burgeoning population.

Why did so much growth happen where there was so little land? Boston was, for many years, the major port in the United States, one of the first entry points for settlers coming to the country. As a port, Boston offered easy access for shipping of all kinds, which in turn spawned numerous industries. A magnet for commerce, Boston soon became a center of education and political organization as well.

Much of the city as it is known today evolved through land reclamation efforts—backfilling the marshlands and reclaiming land from the ocean. A diorama located at the city's John Hancock skyscraper was developed as a visually graphic demonstration of how Boston began as a small waterfront community and then, with landfill projects, grew to its present-day size. Looking at the list of landfill projects is like reviewing a history of Boston's construction technology. Early projects took a long time; equipment was primitive, and the main engine of power was human labor. Those landfill projects include the following:

Table 41.1 The Boston Land Fill Projects[1]

Location	Approximate Acreage of Landfill	Dates
Oliver's Dock	unavailable	1643–1803
West Cove	80	1803–1863
Mill Pond	70	1804–1835
South Cove	86	1806–1843
East Cove (Great Cove)	112	1823–1874
South Boston	714	1836–1988
South Bay	138	1850–1988
Back Bay	580	1857–1894
Charlestown	416	1860–1896
Fenway (Roxbury)	322	1878–1890
East Boston	370	1880–1988
Marine Park	57	1883–1900
Columbus Park	265	1890–1901
Logan Airport	750	1922–1988

Source: Dalia and Stevenson (with permission of the authors).
[1] http://site.www.umb.edu/conne/wendy/#1.

Much of the landfill came from topographical entities that remain in the city, albeit much smaller today. The site of the state capitol, Beacon Hill, for example, was rather steep—high enough, in fact, to serve as the place for a beacon that could be seen for a considerable distance. Earth moved from Beacon Hill filled in the Mill Pond sector. Other hills were trimmed and the earth was used for fill. If hills were not available, anything handy was used: rubbish, rubble from the fire of 1872, even ships taken out of service and filled with rocks, sunk, and shoved against the marshes to make more land (McNichol, 20). "The reduction of the steep hills added more land along the edges of the coves, and also made the hills themselves more habitable" (Howe study guide). In all, Boston gained over 2,000 acres as a result of the landfill process.

Boston's historically tangled streets were laid out long before the advent of automobiles; puckish residents suggest the original road mappers were meandering cows. By the 1950s, traffic in the inner city was extremely congested, especially for north–south travel. Commissioner of Public Works William Callahan pushed through plans for an elevated expressway, the Central Artery, which eventually was constructed, splitting the downtown area from the waterfront. Even with the new artery, traffic continued to be staggering.

Constructed before strict federal standards for interstate highways were developed during the Eisenhower administration, the expressway was plagued by tight turns, entrance ramps without merge lanes, and continually escalating vehicular loads. Designed in the 1950s to carry 75,000 vehicles a day, a few decades later the deteriorating six-lane highway was used by over 200,000 vehicles, resulting in traffic jams of legendary magnitude, along with an accident rate four times higher than the average for urban interstates.

The Central Artery was regarded by many as a social and environmental disaster. Its original construction displaced more than 20,000 people and tore apart venerable residential neighborhoods such as the North End, home of the Old North Church from which the lanterns—"one if by land, two if by sea"—sent Paul Revere on his famous ride. The highway, known by such sobriquets as *the other Green Monster* (for baseball fans, the first Green Monster is the huge wall at the back of Fenway Park, home of the local baseball team) and *the Green Snake* (for the green paint that covered the metalwork), caused a scar through the heart of Boston.

Congress took account of evolving road and highway needs in the United States in the Federal-Aid Highway Act of 1976 (reproduced here as Document I). That measure opened the door to the use of federal funds for roadways in urbanized areas and for highway maintenance and rehabilitation. The law also created new opportunities for Boston-area officials looking for ways to deal with the myriad problems associated with the Central Artery.

CULTURAL CONTEXT

Nine out of 10 American adults drive. Men spend an average of 81 minutes a day behind the wheel, and women 64 minutes a day (Edmondson). This is more time than the average American spends cooking or eating, according to the University of Maryland. It is more than twice as much time as the average parent spends with his or her children.

Between 1969 and 1995, the number of drivers in the United States increased more then three times as fast as the population, and the number of household vehicles increased six times as fast (Edmondson). This increase occurred for many reasons: construction of interstate highways, entry of women into the workforce, growth of suburbs as the only choice for workers seeking affordable housing, neglect of rail systems, and failure to develop alternative commuting options such as buses, protected bikeways, and lanes equipped for PAT (palleted automated transit) systems.

While Boston grew sizably as a result of landfill, that process wrought significant environmental damage with the loss of native salt marshes, according to Dalia and Stevenson. In creating the Back Bay, for example, more than 400 acres of salt marshes were destroyed. Environmental concerns were an important factor in gaining approval for the Central Artery Project.

PLANNING

The eventual solution to Boston's traffic congestion may have been inspired by a promise to a grandmother. Or was it the brew at Jacob Wirth's, a venerable old German *wursthaus* where the floors were strewn with sawdust and the beer was vintage home brew? Two friends were enjoying lunch when one lamented that traffic was going to kill Boston. But how could the roads be widened without further harm to communities and families? "I promised my grandmother when I became a politician," confessed one fellow to the other, "that I'd never allow a family to be displaced again as the result of a political decision." But, then, how could the road be widened? Was it the fine fare or the fine minds that suddenly produced the thought "What if we just put the road underground, you know, depressed it below the current road?"

William Reynolds and Fred Salvucci were the lunchmates. Reynolds had the idea and Salvucci had the motivation and commitment to the political process. Fred Salvucci was one of the progenitors of the project; during his time as state secretary of transportation, extending public transit became his passion. When he heard Reynolds's idea, he moved through the halls of political leadership and local neighborhoods to drum up support for what he knew could become a massive project beyond anybody's imagination. Alan A. Altshuler, Salvucci's successor as state secretary of transportation, took up the cause. U.S. Senator Edward M. Kennedy, Congressman Thomas "Tip" O'Neill Jr., and the late Congressman Joseph Moakley promoted the project in Washington. It was Moakley who quipped that his favorite bird was the crane—the construction crane—because it symbolized all the new jobs for people in his district during the Big Dig.

The proposal that gradually took shape—placing the artery roadway underground and building new connecting bridges and tunnels—promised to solve both traffic and environmental issues at one stroke. To quell resistance from residents and businesses in the area, proponents pledged that no families would be uprooted, that traffic would be kept flowing, and that the land under the old Central Artery, which would be demolished, would be transformed into parks and open space. It was claimed that by speeding up the flow of traffic, the project would produce a 12 percent drop in carbon-monoxide levels in the city. Individuals, businesses, and communities that would be affected by the Big Dig were offered a long list of budget-expanding "mitigations" to make the project more palatable to them.

Business leaders pushed for a third harbor tunnel to improve access to Logan Airport, located across Boston Harbor from the city itself, accessible only by two single-direction tunnels. In their second terms as governor and state secretary of transportation, respectively, Michael Dukakis and Salvucci came up with the strategy of tying the two projects together, thereby creating a single project that would be supported by the business community and by the city of Boston.

Planning for the Big Dig officially began in 1982, with environmental-impact studies starting in 1983. However, the final plank in the plan was almost not laid. Because it qualified for federal funds in its function of connecting the I-90 interstate fully to Boston Logan Airport, the Central Artery project began to draw federal scrutiny, followed by a major obstacle: the president of the United States, Ronald Reagan, opposed the scheme. Advised by his secretary of transportation, Elizabeth Dole, that there was too little federal benefit and that "the cost is not justified on the basis of the transportation benefits to the nation" (McNichol, 37), President Reagan would not sign off on the funding. This resulted in a stalemate between Speaker of the House Tip O'Neill, who favored the project, and the president. When O'Neill retired from his post in 1986, the new Speaker of the House, James Wright, proposed a transportation bill that passed in the Senate but was vetoed by the president on March 27, 1987. A series of reinstatements and vetoes ensued, with Senator Kennedy moving behind the scenes and through the hallways to visit his colleagues. For instance, if the transportation bill did not pass, what might happen to tobacco subsidies for North Carolina? Could Senator Sanford of North Carolina be persuaded to change his mind and support the transportation legislation? The next day, even as President Reagan traveled to Capitol Hill to save his veto, Sanford cast the vote that overrode the presidential veto. The Surface Transportation and Uniform Relocation Assistance Act of 1987 was the result, becoming law in April and assuring the federal funds to begin the Big Dig (McNichol).

Finally, at the beginning of 1991, the Massachusetts secretary of environmental affairs issued a certificate approving the project's final supplemental environmental-impact statement, a 5,000-page document in 12 volumes weighing 44 pounds (20 kilograms); formal endorsement by the U.S. Federal Highway

Administration followed a few months later. The secretary's certificate (reproduced here as Document II) was replete with environmental and quality-of-life concerns, including release of toxins by the excavation and the possibility of disrupting the nests of millions of rats, causing them to roam city streets and alleys in search of new housing. It also called for further review of a planned crossing of the Charles River; indeed, a different design was later adopted. By the time the federal environmental clearances were delivered in 1994, the process had taken seven years, during which time inflation had greatly increased the project's original cost estimates.

In addition to these political and financial difficulties, the project faced several environmental and engineering obstacles. The downtown area through which the tunnels were to be dug was largely landfill, and included existing subway lines and innumerable pipes and utility lines. Before excavation for the tunnels could begin, the utility lines had to be replaced or moved. Tunnel workers encountered many unexpected barriers, ranging from glacial debris to foundations of buried houses and the ships that had been sunk so long ago to reclaim the land.

The project was managed by the Massachusetts Turnpike Authority, with design and construction supervised by a joint venture between Bechtel Corporation and Parsons Brinckerhoff. In the world of macro construction there are only a handful of players; not many firms have the requisite experience and expertise. When bids were submitted in 1985, the committee unanimously selected the joint venture of Bechtel and Parsons Brinckerhoff. Fred Salvucci remarked at the time, "Bechtel and Parsons Brinckerhoff had done 90 percent of the slurry-wall construction in the United States. If either had joined another team, there might have been more choice. By going together, they blew everyone away" (McNichol, 42).

Due to the enormous size of the project—too large for any company to undertake alone—the construction of the Big Dig was broken into dozens of smaller subprojects and let to several contractors. Major heavy-construction contractors on the project included Jay Cashman, Modern Continental, Obayashi, Perini Corporation, Peter Kiewit Sons', J. F. White, and the Slattery division of Skanska USA. Of these, Modern Continental was awarded the greatest gross value of contracts.

BUILDING

The primary requirement was to construct the project while tens of thousands of commuters used the artery and its integrated connectors and off-ramps every day. Demolishing the hectic roadway without seriously restricting traffic flow required considerable preconstruction preparation and numerous state-of-the-art construction techniques. According to Alex Krieger, chairman of Harvard's Department of Urban Planning and Design, one-third of the cost of the project was spent just to keep the city functioning while the work went on.

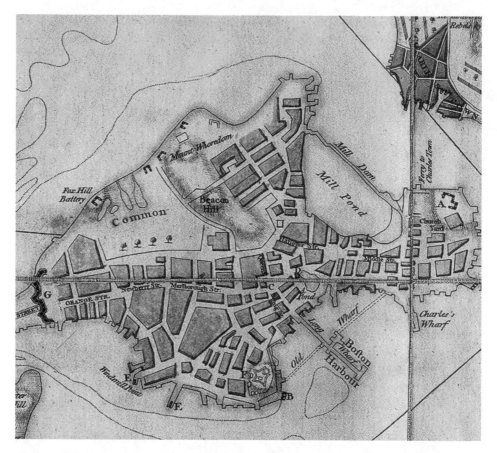

A plan of the city of Boston by Richard Williams, 1775. © The Bodleian Library, University of Oxford.

Because the old elevated highway (which remained in operation throughout the construction process) rested on pylons located throughout the designated dig area, engineers first utilized slurry-wall techniques to create 120-foot-deep concrete walls upon which the highway could rest. These concrete walls also stabilized the sides of the site, preventing cave-ins during the excavation process. The walls were the reason the road above could stay operational while the underground highway was being built: the slurry walls supported the upper structure.

Another challenge was a functioning subway tunnel that crossed the path of the projected underground highway. In order to build new walls past this existing tunnel, it was necessary to dig under the tunnel and build an underground concrete bridge to support the tunnel's weight.

Tunnel jacking was another innovation. The new tunnels were partially built in a jacking pit. Because each tunnel section weighed up to 35,000 tons and went under rail lines and roads, something was needed to keep the tracks and roads from collapsing onto the tunnels. The solution? Embed pipes in the ground (2,400 were placed) and then pump them full with ice-cold liquid. A similar technology had helped cool construction at the Hoover Dam: for that 1930s project, ice water was used. In the case of the Central Artery, the pipes were filled with a subzero calcium chloride solution (Wallenberg). With the ground frozen, the rails and roads could be jacked up and tunnels placed safely beneath them.

One facet of the project required crossing the Charles River, which flows into Boston Harbor. In what form the crossing would be done had been a major source of controversy throughout the design phase of the project. Many environmental advocates preferred a river crossing entirely in tunnels, but this (and 27 other plans) was rejected as too costly. Finally, with the deadline looming to begin construction on a separate project that would connect the Tobin Bridge to the Charles River crossing, Fred Salvucci overrode the objections and chose a plan known as *Scheme Z*. This plan was considered to be reasonably cost-effective, but had the drawback of requiring highway ramps stacked as high as 100 feet (30 meters) immediately adjacent to the Charles River.

The Central Artery/Tunnel Project, in its final form, runs mostly below ground and is about 7.8 miles (12.5 kilometers) long, with approximately 160 lane-miles (260 lane-kilometers) of new and reconstructed highway. There are many components to the project, all built in sequence. Besides the new eight-lane highway replacing the Central Artery, other key components include four major interchanges and two bridges over the Charles River. The first major segment to be completed, in late 1995, was the Ted Williams Tunnel under Boston Harbor to Logan Airport. The tunnel, named after a much-loved Boston baseball player, was another wonder, requiring construction techniques never before used for this type of work.

The Leonard P. Zakim Bunker Hill Bridge was the architectural jewel of the project. It opened in 2002 to rave reviews for its soaring beauty and clever integration of anchor towers reminiscent of its nearby namesake, the Bunker Hill Monument. Created by Swiss architect Christian Menn, the bridge was placed at the north terminus of the project, connecting the underground highway with I-93 and U.S. 1. A distinctive cable-stayed, hybrid bridge (both steel and concrete used in the frame), it was supported by two forked towers connected to the span by cables and girders. It was the widest bridge of its type ever built and the first to use an asymmetrical design. At night the entire structure is swathed in soft blue light.

The Rose Fitzgerald Kennedy Greenway, the final piece of the project, added over 150 acres (60 hectares) of much-sought green space, parks, and walkways to areas near the harbor, turning what was once a river of concrete and exhaust fumes into gardens with public art that grace the city and complete Frederick Law Olmsted's "Emerald Necklace," the series of interconnecting parks that run through several of Boston's contiguous neighborhoods.

Some commentators saw the great efforts taken by proponents of the Big Dig—seeking public input and building a consensus, with the help of mitigation concessions to stakeholders—as a model for how future macro projects in the United States and other countries might negotiate the arduous path to approval. Advocates of future projects will doubtless take a closer look at the lessons of the Big Dig; so too will opponents. Despite efforts to keep Boston open for business during construction, the project was not without major annoyances for those who lived and worked in central Boston, including maddening shifts in traffic patterns, irregular and unpredictable night-time road closings to accommodate construction requirements, and layers of dirt and concrete dust on the furniture and windowsills of every office, apartment, and home in the vicinity.

At the time construction began, the entire project (including the Charles River crossing) was projected to cost $5.8 billion. Eventual cost overruns were so high that the chairman of the Massachusetts Turnpike Authority, James Kerasiotes, was fired in 2000 and his replacement required to commit to a cap on federal contributions of $8.549 billion. Total expenses surpassed $15 billion. Of these expenses, one-third went to mitigation payments and amenities.

IMPORTANCE IN HISTORY

No discussion of spending $15 billion to improve a highway should occur without mentioning the associated concerns of air pollution. According to the former director of the Massachusetts Public Health Association, Laurie Stillman, "Public health costs associated with air pollution include: high rates of school absenteeism, lost work time and wages, rising health insurance costs, lower work productivity, and money spent in direct costs for health care" (Stillman). Stillman questioned the wisdom of allocating so much money to enable people to continue to use cars to get to work. If the same amount of money were spent on improving public transportation, the resulting advantages would be considerable.

Several other issues need to be considered, beginning with construction delays. "Special interest groups, government organizations and individual communities all wanted a piece of the well-funded action. The problem was that for every month that construction was delayed by partisan demands, the price tag rose eighteen million to cover inflation," according to Fred Salvucci (McNichol, 43). Based on the experience in Boston, construction specification plans may need to take these kinds of (inevitable?) delays into financial account in their projections.

It may no longer be possible to estimate construction costs prior to construction; the original estimate for the Big Dig was just one-third of what turned out to be the actual cost. In the future, when cities want to undertake infrastructure development to ease crowding or undertake a macro project, the new metric might need to be one-third for construction and two-thirds to keep the city functioning through the rebuilding.

Another lesson learned from the Big Dig concerns rights-of-way. As with many projects in history, one of the most frequent challenges is not the advance of technology but the acquisition of rights-of-way. From the Roman aqueducts to the New River (a project almost stopped by stalled rights-of-way until the king himself stepped in) to the Eiffel Tower (which has a special clause written into the contract concerning the rights of a protesting abutter), rights-of-way are the major source of delays. And delays inflate cost. In the case of the Central Artery Project, it was possible to build the new road under the existing road because the rights-of-way were already in place. This is true for the entire U.S. interstate road system, and for many road systems in the developed world. Would it then be possible to transform outmoded roads built for single automobiles to more futuristic transport systems such as the palleted automated transit (PAT) system proposed by Professor David Gordon Wilson of MIT? Even if conventional roads remained, one lane in each direction could be devoted to such a PAT system.

The same logic can be applied to railroad rights-of-way. Trains were once an important part of transport connectivity in the United States, but today old trains and even older tracks are rapidly becoming outmoded. Evidence of why old tracks cannot sustain the newer trains becomes immediately apparent when one looks at Acela, America's version of a high-speed train. Acela cannot achieve the high speeds of which it is capable because of the inadequate structure of the tracks on which it must travel. The solution is simple: use existing rights-of-way, tear out the old tracks, and install new ones in the same space using the same permits and hard-won rights-of-way. It is always a long and expensive process to secure rights-of-way, but once they are in hand, they should be used and reused, with each iteration becoming more technologically advanced.

An offshore route for a dredged trench with a tube for high-speed trains running between Boston, New York, and Washington is another viable possibility. The tube could be put in a submerged environment, or placed in new construction along existing rights-of-way up and down the East Coast of the United States. While a land route has curves that slow down a train's speed, an ocean route could be more direct. It would require new permissions and new construction.

As a city that was largely created from reclaimed land, Boston has something in common with Amsterdam. Both prospered due to engineering that created more land. In the future, it may be possible for the world to create more land in locations where land is in dispute, create ports where they are needed, or fashion safer coastlines.

FOR FURTHER REFERENCE

Books and Articles

Aliosi, James A., Jr., and Robert J. Allison. *The Big Dig (New England Remembers)*. Boston: Commonwealth Editions, 2004.

Danigelis, Alyssa. "The Man Behind the Big Dig." *Technology Review*, July/August 2004. To view the article online, see http://www.technologyreview.com/articles/04/07/danigelis0704.asp.

Gewertz, Ken. "A Long View of the Big Dig." *Harvard University Gazette*, September 27, 2001. To view the article online, see http://www.news.harvard.edu/gazette/2001/09.27/11-bigdig.html.

Macaulay, David. *Building Big*. New York: Houghton Mifflin, 2000.

McNichol, Dan. *The Big Dig: The Largest Urban Construction Project in the History of the Modern World*. New York: Silver Lining Books, 2000.

Salvucci, Frederick, and Michael Hintlian. *Digging: The Workers of Boston's Big Dig*. Boston: Commonwealth Editions, 2004.

Seasholes, Nancy. "Landmaking and the Process of Urbanization: The Boston Landmaking Projects, 1630s–1888." PhD diss., Boston University, 1994.

Stillman, Laurie. "Health and the Big Dig." *Boston Globe*, January 19, 2005. To read the article online, see http://www.bostonglobe.com/.

Teale, J., and M. Teale. *Life and Death of the Salt Marsh*. New York: Ballantine Books, 1969.

Tobin, James. *Great Projects: The Epic Story of the Building of America, from the Taming of the Mississippi to the Invention of the Internet*. New York: Free Press, 2001.

Toynbee, Arnold, ed. *Cities of Destiny*. London: Thames and Hudson, 1976.

Tsipis, Yanni K. *Boston's Central Artery: Images of America*. Mount Pleasant, SC: Arcadia Publishing, 2001. See other relevant books by Tsipis: *Building the Mass Pike* (2002); *Building 128* (2003); and *Boston's Bridges* (2004), all published by Arcadia Publishing.

Vanderwarker, Peter. *The Big Dig: Reshaping an American City*. Boston: Little, Brown, 2001.

Wallenberg, Christopher. "Digging Deep." *Panorama*. To read the article online, see http://www.panoramamagazine.com/panoramamagazine/articles/digging_deep.asp.

Wilkie, Richard W., and Jack Tager. *Historical Atlas of Massachusetts*. Amherst: University of Massachusetts Press, 1991.

Wilson, David Gordon. "Guided Transportation Systems: Low-Impact, High-Volume, Fail-Safe Travel." In *Macro-engineering: MIT Brunel Lectures on Global Infrastructure*, edited by Frank P. Davidson, Ernst G. Frankel, and C. Lawrence Meador. Horwood Series in Engineering Science. Chichester, England: Horwood Publishing, 1997.

Internet

For the Big Dig's own Web site, see http://www.masspike.com/bigdig/index.html.

For Wendy Dalia and Robert Stevenson's "Boston and Its Environs," part of "Conservation New England: Past Present & Future," a Web site created by students of Conservation Biology of New England, a graduate course in the Biology Department of the University of Massachusetts, Boston, see http://site.www.umb.edu/conne/wendy/Boston.html.

Edmondson, Brad. "In the Driver's Seat." *American Demographics*, March 1998. http://www.findarticles.com/p/articles/mi_m4021/is_n3_v20/ai_20375004.

For a history of Boston landfills, see Jeffery Howe's "A Digital Archive of American Architecture: A Study Guide for FA 267—From Saltbox to Skyscraper: Architecture in Boston," which includes a moving map where the reader can see the land of Boston being filled in serially, see http://www.bc.edu/bc_org/avp/cas/fnart/fa267/bos_fill2.html.

For further information on the Nationwide Personal Transportation Survey (NPTS), see http://www.fhwa.dot.gov/ohim/1983/1983page.htm.

For further information on the University of Maryland time-use survey, see http://www.webuse.umd.edu/.

Documents of Authorization—I

PL 94-280 (HR 8235)
MAY 5, 1976
90 Stat. 425

An Act to authorize appropriations for the construction of certain highways in accordance with title 23 of the United States Code, and for other purposes.

Be it enacted by the Senate and House of Representatives of the United States of America in Congress assembled,

TITLE I

SHORT TITLE

Sec. 101. This title may be cited as the "Federal-Aid Highway Act of 1976".

REVISION OF AUTHORIZATION FOR APPROPRIATIONS FOR THE INTERSTATE SYSTEM

Sec. 102. (a) Subsection (b) of section 108 of the Federal-Aid Highway Act of 1956, // 23 USC 101 note. // as amended, is amended by striking out "the additional sum of $3,250,000,000 for the fiscal year ending June 30, 1978, and the additional sum of $3,250,000,000 for the fiscal year ending June 30, 1979.", and by inserting in lieu thereof the following: "the additional sum of $3,250,000,000 for the fiscal year ending September 30, 1978, the additional sum of $3,250,000,000 for the fiscal year ending September 30, 1979, the additional sum of $3,625,000,000 for the fiscal year ending September 30, 1980, the additional sum of $3,625,000,000 for the fiscal year ending September 30, 1981, the additional sum of $3,625,000,000 for the fiscal year ending September 30, 1982, the additional sum of $3,625,000,000 for the fiscal year ending September 30, 1983, the additional sum of $3,625,000,000 for the fiscal year ending September 30, 1984, the additional sum of $3,625,000,000 for the fiscal year ending September 30, 1985, the additional sum of $3,625,000,000 for the fiscal year ending September 30, 1986, the additional sum of $3,625,000,000 for the fiscal year ending September 30, 1987, the additional sum of $3,625,000,000 for the fiscal year ending September 30, 1988, the additional sum of $3,625,000,000 for the fiscal year ending

September 30, 1989, and the additional sum of $3,625,000,000 for the fiscal year ending September 30, 1990."

(b) (1) At least 30 percent of the apportionment made to each State for each of the fiscal years ending September 30, 1978, and September 30, 1979, of the sums authorized in subsection (a) of this section shall be expended by such State for projects for the construction of intercity portions (including beltways) which will close essential gaps in the Interstate System and provide a continuous System.

(2) The Secretary of Transportation shall report to Congress before October 1, 1976, on those intercity portions of the Interstate System the construction of which would be needed to close essential gaps in the System.

(3) A State which does not have sufficient projects to meet the 30 percent requirement of paragraph (1) of this subsection may, upon approval of the Secretary of Transportation, be exempt from the requirements of such paragraph to the extent of such inability.

(c) No part of the funds authorized by section 108(b) of the Federal-Aid Highway Act of 1956, // 23 USC 101 note. // as amended, for the Interstate System, shall be obligated for any project for resurfacing, restoring, or rehabilitating any portion of the Interstate System.

AUTHORIZATION OF USE OF COST ESTIMATES FOR APPORTIONMENT OF INTERSTATE FUNDS

Sec. 103. The Secretary of Transportation shall apportion for the fiscal year ending September 30, 1978, the sums authorized to be appropriated for such periods by section 108(b) of the Federal-Aid Highway Act of 1956, as amended, for expenditures on the National System of Interstate and Defense Highways, using the apportionment factors contained in revised table 5 of Committee Print 94-38 of the Committee on Public Works and Transportation of the House of Representatives.

TRANSITION QUARTER AUTHORIZATION

Sec. 104. (a) There is hereby authorized to be appropriated, out of the Highway Trust Fund, $1,637,390,000 for the transition quarter ending September 30, 1976, for those projects authorized by title 23 of the United States Code, the approval of which creates a contractual obligation of the United States for payment out of the Highway Trust Fund of the Federal share of such projects except those authorized by section 142 of such title, and those on the Interstate System (other than as permitted in subsection (b)). Such sums shall be apportioned or allocated on the date of enactment of this Act among the States, as follows:

(1) 60 percent according to the formula established under section 104(b) (1) of title 23, United States Code, as such section is in effect on the day preceding the date of enactment of this Act.

(2) 40 percent in the ratio which the population of each State bears to the total population of all the States shown by the latest available Federal census.

(b) Any State which received less than one-half of 1 percent of the apportionment made under section 104(b) (5) of title 23, United States Code, for the Interstate System for fiscal year 1977 may expend all or any part of its apportionment under this section for projects on the Interstate System in such State.

(c) There is hereby authorized to be appropriated out of the Highway Trust Fund, for the transition quarter ending September 30, 1976, $8,250,000 for forest highways, and $4,000,000 for public lands highways. Such sums shall be apportioned or allocated on the date of enactment of this Act in accordance with section 202 of title 23, United States Code.

(d) There is authorized to be appropriated, out of the Highway Trust Fund, for the transition quarter ending September 30, 1976, $120,000 to the Virgin Islands, $120,000 to Guam, and $120,000 to American Samoa, for projects and programs under section 152, 153, and 402 of title 23, United States Code, such sums shall be apportioned on the date of enactment of this Act in accordance with section 402(c) of title 23, United States Code.

HIGHWAY AUTHORIZATIONS

Sec. 105. (a) For the purpose of carrying out the provisions of title 23, United States Code, the following sums are hereby authorized to be appropriated:

(1) For the Federal-aid primary system in rural areas, including the extensions of the Federal-aid primary system in urban areas, and the priority primary routes, out of the Highway Trust Fund, $1,350,000,000 for the fiscal year ending September 30, 1977, and $1,350,000,000 for the fiscal year ending September 30, 1978. For the Federal-aid secondary system in rural areas, out of the Highway Trust Fund, $400,000,000 for the fiscal year ending September 30, 1977, and $400,000,000 for the fiscal year ending September 30, 1978.

(2) For the Federal-aid urban system, out of the Highway Trust Fund, $800,000,000 for the fiscal year ending September 30, 1977, and $800,000,000 for the fiscal year ending September 30, 1978.

(3) For forest highways, out of the Highway Trust Fund, $33,000,000 for the fiscal year ending September 30, 1977, and $33,000,000 for the fiscal year ending September 30, 1978.

(4) For public lands highways, out of the Highway Trust Fund, $16,000,000 for the fiscal year ending September 30, 1977, and $16,000,000 for the fiscal year ending September 30, 1978.

(5) For forest development roads and trails, $35,000,000 for the three-month period ending September 30, 1976, $140,000,000 for the fiscal year ending September 30, 1977, and $140,000,000 for the fiscal year ending September 30, 1978.

(6) For public lands development roads and trails, $2,500,000 for the three-month period ending September 30, 1976, $10,000,000 for the fiscal year ending September 30, 1977, and $10,000,000 for the fiscal year ending September 30, 1978.

(7) For park roads and trails, $7,500,000 for the three-month period ending September 30, 1976, $30,000,000 for the fiscal year ending September 30, 1977, and $30,000,000 for the fiscal year ending September 30, 1978.

(8) For parkways, $11,250,000 for the three-month period ending September 30, 1976, $45,000,000 for the fiscal year ending September 30, 1977, and $45,000,000 for the fiscal year ending September 30, 1978, except that the entire cost of any parkway project on any Federal-aid system paid under the authorization contained in this paragraph shall be paid from the Highway Trust Fund.

(9) For Indian reservation roads and bridges, $20,750,000 for the three-month period ending September 30, 1976, $83,000,000 for the fiscal year ending September 30, 1977, and $83,000,000 for the fiscal year ending September 30, 1978.

(10) For economic growth center development highways under section 143 of title 23, United States Code, out of the Highway Trust Fund, $50,000,000 for the fiscal year ending September 30, 1977, and $50,000,000 for the fiscal year ending September 30, 1978.

(11) For necessary administrative expenses in carrying out section 131 and section 136 of title 23, United States Code, $375,000 for the three-month period ending September 30, 1976, $1,500,000 for the fiscal year ending September 30, 1977, and $1,500,000 for the fiscal year ending September 30, 1978.

(12) For carrying out section 215(a) of title 23, United States Code—,

(A) for the Virgin Islands, not to exceed $1,250,000 for the three-month period ending September 30, 1976, not to exceed $5,000,000 for the fiscal year ending September 30, 1977, and not to exceed $5,000,000 for the fiscal year ending September 30, 1978.

(B) for Guam, not to exceed $1,250,000 for the three-month period ending September 30, 1976, not to exceed $5,000,000 for the fiscal year ending September 30, 1977, and not to exceed $5,000,000 for the fiscal year ending September 30, 1978.

(C) for American Samoa, not to exceed $250,000 for the three-month period ending September 30, 1976, not to exceed $1,000,000 for the fiscal year ending September 30, 1977, and not to exceed $1,000,000 for the fiscal year ending September 30, 1978. Sum authorized by this paragraph shall be available for obligation at the beginning of the period for which authorized in the same manner and to the same extent as if such sums were apportioned under chapter 1 of title 23, United States Code.

(13) For authorized landscaping, including, but not limited to, the planting of flowers and shrubs indigenous to the area, and for litter removal an additional $25,000,000 for the fiscal year ending September 30, 1977 and $25,000,000 for the fiscal year ending September 30, 1978.

(14) For the Great River Road, $2,500,000 for the three-month period ending September 30, 1976, $100,000,000 for the fiscal year ending September 30, 1977, and $10,000,000 for the fiscal year ending September 30, 1978, for construction or reconstruction of roads not on a Federal-aid highway system; and out

of the Highway Trust Fund, $6,250,000 for the three-month period ending September 30, 1976, $25,000,000 for the fiscal year ending September 30, 1977, and $25,000,000 for the fiscal year ending September 30, 1978, for construction or reconstruction of roads on a Federal-aid highway system.

(15) For control of outdoor advertising under section 131 of title 23, United States Code, $25,000,000 for the fiscal year ending September 30,1977, and $25,000,000 for the fiscal year ending September 30, 1978.

(16) For control of junkyards under section 136 of title 23, United States Code, $15,000,000 for the fiscal year ending September 30, 1977, and $15,000,000 for the fiscal year ending September 30, 1978.

(17) For safer off-system roads under section 219 of title 23, United States Code, $200,000,000 for the fiscal year ending September 30, 1977, and $200,000,000 for the fiscal year ending September 30, 1978.

(18) For access highways under section 155 of title 23, United States Code, $3,750,000 for the three-month period ending September 30, 1976, $15,000,000 for the fiscal year ending September 30, 1977, and $15,000,000 for the fiscal year ending September 30, 1978.

(19) Nothing in the first ten paragraphs or in paragraph (12), (13), (14), (17), or (18) of this section shall be construed to authorize the appropriation of any sums to carry out section 131, 136, or chapter 4 of title 23, United States Code.

(b) (1) For each of the fiscal years 1978 and 1979, no State, including the State of Alaska, shall receive less than one-half of 1 percent of the total apportionment for the Interstate System under section 104(b) (5) of title 23, United States Code. Whenever amounts made available under this subsection for the Interstate System in any State exceed the estimated cost of completing that State's portion of the Interstate System, and exceed the estimated cost of necessary resurfacing, restoration, and rehabilitation of the Interstate System within such State, the excess amount shall be transferred to and added to the amounts last apportioned to such State under paragraphs (1), (2) and (6) of section 104(b) in the ratio which these respective amounts bear to each other in that State, and shall thereafter be available for expenditure in the same manner and to the same extent as the amounts to which they are added. In order to carry out this subsection, there are authorized to be appropriated, out of the Highway Trust Fund, not to exceed $91,000,000 for the fiscal year ending September 30, 1978, and $125,000,000 for the fiscal year ending September 30, 1979.

(2) In addition to funds otherwise authorized, $65,000,000 for the fiscal year ending September 30, 1977, and $65,000,000 for the fiscal year ending September 30, 1978, out of the Highway Trust Fund, are hereby authorized for the purpose of completing projects approved under the urban high density traffic program prior to the enactment of this paragraph. Such sums shall be in addition to sums previously authorized.

(c) (1) In the case of priority primary routes, $50,000,000 of the sum authorized for fiscal year ending September 30, 1977, by the amendment made by subsection (a) (1) of this section, shall not be apportioned. Such $50,000,000 shall be available for obligation on July 1, 1976, in the same manner and to the same extent as sums apportioned for fiscal year 1977 except that such $50,000,000 shall be available for obligation at the discretion of the Secretary of Transportation only for projects of unusually high cost which require long periods of time for their construction. Any part of such $50,000,000 not obligated by such Secretary before October 1, 1977, shall be immediately apportioned in the same manner as funds apportioned on October 1, 1977, for priority primary routes and available for obligation for the same period as such apportionment.

(2) In the case of priority primary routes, $50,000,000 of the sum authorized for the fiscal year ending September 30, 1978, by the amendment made by subsection (a) (1) of this section, shall not be apportioned. Such $50,000,000 of such authorized sum shall be available for obligation on the date of such apportionment, in the same manner and to the same extent as the sums apportioned on such date, except that such $50,000,000 shall be available for obligation at the discretion of the Secretary of Transportation only for projects of unusually high cost which require long periods of time for their construction. Any part of such $50,000,000 not obligated by such Secretary before October 1, 1978, shall be immediately apportioned in the same manner as funds apportioned on October 1, 1978, for such routes, and available for obligation for the same period as such apportionment.

INTERSTATE SYSTEM RESURFACING

Sec. 106. (a) In addition to any other funds authorized for the Interstate System, there is authorized to be appropriated out of the Highway Trust Fund not to exceed $175,000,000 for the fiscal year ending September 30, 1978, and $175,000,000 for the fiscal year ending September 30, 1979. Such sums shall be obligated only for projects for resurfacing, restoring, and rehabilitating those lanes on the Interstate System which have been in use for more than five years and which are not on toll roads.

(b) Paragraph (5) of subsection (b) of section 104 of title 23, United States Code, is amended by inserting "(A) Except as provided in subparagraph (B)—" immediately after "(5)" and by adding at the end of such paragraph the following:

"(B) For resurfacing, restoring, and rehabilitating the Interstate System:

"In the ratio which the lane miles on the Interstate System which have been in use for more than five years (other than those on toll roads) in each State bears to the total of the lane miles on the Interstate System which have been in use for more than five years (other than those on toll roads) in all States."

EXTENSION OF TIME FOR COMPLETION OF SYSTEM

Sec. 107. (a) The second sentence of the second paragraph of section 101(b) of title 23, United States Code, is amended by striking out "twenty-three years" and inserting in lieu thereof "thirty-four years" and by striking out "June 30, 1979", and inserting in lieu thereof "September 30, 1990".

(b) (1) The introductory phrase and the second and third sentences of section 104(b) (5) of title 23, United States Code, are amended by striking out "1979" each place it appears and inserting in lieu thereof at each such place "1990".

(2) The last four sentences of such section 104(b) (5) are amended to read as follows: "Upon the approval by Congress, the Secretary shall use the Federal share of such approved estimate in making the apportionment for the fiscal year ending September 30, 1977. The Secretary shall make the apportionment for the fiscal year ending September 30, 1978, in accordance with section 103 of the Federal-Aid Highway Act of 1976. The Secretary shall make a revised estimate of the cost of completing the then designated Interstate System after taking into account all previous apportionments made under this section in the same manner as stated above, and transmit the same to the Senate and the House of Representatives within ten days subsequent to January 2, 1977. Upon the approval by Congress, the Secretary shall use the Federal share of such approved estimates in making apportionments for the fiscal years ending September 30, 1979, and September 30, 1980. The Secretary shall make a revised estimate of the cost of completing the then designated Interstate System after taking into account all previous apportionments made under this section in the same manner as stated above and transmit the same to the Senate and the House of Representatives within ten days subsequent to January 2, 1979. Upon the approval by Congress, the Secretary shall use the Federal share of such approved estimates in making apportionments for the fiscal years ending September 30, 1981, and September 30, 1982. The Secretary shall make a revised estimate of the cost of completing the then designated Interstate System after taking into account all previous apportionments made under this section in the same manner as stated above and transmit the same to the Senate and the House of Representatives within ten days subsequent to January 2, 1981. Upon the approval by Congress, the Secretary shall use the Federal share of such approved estimates in making apportionments for the fiscal years ending September 30, 1983, and September 30, 1984. The Secretary shall make a revised estimate of the cost of completing the then designated Interstate System after taking into account all previous apportionments made under this section in the same manner as stated above and transmit the same to the Senate and the House of Representatives within ten days subsequent to January 2, 1983. Upon the approval by Congress, the Secretary shall use the Federal share of such approved estimates in making apportionments for the fiscal years ending September 30, 1985, and September 30, 1986. The Secretary shall make a revised estimate of the cost of completing the then designated Interstate System after taking into account all previous apportionments made under this section in the same manner as stated above and transmit the same to the Senate and the House of Representatives within ten days subsequent to January 2, 1985. Upon the approval by Congress, the Secretary shall use the Federal share of such approved estimates in making apportionments for the

fiscal years ending September 30, 1987, and September 30, 1988. The Secretary shall make a revised estimate of the cost of completing the then designated Interstate System after taking into account all previous apportionments made under this section in the same manner as stated above and transmit the same to the Senate and the House of Representatives within ten days subsequent to January 2, 1987. Upon the approval by Congress, the Secretary shall use the Federal share of such approved estimates in making apportionments for the fiscal years ending September 30, 1989, and September 30, 1990. Whenever the Secretary, pursuant to this subsection, requests and receives estimates of cost from the State highway departments, he shall furnish copies of such estimates at the same time to the Senate and the House of Representatives."

DEFINITIONS

Sec. 108. (a) Subsection (a) of section 101 of title 23, United States Code, is amended as follows:

(1) The definition of the term "construction" is amended by inserting immediately after "Commerce", the following: "resurfacing, restoration, and rehabilitation,"

(2) The definition of the term "urban area" is amended by striking out the period at the end thereof and inserting in lieu thereof a comma and the following: "except in the case of cities in the State of Maine and in the State of New Hampshire."

(b) Section 101(a) of title 23, United States Code, is amended by adding the following definition after "public lands highways"

"The term 'public road' means any road or street under the jurisdiction of and maintained by a public authority and open to public travel."

ELIGIBILITY FOR WITHDRAWAL

Sec. 109. (a) The second sentence of paragraph (2) of subsection (e) of section 103 of title 23, United States Code, is amended by striking out "prior to the enactment of this paragraph".

(b) Section 103(e) of title 23, United States Code, is amended by adding the following new paragraph at the end thereof:

"(5) Interstate mileage authorized for any State and withdrawn and transferred under the provisions of paragraph (2) of this subsection after the date of enactment of the Federal-Aid Highway Act of 1976, must be constructed by the State receiving such mileage as part of its Interstate System. Any State receiving such transfer of mileage may not, with respect to that transfer, avail itself of the optional use of Interstate funds under the second sentence of paragraph (4) of this subsection."

INTERSTATE SYSTEM

Sec. 110. (a) Section 103(e) (4) of title 23, United States Code, is amended to read as follows:

"(4) Upon the joint request of a State Governor and the local governments concerned, the Secretary may withdraw his approval of any route or portion thereof on the Interstate System which is within an urbanized area or which passes through and connects urbanized areas within a State and which was selected and approved in accordance with this title, if he determines that such route or portion thereof is not essential to completion of a unified and connected Interstate System and if he receives assurance that the State does not intend to construct a toll road in the traffic corridor which would be served by the route or portion thereof. When the Secretary withdraws his approval under this paragraph, a sum equal to the Federal share of the cost to complete the withdrawn route or portion thereof, as that cost is included in the latest Interstate System cost estimate approved by Congress, subject to increase or decrease, as determined by the Secretary based on changes in construction cost of the withdrawn route or portion thereof as of the date of enactment of the Federal-Aid Highway Act of 1976 or the date of approval of each substitute project under this paragraph, whichever is later, and in accordance with the design of the route or portion thereof that is the basis of the latest cost estimate, shall be available to the Secretary to incur obligations for the Federal share of either public mass transit projects involving the construction of fixed rail facilities or the purchases of passenger equipment including rolling stock, for any mode of mass transit, or both, or projects authorized under any highway assistance program under section 103 of this title; or both, which will serve the urbanized area and the connecting non-urbanized area corridor from which the Interstate route or portion thereof was withdrawn, which are selected by the responsible local officials of the urbanized area or area to be served, and which are submitted by the Governor of the State in which the withdrawn route was located. Approval by the Secretary of the plans, specifications, and estimates for a substitute project shall be deemed to be a contractual obligation of the Federal Government. The Federal share of the substitute projects shall be determined in accordance with the provisions of section 120 of this title applicable to the highway program of which the substitute project is a part, except that in the case of mass transit projects, the Federal share shall be that specified in section 4 of the Urban Mass Transportation Act of 1964, // 49 USC 1603. // as amended. The sums available for obligation shall remain available until obligated. The sums obligated for mass transit projects shall become part of, and be administered through, the Urban Mass Transportation Fund. There are authorized to be appropriated for liquidation of the obligations incurred under this paragraph such sums as may be necessary out of the general fund of the Treasury. Unobligated apportionments for the Interstate System in any State where a withdrawal is approved under this paragraph shall, on the date of such approval, be reduced in the proportion that the Federal share of the cost of the withdrawn route or portion thereof bears to the Federal share of the total cost of all Interstate routes in that State as reflected in the latest cost estimate approved by the Congress. In any State where the withdrawal of an Interstate route or portion thereof has been approved under section 103(e) (4) of this title prior to the date of enactment of the Federal-Aid Highway Act of 1976, the unobligated apportionments for the Interstate System in that State on the date of enactment of the Federal-Aid Highway Act of 1976 shall be reduced in the proportion that the Federal share of the cost to complete such route or portion thereof, as shown on the latest cost estimate approved by Congress prior to such approval of withdrawal, bears to the Federal share of the cost of all Interstate routes in that State, as shown on such cost estimate, except that the amount of such proportional reduction shall be credited with the amount of any reduction in such State's Interstate apportionment which was attributable to the Federal share of any substitute project approved under this paragraph prior to enactment of such Federal-Aid Highway Act. Funds available for expenditure to carry out the purposes of this paragraph shall be supplementary to and not in substitution for funds authorized and available for obligation pursuant to the Urban Mass Transportation Act of 1964, // 49 USC 1601 note. // as amended. The provisions of this paragraph as amended by the Federal-Aid Highway Act of 1976, shall be effective as of August 13, 1973."

(b) Section 103(e) (4) of title 23, United States Code, is further amended by adding the following sentence at the end thereof: "In the event a withdrawal of approval is accepted pursuant to this section, the State shall not be required to refund to the Highway Trust Fund any sums previously paid to the State for the withdrawn route or portion of the Interstate System as long as said sums were applied to a transportation project permissible under this title."

ROUTE WITHDRAWALS

Sec. 111. (a) The existing fourth sentence of paragraph (2) of subsection (e) of section 103 of title 23, United States Code, is amended by striking out "increased or decreased," and all that follows down through and including the period at the end thereof and inserting in lieu thereof the following: "or if the cost of any such withdrawn route was not included in such 1972 Interstate System cost estimate, the cost of such withdrawn route as set forth in the last Interstate System cost estimate before such 1972 cost estimate which was approved by Congress and which included the cost of such withdrawn route, increased or decreased, as the case may be, as determined by the Secretary, based on changes in construction costs of such route or portion thereof, which, (i) in the case of a withdrawn route the cost of which was not included in the 1972 cost estimate but in an earlier cost estimate, have occurred between such earlier cost estimate and the date of enactment of the Federal-Aid Highway Act of 1976, and (ii) in the case of a withdrawn route the cost of which was included in the 1972 cost estimate, have occurred between the 1972 cost estimate and the date of enactment of the Federal-Aid Highway Act of 1976, or the date of withdrawal of approval, whichever date is later, and in each case costs shall be based on that design of such route or portion thereof which is the basis of the applicable cost estimate."

(b) The amendment made by subsection (a) of this section shall be applicable to each route on the Interstate System approval of which was withdrawn or is hereafter withdrawn by the Secretary of Transportation in accordance with the provisions of section 103(e) (2) of title 23, United States Code, including any route on the Interstate System approval of which was withdrawn by the Secretary of Transportation in accordance with the provisions of title 23, United States Code, on August 30, 1965, for the purpose of designating an alternative route.

APPORTIONMENTS

Sec. 112. (a) Section 104(b) of title 23, United States Code, is amended by striking "On or before January 1 next preceding the commencement of each fiscal year, except as provided in paragraphs (4) and (5) of this subsection," and inserting in lieu thereof "On October 1 of each fiscal year except as provided in paragraphs (4) and (5) of this subsection,"

(b) Section 104(b) (1) of title 23, United States Code, is amended to read as follows:

"(1) For the Federal-aid primary system (including extensions in urban areas and priority primary routes)—,

"Two-thirds according to the following formula: one-third in the ratio which the area of each State bears to the total area of all the States, one-third in the ratio which the population of rural areas of each State bears to the total population of rural areas of all the States as shown by the latest available Federal census, and one-third in the ratio which the mileage of rural delivery routes and intercity mail routes where service is performed by motor vehicles in each State bear to the total mileage of rural delivery and intercity mail routes where service is performed by motor vehicles, as shown by a certificate of the Postmaster General, which he is directed to make and furnish annually to the Secretary; and one-third as follows: in the ratio which the population in urban areas in each State bears to the total population in urban areas in all the States as shown by the latest Federal census. No State (other than the District of Columbia) shall receive less than one-half of 1 percent of each year's apportionment."

(c) Section 104(b) (3) of title 23, United States Code, is repealed.

(d) Section 104(e) of title 23, United States Code, is amended to read as follows:

"(e) On October 1 of each fiscal year the Secretary shall certify to each of the State highway departments the sums which he has apportioned hereunder (other than under subsection (b) (5) of this section) to each State for such fiscal year, and also the sums which he has deducted for administration and research pursuant to subsection (a) of this section. On October 1 of the year preceding the fiscal year for which authorized, the Secretary shall certify to each of the State highway departments the sums which he has apportioned under subsection (b) (5) of this section to each State for such fiscal year, and also the sums which he has deducted for administration and research pursuant to subsection (a) of this section. To permit the States to develop adequate plans for the utilization of apportioned sums, the Secretary shall advise each State of the amount that will be apportioned each year under this section not later than ninety days before the beginning of the fiscal year for which the sums to be apportioned are authorized, except that in the case of the Interstate System the Secretary shall advise each State ninety days prior to the apportionment of such funds."

(e) Section 104(f) (1) of title 23, United States Code, is amended by striking out "On or before January 1 next preceding the commencement" and inserting in lieu thereof "On October 1". Section 104(f) (1) is further amended by striking out the period at the end thereof and inserting in lieu thereof a comma and the following: "except that in the case of funds authorized for apportionment on the Interstate System, the Secretary shall set aside that portion of such funds (subject to the overall limitation of one-half of 1 percent) on October 1 of the year next preceding the fiscal year for which such funds are authorized for such System."

(f) Section 104(f) (3) of title 23, United States Code, is amended by striking out the period at the end of the first sentence and inserting in lieu thereof ", except that States receiving the minimum apportionment under paragraph (2) may, in addition, subject to the approval of the Secretary, use the funds apportioned to finance transportation planning outside of urbanized areas."

(g) Section 104(b) (5) of title 23, United States Code, is amended by striking out "a date as far in advance of the beginning of the fiscal year for which authorized as practicable but in no case more than eighteen months prior to the beginning of the fiscal year for which authorized." and inserting in lieu thereof the following: "October 1 of the year preceding the fiscal year for which authorized."

(h) Notwithstanding any other provision of this Act, including any amendments made by this Act, funds authorized by this Act (other than for the Interstate System) for the transition quarter ending September 30, 1976, and for the fiscal year ending September 30, 1977, shall be apportioned on July 1, 1976, except as otherwise provided in section 104.

TRANSFERABILITY

Sec. 113. (a) Subsections (c) and (d) of section 104 of title 23, United States Code, are amended to read as follows:

"(c) (1) Subject to subsection (d), the amount apportioned in any fiscal year, commencing with the apportionment of funds authorized to be appropriated under subsection (a) of section 102 of the Federal-Aid Highway Act of 1956 (70 Stat. 374), to each State in accordance with paragraph (1) or (2) of subsection

(b) of this section may be transferred from the apportionment under one paragraph to the apportionment under the other paragraph if such a transfer is requested by the State highway department and is approved by the Governor of such State and the Secretary as being in the public interest.

"(2) Subject to subsection (d), the amount apportioned in any fiscal year to each State in accordance with paragraph (1) or (6) of subsection (b) of this section may be transferred from the apportionment under one paragraph to the apportionment under the other paragraph if such transfer is requested by the State highway department and is approved by the Governor of such State and the Secretary as being in the public interest. Funds apportioned in accordance with paragraph (6) of subsection (b) of this section shall not be transferred from their allocation to any urbanized area of two hundred thousand population or more under section 150 of this title, without the approval of the local officials of such urbanized area.

"(d) Each transfer of apportionments under subsection (c) of this section shall be subject to the following conditions:

"(1) In the case of transfers under paragraph (1), the total of all transfers during any fiscal year to any apportionment shall not increase the original amount of such apportionment for such fiscal year by more than 40 percent. Not more than 40 percent of the original amount of an apportionment for any fiscal year shall be transferred to other apportionments.

"(2) In the case of transfers under paragraph (2), the total of all transfers during any fiscal year to any apportionment shall not increase the original amount of such apportionment for such fiscal year by more than 20 percent. Not more than 20 percent of the original amount of an apportionment for any fiscal year shall be transferred to other apportionments.

"(3) No transfer shall be made from an apportionment during any fiscal year if during such fiscal year a transfer has been made to such apportionment.

"(4) No transfer shall be made to an apportionment during any fiscal year if during such fiscal year a transfer has been made from such apportionment."

(b) The amendment made by subsection (a) of this section shall take effect on July 1, 1976, and shall be applicable with respect to funds authorized for the fiscal year ending September 30, 1977, and for subsequent fiscal years. With respect to the fiscal year 1976 and earlier fiscal years, the provisions of subsections (c) and (d) of section 104 of title 23, United States Code, as in effect on June 30, 1976, shall remain applicable to funds authorized for such years.

CONSTRUCTION ESTIMATES

Sec. 114. Section 106(c) of title 23, United States Code, is amended to read as follows:

"(c) Items included in any such estimate for construction engineering shall not exceed 10 percent of the total estimated cost of a project financed with Federal-aid highway funds, after excluding from such total estimate cost, the estimated costs of rights-of-way, preliminary engineering, and construction engineering. However, this limitation shall be 15 percent in any State with respect to which the Secretary finds such higher limitation to be necessary."

ADVANCE ACQUISITION OF RIGHTS-OF-WAY

Sec. 115. (a) Paragraph (2) of subsection (c) of section 108 of title 23, United States Code, is amended by striking out "made pursuant to section 133 or chapter 5 of this title."

(b) Section 108(a) of title 23, United States Code, is amended by inserting after "request is made" the words "unless a longer period is determined to be reasonable by the Secretary" in the last sentence.

(c) Section 108(c) (3) of title 23, United States Code, is amended by inserting "or later" following "earlier" in the first sentence.

CERTIFICATION ACCEPTANCE

Sec. 116. (a) Subsection (a) of section 117 of title 23, United States Code, is amended by striking out "establishing requirements at least equivalent to those contained in, or issued pursuant to, this title." and inserting in lieu thereof "which will accomplish the policies and objectives contained in or issued pursuant to this title."

(b) Section 117 of title 23 of the United States Code is amended by adding at the end thereof the following new subsection:

"(f) (1) In the case of the Federal-aid secondary system, in lieu of discharging his responsibilities in accordance with subsections (a) through (d) of this section, the Secretary may, upon the request of any State highway department, discharge his responsibility relative to the plans, specifications, estimates, surveys, contract awards, design, inspection, and construction of all projects on the Federal-aid Secondary system by his receiving and approving a certified statement by the State highway department setting forth that the plans, design, and construction for each such project are in accord with those standards and procedures which (A) were adopted by such State highway department, (B) were applicable to projects in this category, and (C) were approved by him.

"(2) The Secretary shall not approve such standards and procedures unless they are in accordance with the provisions of subsection (b) of section 105, subsection (b) of section 106, and subsection (c) of section 109, of this title.

"(3) Paragraphs (1) and (2) of this subsection shall not be construed to relieve the Secretary of his obligation to make a final inspection of each project after construction and to require an adequate showing of the estimated cost of construction and the actual cost of construction."

AVAILABILITY

Sec. 117. (a) Subsection (b) of section 118 of title 23, United States Code, is amended to read as follows:

"(b) Sums apportioned to each Federal-aid system (other than the Interstate System) shall continue available for expenditure in that State for the appropriate Federal-aid system or part thereof (other than the Interstate System) for a period of three years after the close of the fiscal year for which such sums are authorized and any amounts so apportioned remaining unexpended at the end of such period shall lapse. Sums apportioned to the Interstate System shall continue available for expenditure in that State for the Interstate System for a period of two years after the close of the fiscal year for which such sums are authorized. Any amount apportioned to the States for the Interstate System under subsection (b) (5) (A) of section 104 of this title remaining unexpended at the end of the period during which it is available under this section shall lapse and shall immediately be reapportioned among the other States in accordance with the provisions of subsection (b) (5) (A) of section 104 of this title. Any amount apportioned to the States for the Interstate System under subsection (b) (5) (B) of section 104 of this title remaining unexpended at the end of the period of its availability shall lapse. Sums apportioned to a Federal-aid system for any fiscal year shall be deemed to be expended if a sum equal to the total of the sums apportioned to the State for such fiscal year and previous fiscal years is obligated. Any Federal-aid highway funds released by the payment of the final voucher or by the modification of the formal project agreement shall be credited to the same class of funds, primary, secondary, urban, or interstate, previously apportioned to the State and be immediately available for expenditure.".

(b) (1) The first sentence of section 203 of title 23, United States Code, is amended by striking out "or a date not earlier than one year preceding the beginning" and inserting in lieu thereof "or on October 1,".

(2) The second sentence of such section 203 is amended by striking out "two years" and inserting in lieu thereof "three years".

(c) The funds authorized by section 104 of this Act and all funds authorized by titles I and II of this Act for the transition quarter ending September 30, 1976, shall, for the purposes of the application of sections 118 and 203 of title 23, United States Code, remain available for expenditure for the same period as funds authorized by this Act for the fiscal year ending September 30, 1977.

PAYMENT TO STATES FOR CONSTRUCTION

Sec. 118. (a) Section 121(d) of title 23, United States Code, is amended to read as follows:

"(d) In making payments pursuant to this section, the Secretary shall be bound by the limitations with respect to the permissible amounts of such payments contained in sections 120 and 130 of this title. Payments for construction engineering on any project financed with Federal-aid highway funds shall not exceed 10 percent of the Federal share of the cost of construction of such project after excluding from the cost of construction the costs of rights-of-way, preliminary engineering, and construction engineering. However, this limitation shall be 15 percent in any State with respect to which the Secretary finds such higher limitation to be necessary."

EMERGENCY RELIEF

Sec. 119. (a) Section 125(a) of title 23, United States Code, is amended—,

(1) by striking out "June 30, 1972," and inserting in lieu thereof "June 30, 1972, and ending before June 1, 1976,"

(2) by striking out "June 30, 1973," and inserting in lieu thereof "June 30, 1973, to carry out the provisions of this section, and not more than $25,000,000 for the three-month period beginning July 1, 1976, and ending September 30, 1976, is authorized to be expended to carry out the provisions of this section, and not more than $100,000,000 is authorized to be expended in any one fiscal year commencing after September 30, 1976,"; and

(3) by adding before the last sentence the following new sentence: "For the purposes of this section the period beginning July 1, 1976, and ending September 30, 1976, shall be deemed to be a part of the fiscal year ending September 30, 1977."

(b) The second sentence of section 125(b) of such title is amended by striking out the period and inserting in lieu thereof the following: ", except that if the President has declared such emergency to be a major disaster for the purposes of the Disaster Relief Act of 1974 (Public Law 93-288) 42 USC 5121 note. Concurrence of the Secretary is not required."

BUS WIDTHS

Sec. 120. Section 127 of title 23, United States Code is amended by adding at the end thereof the following new sentence: "Notwithstanding any limitation relating to vehicle widths contained in this section, a State may permit any bus having a width of 102 inches or less to operate on any lane of 12 feet or more in width on the Interstate System."

FERRY OPERATIONS

Sec. 121. The first sentence of paragraph (5) of subsection (g) of section 129 of title 23, United States Code, is amended by inserting after "Hawaii" the following: "and the islands which comprise the Commonwealth of Puerto Rico". The second sentence of such paragraph (5) is amended by inserting after "Hawaii" the following: "and operations between the islands which comprise the Commonwealth of Puerto Rico."

CONTROL OF OUTDOOR ADVERTISING

Sec. 122. (a) Subsection (f) of section 131 of title 23, United States Code, is amended by inserting the following after the first sentence: "The Secretary may also, in consultation with the States, provide within the rights-of-way of the primary system for areas in which signs, displays, and devices giving specific information in the interest of the traveling public may be erected and maintained".

(b) Section 131 of title 23, United States Code, is amended by adding at the end thereof the following new subsections:

"(o) The Secretary may approve the request of a State to permit retention in specific areas defined by such State of directional signs, displays, and devices lawfully erected under State law in force at the time of their erection which do not conform to the requirements of subsection (c), where such signs, displays, and devices are in existence on the date of enactment of this subsection and where the State demonstrates that such signs, displays, and devices (1) provide directional information about goods and services in the interest of the traveling public, and (2) are such that removal would work a substantial economic hardship in such defined area.

"(p) In the case of any sign, display, or device required to be removed under this section prior to the date of enactment of the Federal-Aid Highway Act of 1974, which sign, display, or device was after its removal lawfully relocated and which as a result of the amendments made to this section by such Act is required to be removed, the United States shall pay 100 percent of the just compensation for such removal (including all relocation costs).

"(q) (1) During the implementation of State laws enacted to comply with this section, the Secretary shall encourage and assist the States to develop sign controls and programs which will assure that necessary directional information about facilities providing goods and services in the interest of the traveling public will continue to be available to motorists. To this end the Secretary shall restudy and revise as appropriate existing standards for directional signs authorized under subsections 131(c) (1) and 131(f) to develop signs which are functional and esthetically compatible with their surroundings. He shall employ the resources of other Federal departments and agencies, including the National Endowment for the Arts, and employ maximum participation of private industry in the development of standards and systems of signs developed for those purposes.

"(2) Among other things the Secretary shall encourage States to adopt programs to assure that removal of signs providing necessary directional information, which also were providing directional information on June 1, 1972, about facilities in the interest of the traveling public, be deferred until all other nonconforming signs are removed."

(c) Section 131(i) of title 23, United States Code, is amended to read as follows:

"(i) In order to provide information in the specific interest of the traveling public, the State highway departments are authorized to maintain maps and to permit information directories and advertising pamphlets to be made available at safety rest areas. Subject to the approval of the Secretary, a State may also establish information centers at safety rest areas and other travel information systems within the rights-of-way for the purpose of informing the public of places of interest within the State and providing such other information as a State may consider desirable. The Federal share of the cost of establishing such an information center or travel information system shall be that which is provided in section 120 for a highway project on that Federal-aid system to be served by such center or system."

TRAFFIC OPERATIONS IMPROVEMENT PROGRAMS

Sec. 123. (a) Section 135 of title 23, United States Code, is amended to read as follows:

"Sec. 135. Traffic operations improvements programs.

"(a) The Congress hereby finds and declares it to be in the national interest that each State shall have a continuing program designed to reduce traffic congestion and facilitate the flow of traffic.

"(b) The Secretary may approve under this section any project for improvements on any public road which project will directly facilitate and control traffic flow on any of the Federal-aid systems."

(b) The analysis of chapter 1 is amended by striking out: "135. Urban area traffic operations improvement programs." and inserting in lieu thereof: "135. Traffic operations improvement programs."

PRESERVATION OF PARKLANDS

Sec. 124. Section 138 of title 23, United States Code, is amended by adding a new sentence at the end thereof to read as follows: "In carrying out the national policy declared in this section the Secretary, in cooperation with the Secretary of the Interior and appropriate State and local officials, is authorized to conduct studies as to the most feasible Federal-aid routes for the movement of motor vehicular traffic through or around national parks so as to best serve the needs of the traveling public while preserving the natural beauty of these areas."

ADDITIONS TO INTERSTATE SYSTEM

Sec. 125. Section 139(b) of title 23, United States Code, is amended by striking "(d)" the two places it appears and inserting in lieu thereof "(e)."

EQUAL EMPLOYMENT OPPORTUNITY

Sec. 126. The second sentence of subsection (b) of section 140, title 23, United States Code, is amended to read as follows: "Whenever apportionments are made under section 104(b) of this title, the Secretary shall deduct such sums as he may deem necessary, not to exceed $2,500,000 for the transition quarter ending September 30, 1976, and not to exceed $10,000,000 per fiscal year, for the administration of this subsection."

PUBLIC TRANSPORTATION

Sec. 127. (a) Section 142(a) (1) of title 23, United States Code, is amended by adding at the end thereof the following new sentence: "If fees are charged for the use of any parking facility constructed under this section, the rate thereof shall not be in excess of that required for maintenance and operation of the facility (including compensation to any person for operating the facility)."

(b) Section 142(e) (3) of title 23, United States Code, is amended by striking out "section." and inserting in lieu thereof "title."

SPECIAL URBAN HIGH DENSITY

Sec. 128. (a) Section 146 of title 23, United States Code, is repealed.

(b) The analysis of chapter 1 of title 23, United States Code, is amended by striking out:

"146. Special urban high density traffic programs." and inserting in lieu thereof:

"146. Repealed."

RURAL BUS DEMONSTRATION

Sec. 129. Section 147(a) of the Federal-Aid Highway Act of 1973, as amended, is amended by adding after the first sentence a new sentence as follows: "Such sums shall remain available for a period of two years after the close of the fiscal year for which such sums are authorized."

PRIORITY PRIMARY

Sec. 130. Section 147(b) of title 23, United States Code, is amended to read as follows:

"(b) The Federal share of any project on a priority primary route shall be that provided in section 120(a) of this title. All provisions of this title applicable to the Federal-aid primary system shall be applicable to the priority primary routes selected under this section."

DEFINING STATE

Sec. 131. Section 152 and section 153 of title 23, United States code, are amended by adding at the end of each such section the following new subsection:

"(f) For the purposes of this section the term 'State' shall have the meaning given it in section 401 of this title."

HIGHWAYS CROSSING FEDERAL PROJECTS

Sec. 132. (a) Chapter I of title 23, United States Code, is amended by adding at the end thereof the following new section:

"Sec. 156. Highways crossing Federal projects

"(a) The Secretary is authorized to construct and to reconstruct any public highway or highway bridge across any Federal public works project, notwithstanding any other provision of law, where there has been a substantial change in the requirements and costs of such highway or bridge since the public works project was authorized, and where such increased costs would work an undue hardship upon any one State. No such highway or bridge shall be constructed or reconstructed under authority of this section until the State shall agree that upon completion of such construction or reconstruction it will accept ownership to such highway or bridge and will thereafter operate and maintain such highway or bridge.

"(b) There is hereby authorized to be appropriated not to exceed $100,000,000 to carry out this section. Amounts authorized by this subsection shall be available for the fiscal year in which appropriated and for two succeeding fiscal years."

(b) The analysis of chapter I of title 23 of the United States Code is amended by adding at the end thereof the following:

"156. Highways crossing Federal projects."

APPORTIONMENTS OR ALLOCATIONS

Sec. 133. Section 202(a) of title 23, United States Code, is amended by striking "On or before January 1 next preceding the commencement" and inserting in lieu thereof "On October 1"

BICYCLE TRANSPORTATION AND PEDESTRIAN WALKWAYS

Sec. 134. Section 217(e) of title 23, United States Code, is amended by striking out "$40,000,000" and inserting in lieu thereof "$45,000,000", and by striking out "$2,000,000" and inserting in lieu thereof "$2,500,000".

SAFER OFF-SYSTEM ROADS

Sec. 135. (a) Section 219 of title 23 of the United States Code, is amended to read as follows:

"Sec. 219. Safer off-system roads.

"(a) The Secretary is authorized to make grants to States for projects for the construction, reconstruction, and improvement of any off-system road, including, but not limited to, the correction of safety hazards, the replacement of bridges, the elimination of high-hazard locations and roadside obstacles.

"(b) On October 1 of each fiscal year the Secretary shall apportion the sums authorized to be appropriated to carry out this section among the several States as follows:

"(1) Two-thirds according to the following formula—,

"(A) one-third in the ratio which the area of each State bears to the total area of all States;

"(B) one-third in the ratio which the population of rural areas of each State bears to the total population of rural areas of all the States; and

"(C) one-third in the ratio in which the off-system road mileage of each State bears to the total off-system road mileage of all the States. Off-system road mileage as used in this subsection shall be determined as of the end of the calendar year preceding the year in which the funds are apportioned and shall be certified to by the Governor of the State and subject to approval by the Secretary.

"(2) One-third in the ratio which the population in urban areas in all the States as shown by the latest Federal census.

"(c) Sums apportioned to a State under this section shall be made available for obligation throughout such State on a fair and equitable basis.

"(d) In any State wherein the State is without legal authority to construct or maintain a project under this section, such State shall enter into a formal agreement for such construction or maintenance with the appropriate local officials of the county or municipality in which such project is located.

"(e) Sums apportioned under this section and programs and projects under this section shall be subject to all of the provisions of chapter 1 of this title applicable to highways on the Federal-aid secondary system except the formula for apportionment, the requirement that these roads be on the Federal-aid system, and those other provisions determined by the Secretary to be inconsistent with this section. The Secretary is not authorized to determine as inconsistent with this section any provision relating to the obligation and availability of funds.

"(f) As used in this section, the terms 'off-system road' means any toll-free road (including bridges), which road is not on any Federal-aid system and which is under the jurisdiction of and maintained by a public authority and open to public travel."

(b) The analysis of chapter 1 of title 23 of the United States Code is amended by striking out "219. Off-system roads." and inserting in lieu thereof the following: "219. Safer off-system roads."

(c) Section 405 of title 23 of the United States Code is hereby repealed.

(d) The analysis of chapter 4 of title 23 of the United States Code is amended by striking out "405. Federal-aid safer roads demonstration program." and inserting in lieu thereof the following: "405. Repealed."

LANDSCAPING AND SCENIC ENHANCEMENT

Sec. 136. (a) Section 319 of title 23, United States Code, is amended to read as follows:

"Sec. 319. Landscaping and scenic enhancement.

"The Secretary may approve as a part of the construction of Federal-aid highways the costs of landscape and roadside development, including acquisition and development of publicly owned and controlled rest and recreation areas and sanitary and other facilities reasonably necessary to accommodate the traveling public, and for acquisition of interests in and improvement of strips of land necessary for the restoration, preservation, and enhancement of scenic beauty adjacent to such highways."

(b) All sums authorized to be appropriated to carry out section 319(b) of title 23, United States Code, as in effect immediately before the date of enactment of this section shall continue to be available for appropriation, obligation, and expenditure in accordance with such section 319(b), notwithstanding the amendment made by the subsection (a) of this section.

BRIDGES ON FEDERAL DAMS

Sec. 137. (a) Section 320(d) of title 23, United States Code, is amended by striking out "$27,761,000" and inserting in lieu thereof "$50,000,000".

(b) Sums appropriated or expended under authority of the increased authorization established by the amendment made by subsection (a) of this section shall be appropriated out of the Highway Trust Fund for the fiscal year ending September 30, 1977, and for subsequent fiscal years.

OVERSEAS HIGHWAY

Sec. 138. Subsection (b) of section 118 of the Federal-Aid Highway Amendments of 1974 (Public Law 93-643) is amended—,

(1) by striking out "1975, and" and inserting in lieu thereof "1975."; and

(2) by striking out "can be obligated." and inserting in lieu thereof "$8,750,000 for the three-month period ending September 30, 1976, $35,000,000 for the fiscal year ending September 30, 1977, and $35,000,000 for the fiscal year ending September 30, 1978, can be obligated."

TECHNICAL AMENDMENTS

Sec. 139. (a) The analysis of chapter I of title 23, United States Code, is amended by striking out "111. Use of an access to rights-of-way—Interstate System." and inserting in lieu thereof the following: "111. Agreements relating to use of and access to rights-of-way—Interstate System."

(b) The analysis of chapter I of title 23, United States Code, is amended by striking out "119. Administration of Federal-aid for highways in Alaska." and inserting in lieu thereof the following: "119. Repealed."

(c) The analysis of chapter I of title 23, United States Code, is amended by striking out 133. Relocation assistance." and inserting in lieu thereof the following:"133. Repealed."

DEMONSTRATION PROJECTS—RAILROAD HIGHWAY CROSSINGS

Sec. 140. (a) Section 163 of the Federal-Aid Highway Act of 1973 (Public Law 93-87) is amended by inserting immediately after subsection (h) the following new subsections:

"(i) The Secretary of Transportation shall carry out a demonstration project in Metairie, Jefferson Parish, Louisiana, for the relocation or grade separation of rail lines whichever he deems most feasible in order to eliminate certain grade level railroad highway crossings.

"(j) The Secretary of Transportation shall enter into such arrangements as may be necessary to carry out a demonstration project in Augusta, Georgia, for the relocation of railroad lines and for the purpose of eliminating highway railroad grade crossings.

"(k) The Secretary of Transportation shall enter into such arrangements as may be necessary to carry out a demonstration project in Pine Bluff, Arkansas, for the relocation of railroad lines for the purpose of eliminating highway railroad grade crossings.

"(1) The Secretary of Transportation shall carry out a demonstration project in Sherman, Texas, for the relocation of rail lines in order to eliminate the ground level railroad crossing at the crossing of the Southern Pacific and Frisco Railroads with Grand Avenue-Roberts Road."

(b) Existing subsections (i), (j), (k), and (l) of section 163 of the Federal-Aid Highway Act of 1973 are relettered as (m), (n), (o), and (p), respectively, including any references to such subsections.

(c) Subsection (m) (as relettered by subsection (b) of this section) of section 163 of the Federal-Aid Highway Act of 1973 is amended by striking out the period at the end thereof and inserting in lieu thereof a comma and the following: "except that in the case of projects authorized by subsections (i), (j), (k), and (l), the Federal share payable on account of such projects shall not exceed 70 percent and the remaining costs of such projects shall be paid by the State or local governments."

(d) Subsection (o) (as relettered by subsection (b) of this section) of section 163 of the Federal-Aid Highway Act of 1973 is amended by striking out "1976, except that" and inserting in lieu thereof the following: "1976, $6,250,000, for the period beginning July 1, 1976, and ending September 30, 1977, and $51,400,00 for the fiscal year ending September 30, 1978, except that not more than."

(e) Paragraph (2) of subsection (a) of section 163 of the Federal-Aid Highway Act of 1973 is amended by striking out "an engineering and feasibility study for."

(f) Section 302 of the National Mass Transportation Assistance Act of 1974 (Public Law 93-503) is amended by striking out "4,000,000, except that" and inserting in lieu thereof "4,000,000, except that not more than"

ACCELERATION OF PROJECTS

Sec. 141. The Secretary of Transportation shall carry out a project to demonstrate the feasibility of reducing the time required from the time of request for project approval through the completion of construction of highway projects in areas that, as a result of recent or imminent change, including but not limited to change in population or traffic flow resulting from the construction of Federal projects, show a need to construct such projects to relieve such areas from the impact of such change. There is authorized to be appropriated out of the Highway Trust Fund to carry out such project not to exceed $5,000,000.

MULTIMODAL CONCEPT

Sec. 142. Section 143 of the Federal-Aid Highway Act of 1973 is amended by inserting "(a)" immediately following "Sec. 143." and by adding the following new subsection at the end thereof:

"(b) The Secretary of Transportation is authorized and directed to study the feasibility of developing a multimodal concept along the route described in paragraph (1) of subsection (a) of this section, which study shall include an analysis of the environmental impact of such multimodal concept. The Secretary shall report to Congress the results of such a study not later than July 1, 1977."

CARPOOL DEMONSTRATION PROJECTS

Sec. 143. Section 3 of the Emergency Highway Energy Conservation Act, as amended (87 Stat. 1047, 88 Stat. 2289), is amended as follows:

(1) Subsection (a) is amended by adding at the end thereof the following: "For the purposes of this section, the term 'carpool' includes a vanpool."

(2) Subsection (c) is amended by inserting after "such measures as" the words "providing carpooling opportunities to the elderly and the handicapped," and by inserting after "opportunities," the words "acquiring vehicles appropriate for carpool use,"

(3) Subsection (d) is amended by striking out "(3) and (6)" from the first sentence, and inserting in lieu thereof "(1) and (6)" and by striking out the second sentence.

USE OF TOLL RECEIPTS FOR HIGHWAY AND RAIL CROSSINGS

Sec. 144. Section 2 of the Act entitled "An Act granting the consent of Congress to the State of California to construct, maintain, and operate a bridge across the Bay of San Francisco from the Rincon Hill district in San Francisco by way of Goat Island to Oakland", approved February 20, 1931, is amended as follows:

(1) Subsection (a) is amended by striking out "heretofore enacted." and inserting in lieu thereof a period.

(2) The first sentence in subsection (b) is amended by striking out "of not to exceed two additional highway crossings and one rail transit crossing across the Bay of San Francisco and their approaches." and inserting in lieu thereof "(1) not to exceed two additional highway crossings and one rail transit crossing across the Bay of San Francisco and their approaches, and (2) any public transportation system in the vicinity of any toll bridge in the San Francisco Bay Area. Such tolls may also be used to pay the cost of constructing new approaches to the Richmond-San Rafael Bridge in the San Francisco Bay Area."

(3) The existing third sentence in subsection (b) which begins "After" is repealed.

EXTENSION OF REPAYMENT

Sec. 145. The first sentence of section 2 of Public Law 94-30 is amended by striking out "before January 1, 1977." and inserting in lieu thereof "January 1, 1979, at a rate of 20 percent by January 1, 1977, 30 percent by January 1, 1978, and 50 percent by January 1, 1979. If a State fails to make any repayment in accordance with the preceding sentence, the entire unpaid balance shall immediately become due and payable."

TRAFFIC CONTROL SIGNALIZATION DEMONSTRATION PROJECTS

Sec. 146. (a) The Secretary of Transportation is authorized to carry out traffic control signalization demonstration projects designed to demonstrate through the use of technology not now in general use the increased capacity of existing highways, the conservation of fuel, the decrease in traffic congestion, the improvement in air and noise quality, the furtherance of highway safety, giving priority to those projects providing coordinated signalization of two or more intersections. Such projects can be carried out on any highway whether on or off a Federal-aid system.

(b) There is authorized to be appropriated to carry out this section of the Highway Trust Fund, not to exceed $40,000,000 for the fiscal year ending September 30, 1978.

(c) Each participating State shall report to the Secretary of Transportation not later than September 30, 1977, and not later than September 30 of each year thereafter, on the progress being made in implementing this section and the effectiveness of the improvements made under it. Each report shall include an analysis and evaluation of the benefits resulting from such projects comparing an adequate time period before and after treatment in order to properly assess the benefits occurring from such traffic control signalization. The Secretary of Transportation shall submit a report to the Congress not later than January 1, 1978, on the progress being made in implementing this section and an evaluation of the benefits resulting therefrom.

ACCESS RAMPS TO PUBLIC BOAT LAUNCHING AREAS

Sec. 147. Funds apportioned to States under subsection (b) (1), (b) (2), and (b) (6) of section 104 of title 23, United States Code, may be used upon the application of the State and the approval of the Secretary of Transportation for construction of access ramps from bridges under construction of which are being reconstructed, replaced, repaired, or otherwise altered on the Federal-aid primary, secondary, or urban system to public boat launching areas adjacent to such bridges. Approval of the Secretary shall be in accordance with guidelines developed jointly by the Secretary of Transportation and the Secretary of the Interior.

DEMONSTRATION PROJECT

Sec. 148. The Secretary of Transportation, acting pursuant to his authority under section 6 of the Urban Mass Transportation Act of 1964, shall conduct a demonstration project in urban mass transportation for design, improvement, modification, and urban deployment of the Automated Guideway Transit system now in operation at the Dallas/Fort Worth Regional Airport. There is authorized to be appropriated to carry out this section $7,000,000 for the fiscal year ending September 30, 1977.

URBAN SYSTEM STUDY

Sec. 149. The Secretary of Transportation is authorized and directed to conduct a study of the various factors involved in the planning, selection, programming, and implementation of Federal-aid urban system routes which shall include but not be limited to the following:

(1) An analysis of the various types of organizations now in being which carry out the planning process required by section 134 of title 23, United States Code. Such analysis shall include but not be limited to the degree of representation of various governmental units within the urbanized area, the organizational structure, size and caliber of staff, authority provided to the organization under State and local law, and relation to State governmental entities.

(2) The status of jurisdiction over roads on the Federal-aid urban system (State, county, city, or other local body having control).

(3) Programming responsibilities under local and State laws with respect to the Federal-aid urban system.

(4) The authority for and capability of local units of government to carry out the necessary steps to process a highway project through and including the plan, specification, and estimate requirement of section 106 of title 23, United States Code, and final construction.

Such study shall be carried out in cooperation with State, county, city, and local organizations which the Secretary deems appropriate. The study shall be submitted to the Congress within six month of enactment of this section.

INTERSTATE FUNDING STUDY

Sec. 150. (a) The Secretary of Transportation is hereby directed to undertake a complete study of the financing of completion of the Interstate Highway System. Such study should identify and analyze optional financing methods including State bonding authority under which the Secretary contracts to reimburse the States for up to 90 percent of the principal and interest on such bonds. The Secretary shall report to the Congress not later than nine months after the date of enactment of this Act the results of the study.

(b) Within one year of the date of enactment of this Act, the Secretary shall submit to the Congress his recommendations regarding the need to provide Federal financial assistance for resurfacing, restoration, and rehabilitation of routes on the Interstate System. In arriving at his recommendations, he shall conduct a full and complete study in cooperation and in consultation with the States of alternative means of assuring that the high level of transportation service provided by the Interstate System is maintained. The results of the study shall accompany the Secretary's recommendations. The study shall include an estimate of the cost of implementing any recommended programs as well as an analysis of alternative methods of apportioning such Federal assistance among the States.

ALASKAN ROADS STUDY

Sec. 151. (a) The Secretary of Transportation is authorized to undertake an investigation and study to determine the cost of, and the responsibility for, repairing the damage to Alaska highways that has been or will be caused by heavy truck traffic during construction of the trans-Alaska pipeline and to restore them to proper standards when construction is complete. The Secretary of Transportation shall report his initial findings to the Congress on or before September 30, 1976, and his final conclusions on rebuilding costs no later than three months after completion of pipeline construction.

(b) There is hereby authorized to be appropriated, out of any money in the Treasury not otherwise appropriated, to be available until expended, the sum of $200,000 for the purpose of making the study authorized by subsection (a) of this section.

GLENWOOD CANYON HIGHWAY CONSTRUCTION

Sec. 152. Notwithstanding section 109(b) of title 23 of the United States Code, the Secretary of Transportation is authorized, upon application of the Governor of the State, to approve construction of that section or portions thereof of Interstate Route 70 from a point three miles east of Dotsero, Colorado, westerly to No-Name Interchange, approximately 2.3 miles east of Glenwood Springs, Colorado, approximately 17.5 miles in length, to provide for variations from the number of lanes and other requirements of said section 109(b) in accordance with geometric and construction standards whether or not in conformance with said section 109(b) which the Secretary determines are necessary for the safety of the traveling public, for the protection of the environment, and for preservation of the scenic and historic values of the Glenwood Canyon. The Secretary shall not approve any project for construction under this section unless he shall first have determined that such variations will not result in creation of safety hazards and that there is no reasonable alternative to such project.

STUDY OF HIGHWAY NEEDS TO SOLVE ENERGY PROBLEMS

Sec. 153. (a) The Secretary of Transportation shall make an investigation and study for the purpose of determining the need for special Federal assistance in the construction or reconstruction of highways on the Federal-aid system necessary for the transportation of coal or other uses in order to promote the solution of the Nation's energy problems. Such study shall include appropriate consultations with the Secretary of the Interior, the Administrator of the Federal Energy Administration, and other appropriate Federal and State officials.

(b) The Secretary shall report the results of such investigation and study together with his recommendations, to the Congress not later than one year after the date of enactment of this Act.

(c) In order to carry out the study, the Secretary is authorized to use such funds as are available to him for such purposes under section 104(a) of title 23, United States Code.

ESTABLISHMENT OF COMMISSION

Sec. 154. (a) (1) There is hereby established a Commission to be known as the National Transportation Policy Study Commission, hereinafter referred to as the "Commission". (2) The Commission shall make a full and complete investigation and study of the transportation needs and of resources, requirements, and policies of the United States to meet such expected needs. It shall take into consideration all reports on National Transportation Policy which have been submitted to the Congress including but not limited to the National Transportation Reports of 1972 and 1974. It shall evaluate the relative merits of all modes of transportation in meeting our transportation needs. Based on such study, it shall recommend those policies which are most likely to insure that adequate transportation systems are in place which will meet the needs for safe and efficient movement of goods and people.

(b) Such Commission shall be comprised of 19 members as follows:

(A) Six members appointed by the President of the Senate from the membership of the Committee on Public Works, Committee on Commerce, and Committee on Banking, Housing and Urban Affairs of the United States Senate;

(B) five members appointed by the Speaker of the House of Representatives from the membership of the Committee on Public Works and Transportation and one member appointed by the Speaker from the membership of the Committee on Interstate and Foreign Commerce; and

(C) seven members of the public appointed by the President.

(c) The Commission shall not later than December 31, 1978 submit to the President and the Congress its final report including its findings and recommendations. The Commission shall cease to exist six months after submission of such report. All records and papers of the Commission shall thereupon be delivered to the Administrator of General Services for deposit in the Archives of the United States.

(d) Such report shall include the Commission's findings and recommendations with respect to ,

(A) the Nation's transportation needs both national and regional, through the year 2000;

(B) the ability of our current transportation systems to meet the projected needs;

(C) the proper mix of highway, rail, waterway, pipeline, and air transportation systems to meet anticipated needs;

(D) the energy requirements and availability of energy to meet anticipated needs;

(E) the existing policies and programs of the Federal government which affect the development of our national transportation systems; and

(F) the new policies required to develop balanced national transportation systems which meet projected needs.

(e) (1) The Chairman of the Commission, who shall be elected by the Commission from among its members, shall request the head of each Federal department or agency which has an interest in or a responsibility with respect to a national transportation policy to appoint, and the head of such department or agency shall appoint a liaison officer who shall work closely with the Commission and its staff in matters pertaining to this section. Such departments and agencies shall include, but not be limited to, the Department of Transportation, the Federal Highway Administration, the Federal Railroad Administration, the Urban Mass Transportation Administration, the Federal Aviation Administration, the Interstate Commerce Commission, the Civil Aeronautics Board, and the U.S. Army Corps of Engineers.

(2) In carrying out its duties the Commission shall seek the advice of various groups interested in national transportation policy including, but not limited to, State and local governments, public and private organizations working in the fields of transportation and safety, industry, education, and labor.

(f) (1) The Commission or, on authorization of the Commission, any Committee of two or more members may, for the purpose of carrying out the provisions of this section, hold such hearings and sit and act at such times and places as the Commission or such authorized committee may deem advisable.

(2) The Commission is authorized to secure from any department, agency, or individual instrumentality of the executive branch of the Government any information it deems necessary to carry out the functions under this section and each department, agency, and instrumentality is authorized and directed to furnish such information to the Commission upon request made by the Chairman.

(g) (1) Members of Congress who are members of the Commission shall serve without compensation in addition to that received for their services as Members of Congress; but they shall be reimbursed for travel, per diem in accordance with the Rules of the House of Representatives or subsistence, and other necessary expenses incurred by them in the performance of the duties vested in the Commission.

(2) Members of the Commission, except Members of Congress, shall each receive compensation at a rate not in excess of the maximum rate of pay for GS-18, as provided in the General Schedule under section 5332 of title 5, United States Code, and shall be entitled to reimbursement for travel expenses, per diem in accordance with the Rules of the House of Representatives or subsistence and other necessary expenses incurred by them in performance of duties while serving as a Commission member.

(h) (1) The Commission is authorized to appoint and fix the compensation of a staff director, and such additional personnel as may be necessary to enable it to carry out its functions. The Director and personnel may be appointed without regard to the provisions of title 5, United States Code, covering appointments in the competitive service, and may be paid without regard to the provisions of chapter 51 and subchapter III of chapter 53 of such title relating to classification and General Schedule pay rates. Any Federal employees subject to the civil service laws and regulations who may be employed by the Commission shall retain civil service status without interruption or loss of status or privilege. In no event shall any employee other than the staff director receive as compensation an amount in excess of the maximum rate for GS-18 of the General Schedule under section 5332 of title 5, United States Code. In addition, the Commission is authorized to obtain the services of experts and consultants in accordance with section 3109 of title 5, United States Code, but at rates not to exceed the maximum rate of pay for grade GS-18, as provided in the General Schedule under section 5332 of title 5, United States Code.

(2) The staff director shall be compensated at a Level 2 of the Executive Schedule in subchapter II of chapter 53 of title 5, United States Code.

(i) The Commission is authorized to enter into contracts or agreements for studies and surveys with public and private organizations and, if necessary, to transfer funds to Federal agencies from sums appropriated pursuant to this section to carry out such of its duties as the Commission determines can best be carried out in that manner.

(j) Any vacancy which may occur on the Commission shall not affect its powers or functions but shall be filled in the same manner in which the original appointment was made.

(k) There are hereby authorized to be appropriated not to exceed $15,000,000 to carry out this section. Funds appropriated under this section shall be available to the Commission until expended.

LIMITATIONS

Sec. 155. To the extent that any section of this Act provides new or increased authority to enter into contracts under which outlays will be made from funds other than the Highway Trust Fund, such new or increased authority shall be effective for any fiscal year only in such amounts as are provided in appropriations Acts.

TITLE II

SHORT TITLE

Sec. 201. This title may be cited as the "Highway Safety Act of 1976". // 23 USC 401 note//.

HIGHWAY SAFETY

Sec. 202. The following sums are hereby authorized to be appropriated:

(1) For carrying out section 402 of title 23, United States Code (relating to highway safety programs), by the National Highway Traffic Safety Administration, out of the Highway Trust Fund, $122,000,000 for the fiscal year ending September 30, 1977, and $137,000,000 for the fiscal year ending September 30, 1978.

(2) For carrying out section 403 of title 23, United States Code (relating to highway safety research and development), by the National Highway Traffic Safety Administration, out of the Highway Trust Fund, $10,000,000 for the three-month period ending September 30, 1976, $40,000,000 for the fiscal year ending September 30, 1977, and $50,000,000 for the fiscal year ending September 30, 1978.

(3) For carrying out section 402 of title 23, United States Code (relating to highway safety programs), by the Federal Highway Administration, out of the Highway Trust Fund, $25,000,000 for the fiscal year ending September 30, 1977, and $25,000,000 for the fiscal year ending September 30, 1978.

(4) For carrying out sections 307(a) and 403 of title 23, United States Code (relating to highway safety research and development), by the Federal Highway Administration, out of the Highway Trust Fund, $2,500,000 for the three-month period ending September 30, 1976, $10,000,000 for the fiscal year ending September 30, 1977, and $10,000,000 for the fiscal year ending September 30, 1978.

(5) For bridge reconstruction and replacement under section 144 of title 23, United States Code, out of the Highway Trust Fund, $180,000,000 for the fiscal year ending September 30, 1977, and $180,000,000 for the fiscal year ending September 30, 1978.

(6) For carrying out section 151 of title 23, United States Code (relating to pavement marking), out of the Highway Trust Fund, $50,000,000 for the fiscal year ending September 30, 1977, and $50,000,000 for the fiscal year ending September 30, 1978.

(7) For projects for high-hazard locations under section 152 of title 23, United States Code, and for the elimination of roadside obstacles under section 153 of title 23, United States Code, out of the Highway Trust Fund, $125,000,000 for the fiscal year ending September 30, 1977, and $125,000,000 for the fiscal year ending September 30, 1978.

(8) For carrying out subsection (j) (2) of section 402 of title 23, United States Code (relating to incentives for the reduction of the rate of traffic fatalities), out of the Highway Trust Fund, $1,875,000 for the three-month period ending September 30, 1976, $7,500,000 for the fiscal year ending September 30, 1977, and $7,500,000 for the fiscal year ending September 30, 1978.

(9) For carrying out subsection (j) (3) of section 402 of title 23, United States Code (relating for incentives for reduction of actual traffic fatalities), out of the Highway Trust Fund, $1,875,000 for the three-month period ending September 30, 1976, $7,500,000 for the fiscal year ending September 30, 1977, and $7,500,000 for the fiscal year ending September 30, 1978.

RAIL-HIGHWAY CROSSINGS

Sec. 203 (a) Subsections (b) and (c) of section 203 of the Highway Safety Act of 1973 (Public Law 93-87) are hereby amended to read as follows:

"(b) (1) In addition to funds which may be otherwise available to carry out section 130 of title 23, United States Code, there is authorized to be appropriated out of the Highway Trust Fund for projects for the elimination of hazards of railway-highway crossings, $25,000,000 for the fiscal year ending June 30, 1974, $75,000,000 for the fiscal year ending June 30, 1975, $75,000,000 for the fiscal year ending June 30, 1976, $125,000,000 for the fiscal year ending

September 30, 1977, and $125,000,000 for the fiscal year ending September 30, 1978. At least half of the funds authorized and expended under this section shall be available for the installation of protective devices at railway-highway crossings. Sums authorized to be appropriated by this subsection shall be available for obligation in the same manner as funds apportioned under chapter 1 of title 23, United States Code.

"(2) Funds authorized by the subsection shall be available solely for expenditure for projects on any Federal-aid system (other than the Interstate System).

"(c) There is authorized to be appropriated for projects for the elimination of hazards of railway-highway crossings on roads other than those on any Federal-aid system $18,750,000 for the three-month period ending September 30, 1976, $75,000,000 for the fiscal year ending September 30, 1977, and $75,000,000 for the fiscal year ending September 30, 1978. Sums apportioned under this section for projects under this subsection shall be subject to all of the provisions of chapter 1 of title 23, United States Code, applicable to highways on the Federal-aid system, except the formula for apportionment, the requirement that these roads be on the Federal-aid system, and those other provisions determined by the Secretary to be inconsistent with this section."

(b) Subsection (d) of section 203 of the Highway Safety Act of 1973 is amended by adding immediately before the first sentence thereof the following new sentence: "50 percent of the funds made available in accordance with subsection (b) shall be apportioned to the States in the same manner as sums authorized to be appropriated under subsection (a) (1) of section 104 of the Federal-aid Highway Act of 1973 and 50 percent of the funds made available in accordance with subsection (b) shall be apportioned to the States in the same manner as sums authorized to be appropriated under subsection (a) (2) of section 104 of the Federal-aid Highway Act of 1973."

INCENTIVE SAFETY GRANTS

Sec. 204. Subsection (j) (3) of section 402 of title 23, United States Code, is hereby amended to read as follows:

"(3) In addition to other grants authorized by this section, the Secretary may make additional incentive grants to those States which have significantly reduced the actual number of traffic fatalities during the calendar year immediately preceding the fiscal year for which such incentive funds are authorized compared to the average of the actual number of traffic fatalities for the four calendar year period preceding such calendar year. Such incentive grants shall be made in accordance with criteria which the Secretary shall establish and publish. Such grants may only be used by recipient States to further the purposes of this chapter. Such grants shall be in addition to other funds authorized by this section.

"(4) No State shall receive from funds authorized for any fiscal year or period by this subsection incentive grants under paragraph (1) of this subsection which exceed an amount equal to 25 percent of the amount apportioned to such State under this section for such fiscal year or period. No State shall receive from funds authorized for any fiscal year or period by this subsection incentive awards under paragraph (2) of this subsection which exceed an amount equal to 25 percent of the amount apportioned to such State under this section for such fiscal year or period. No State shall receive from funds authorized for any fiscal year or period by this subsection incentive awards under paragraph (3) of this subsection which exceed an amount equal to 25 percent of the amount apportioned to such State under this section for such fiscal year or period.

"(5) Notwithstanding subsection (c) of this section, no part of the sums authorized by this subsection shall be apportioned as provided in such subsection. Sums authorized by this subsection shall be available for obligation in the same manner and to the same extent as if such funds were apportioned under subsection (c) of this section."

SCHOOL BUS DRIVER TRAINING

Sec. 205. The second subsection (b) of section 406 of title 23, United States Code (relating to authorizations), is relettered as subsection (c), including all references thereto, and the second sentence of such relettered subsection (c) is amended to read as follows: "Not less than $7,000,000 of the sums authorized to carry out section 402 of this title for each of the fiscal years 1977 and 1978 shall be obligated to carry out this section. All sums authorized to carry out this section shall be apportioned among the States in accordance with the formula established under subsection (c) of section 402 of this title, and shall be available for obligation in the same manner and to the same extent as if such funds were apportioned under such subsection (c)."

TRANSFERABILITY

Sec. 206. (a) The first sentence of subsection (g) of section 104 of title 23, United States Code, is amended by striking out "30 percent" and inserting in lieu thereof "40 percent".

(b) The second sentence of such subsection (g) is amended to read as follows: "The Secretary may approve the transfer of 100 percent of the apportionment under one such section to the apportionment under any other of such sections if such transfer is requested by the State highway department, and is approved by the Secretary as being in the public interest, if he has received satisfactory assurances from such State highway department that the purposes of the program from which such funds are to be transferred have been met."

(c) Subsection (g) of section 104 of title 23, United States Code, is further amended by adding at the end thereof the following new sentences: "All or any part of the funds apportioned in any fiscal year to a State in accordance with section 203(d) of the Highway Safety Act of 1973 from funds authorized in section 203(c) of such Act, may be transferred from that apportionment to the apportionment made under section 219 of this title if such transfer is requested by the State highway department and is approved by the Secretary after he has received satisfactory assurances from such department that the purposes of such section 203 have been met. Nothing in this subsection authorizes the transfer of any amount apportioned from the Highway Trust Fund to any apportionment the funds for which were not from the Highway Trust Fund, and nothing in this subsection authorizes the transfer of any amount apportioned from funds not from the Highway Trust Fund to any apportionment the funds for which were from the Highway Trust Fund."

PAVEMENT MARKING PROGRAM

Sec. 207. (a) Subsection (c) of section 151 of title 23, United States Code, is amended by striking out "and which are" and all that follows down through and including "Federal-aid system."

(b) Subsection (g) of such section 151 is amended by adding at the end thereof the following: "No State shall submit any such report to the Secretary for any year after the second year following completion of the pavement marking program in that State, and the Secretary shall not submit any such report to Congress after the first year following the completion of the pavement marking program in all States."

HIGHWAY SAFETY PROGRAMS

Sec. 208. (a) The last three sentences of subsection (c) of section 402 of title 23, United States Code, are amended to read as follows: "For the purpose of the seventh sentence of this subsection, a highway safety program approved by the Secretary shall not include any requirement that a State implement such a program by adopting or enforcing any law, rule, or regulation based on a standard promulgated by the Secretary under this section requiring any motorcycle operator eighteen years of age or older or passenger eighteen years of age or older to wear a safety helmet when operating or riding a motorcycle on the streets and highways of the State. Implementation of a highway safety program under this section shall not be construed to require the Secretary to require compliance with every uniform standard, or with every element of every uniform standard, in every State."

(b) The Secretary of Transportation shall, in cooperation with the States, conduct an evaluation of the adequacy and appropriateness of all uniform safety standards established under section 402 of title 23 of the United States Code which are in effect on the date of enactment of this Act. The Secretary shall report his findings, together with his recommendations, including but not limited to, the need for revision or consolidation of existing standards and the establishment of new standards, to Congress on or before July 1, 1977. Until such report is submitted, the Secretary shall not, pursuant to subsection (c) of section 402 of title 23, United States Code, withhold any apportionment or any funds apportioned to any State because such State is failing to implement a highway safety program approved by the Secretary in accordance with such section 402.

NATIONAL HIGHWAY SAFETY ADVISORY COMMITTEE

Sec. 209. Section 404(a) (1) of title 23, United States Code, is amended by deleting "who shall be Chairman," from the first sentence thereof, and by adding immediately after such first sentence the following: "The Secretary shall select the Chairman of the Committee from among the Committee members."

STEERING AXLE STUDY

Sec. 210. The Secretary of Transportation is directed to conduct an investigation into the relationship between the gross load on front steering axles of truck tractors and the safety of operation of vehicle combinations of which such truck tractors are a part. Such investigation shall be conducted in cooperation with representatives of (A) manufacturers of truck tractors and related equipment, (B) labor, and (C) users of such equipment. The Secretary shall report the results of such study to the Congress not later than July 1, 1977.

SAFETY PROGRAM APPORTIONMENT

Sec. 211. The sixth sentence of section 402(c) of title 23, United States Code, is amended by deleting the period at the end and adding the following: " except that the apportionments to the Virgin Islands, Guam, and American Samoa shall not be less than one-third of 1 percent of the total apportionment."

PENALTY

Sec. 212. Section 402(c) of title 23, United States Code, is amended by adding at the end thereof the following: "Funds apportioned under this section to any State, that does not have a highway safety program approved by the Secretary or that is not implementing an approved program, shall be reduced by amounts equal to not less than 50 percent of the amounts that would otherwise be apportioned to the State under this section, until such time as the Secretary approves such program or determines that the State is implementing an approved program, as appropriate. The Secretary shall consider the gravity of the State's failure to have or implement an approved program in determining the amount of the reduction. The Secretary shall promptly apportion to the State the funds withheld from its apportionment if he approves the State's highway safety program or determines that the State has begun implementing an approved program, as appropriate, prior to the end of the fiscal year for which the funds were withheld. If the Secretary determines that the State did not correct its failure within such period, the Secretary shall reapportion the withheld funds to the other States in accordance with the formula specified in this subsection not later than 30 days after such determination."

LIMITATIONS

Sec. 213. To the extent that any section of this title provides new or increased authority to enter into contracts under which outlays will be made from funds other than the Highway Trust Fund, such new or increased authority shall be effective for any fiscal year only in such amounts as are provided in appropriations Acts.

TITLE III

EXTENSION OF HIGHWAY TRUST FUND AND CERTAIN RELATED PROVISIONS

SEC. 301. HIGHWAY TRUST FUND.

(a) Subsections (c) and (f) of section 209 of the Highway Revenue Act of 1956 (relating to the Highway Trust Fund; 23 U.S.C. 120 note) are amended—,

(1) by striking out "1977" each place it appears and inserting in lieu thereof "1979"; and

(2) by striking out "1978" each place it appears and inserting in lieu thereof "1980".

(b) Subsection (e) (1) of section 209 of such Act is amended by striking out "June 30, 1978" and inserting in lieu thereof "September 30, 1980".

SEC. 302. TRANSFER FROM LAND AND WATER CONSERVATION FUND.

Subsection (b) of section 201 of the Land and Water Conservation Fund Act of 1965 (16 U.S.C. 4601-11) is amended—,

(1) by striking out "1977" and inserting in lieu thereof "1979"; and

(2) by striking out "1978" each place it appears and inserting in lieu thereof "1980".

SEC. 303. POSTPONEMENT OF CERTAIN EXCISE TAX REDUCTIONS.

(a) The following provisions of the Internal Revenue Code of 1954 are amended by striking out "1977" each place it appears and inserting in lieu thereof "1979":

(1) Section 4041(c) (3) (relating to rate of tax on fuel for noncommercial aviation).

(2) Section 4041(e) (relating to rate reduction).

(3) Section 4061(a) (1) (relating to imposition of tax on trucks, buses, etc.).

(4) Section 4061(b) (1) (relating to imposition of tax on parts and accessories).

(5) Section 4071(d) (relating to imposition of tax on tires and tubes).

(6) Section 4081(b) (relating to imposition of tax on gasoline).

(7) Section 4481(a) (relating to imposition of tax on use of highway motor vehicles).

(8) Section 4481(e) (relating to period tax in effect).

(9) Section 4482(c) (4) (defining taxable period).

(10) Section 6156(e) (2) (relating to installment payments of tax on use of highway motor vehicles).

(11) Section 6421(h) (relating to tax on gasoline used for certain nonhighway purposes or by local transit systems).

(b) Section 6412(a) (2) of such Code (relating to floor stocks refunds) is amended—,

(1) by striking out "1977" each place it appears and inserting in lieu thereof "1979"; and

(2) by striking out "1978" each place it appears and inserting in lieu thereof "1980".

LEGISLATIVE HISTORY

HOUSE REPORTS: No. 94-716 (Comm. on Public Works and Transportation) and No. 94-1017 Comm. of Conference).

SENATE REPORTS: No. 94-485 accompanying S. 2711 (Comm. of Public Works) and No. 94-741 (Comm. of Conference).

CONGRESSIONAL RECORD: Vol. 121 (1975): Dec. 11, 12, S. 2711 considered and passed Senate.

Dec. 18, considered and passed House.

Vol. 122 (1976): Jan. 19, considered and passed Senate, amended, in lieu of S. 2711.

Apr. 13, Senate and House agreed to conference report.

Approved May 5, 1976.

PL 94-280

From Public Law 94-280 (HR 8235), 90 Stat. 425, May 5, 1976.

Documents of Authorization—II

CERTIFICATE OF THE SECRETARY OF ENVIRONMENTAL AFFAIRS ON THE FINAL SUPPLEMENTAL ENVIRONMENTAL IMPACT REPORT

PROJECT NAME: CENTRAL ARTERY/THT
PROJECT LOCATION: BOSTON
EOEA NUMBER: 4324
PROJECT PROPONENT: DPW
DATE NOTICED IN MONITOR: NOVEMBER 25, 1990

The Secretary of Environmental Affairs herein issues a statement that the Final Supplemental Environmental Impact Report submitted on the Central Artery/Third Harbor Tunnel project adequately and properly complies with the Massachusetts Environmental Policy Act (G.L., c. 30, s. 61-62H) and with its implementing regulations (301 CMR 11.00).

Introduction

The Central Artery/Third Harbor Tunnel project (CA/THT) is far and away the largest and most expensive public works project ever undertaken by the Commonwealth of Massachusetts. It will have extraordinary and long-lasting impacts on the cities of Boston and Cambridge, the metropolitan area, eastern Massachusetts, and indeed millions of Americans travelling in the northeast corridor. The Project also has extraordinary potential to improve air quality, improve our ability to commute in and out of and through the city (whether by car, train, bus, or rapid transit), improve recreational opportunities in Cambridge, Boston, and the Harbor, and improve the visual character of downtown Boston and the banks of the Charles River. The project, and the environmental benefits it affords, provide a unique opportunity to make substantial improvements in the quality of life for the citizens of the Boston metropolitan area. This opportunity must not be missed.

The state environmental review process for this project has a lengthy history, beginning in January 1982, with the filing of an Environmental Notification Form by the Department of Public Works (DPW) for a project known as the "Third Harbor Tunnel." Exactly five years ago, on January 2, 1986, the then Secretary of Environmental Affairs found the FEIS/R submitted on the project, which by that time included a depressed Central Artery, to be adequate. In the intervening five years, the DPW has proposed a number of changes to this project, most notably a change in the plan for crossing the Charles River. As I noted in my Certificate on the SDEIR/S in August of 1990, the document now under review is substantially updated from the 1986 FEIS/R, both in terms of technical information presented and refinements and changes to the proposed project. Moreover, and not surprisingly for a project of this size and complexity, many of the issues identified in previous comments and Secretarial Certificates are still the subject of debate, despite the considerable progress that has been made.

I am now faced with certifying the adequacy or inadequacy of the information contained in the environmental document before me. As often happens with large, complex projects, several aspects of the project have become extremely controversial in recent weeks. The attention focused on these issues should not be allowed to detract from the essential purpose of this project: to fix a woefully inadequate highway—a highway that is plagued by hours of congestion resulting in the discharge of tons of toxic pollutants each day to the air we breathe. The current road is the most dangerous and congested stretch of interstate highway in the country. It contributes significantly to Massachusetts' inability to meet state and federal air quality standards. Millions of commuters can attest that it is woefully inadequate for the nearly 200,000 cars it carries each day. We are clearly destined to be choked in further gridlock and bad air if improvements are not made now. The twin benefits of this project—to ease the burden on commuters and improve air quality—are more than laudable; they are essential for the environmental and economic well being of the region, and for the safety of anyone travelling on the current roadway.

As I will discuss in more detail below, the CA/THT project cannot accomplish the goals of cleaner air and an improved transportation system on its own. Fortunately, the transportation agencies of this Commonwealth have made tremendous strides in recent years towards the improvement of mass transit. The MBTA rapid transit and commuter rail systems, though not without problems, nevertheless provide reliable, efficient and reasonably priced service to hundreds of thousands of commuters each day. The Commonwealth's clear support for mass transit is illustrated by the leadership of our state's senior transportation officials over most of the last 20 years. They have brought about the defeat of the proposed Southwest Corridor highway in favor of a relocated, upgraded Orange Line, the extension of the Red Line to Alewife, and additional capital investments in equipment and station improvements that outpace those made anywhere else in America.

But transportation and environmental leaders alike recognize that more progress must be made. This project affords a tremendous opportunity to make this progress by coupling the necessary highway improvements with significant commitments to even better, more accessible mass transit. This must include an expanded mass transit system and fare policies that make the system an attractive alternative to the automobile. The project must also include a comprehensive High Occupancy Vehicle (HOV) system built into the road design that is accompanied by an ambitious program to encourage HOV use, and aggressive strategies, such as parking freezes, that will discourage the introduction of automobiles into Cambridge, Boston and Logan Airport.

The benefits of this project go beyond highway improvements by offering enormous potential to improve the visual character and recreational opportunities of the communities through which this project will be built. The 27 acres created in downtown Boston by the depression of the Central Artery that will reconnect the Harbor to its City, the City to its boulevards and its people to both, is an urban dream come true. This new land creates an unprecedented opportunity to improve the quality of life in Boston. The ugly green wall of highway that cuts like a scar through the heart of our downtown will be replaced by acres of open space and parkland. New housing and recreational opportunities will be created for the people of the North End and Chinatown. A Winter Garden of potentially national cultural importance will flower in the heart of the downtown. And a tree-lined boulevard divided by wide swaths of parks and open space—a Commonwealth Avenue of tomorrow—will replace what is now a steel shrouded strip of macadam—a no man's land of noise, dirt and concrete—with a gracious green border between the bustle of downtown and the waters of Boston Harbor. Spectacle Island—today a leaking environmental nightmare off the shores of South Boston—will be transformed into a spectacular harbor island park and bring additional and much needed recreational opportunities to city dwellers. The new crossing of the Charles River will lead to vastly improved public access to the riverbanks and the long hoped for extension of the Esplanade, making it possible to walk or bike continuously from Brighton to the Back Bay, and around the waterfront to the shores of Charlestown and Dorchester.

The construction period may feel like open heart surgery but it will work like plastic surgery and when it is completed the City of Boston will have a bright new look. So too will this project brighten the prospects for our economy. It is because there are so many good and desirable aspects to this project that the project as a whole enjoys nearly unanimous support from state and local agencies and elected officials, neighborhood groups, environmental organizations, business and trade interests, and private citizens. This project must move forward, and this Certificate should be seen as a message that the Commonwealth's environmental officials are supportive of it.

The Final SEIR before me has discussed the project in exhaustive detail. There are still issues to be analyzed, discussed, debated, and decided. The FSEIR has not answered every comment to the complete satisfaction of the commenter. No environmental review document on a project of this size can address all of the concerns that will arise during the design, permitting, and construction process. The FSEIR has provided a vast amount of information about the project, its potential environmental impacts, and the DPW's plans for mitigating environmental harm. It is therefore time to move on to the next phase of the project.

Comments on the FSEIR

As I have already indicated, this is a vast and complex project that raises a myriad of substantive issues. The public and agency comments submitted on this report have been of exceptionally high quality and have been of great assistance to me and my staff in our review of the document. This Certificate addresses the most important of these hundreds of issues. But given the complexity of this project, it is not possible to address directly in this Certificate all the issues on which I have received comment. However, my silence on any specific issue of concern should not be taken as a judgment that the issue is not worthy of further discussion. I expect the DPW to work diligently with all interested parties throughout the design, permitting and construction phases of this project to respond to and address, to the greatest extent possible, all legitimate concerns.

What follows are my comments on the most critical of the issues before me.

The Charles River Crossing

The current I-93/Route 1 interchange, where seven lanes now must squeeze into three, is the most dangerous bottleneck in the United States Interstate Highway system, with an accident rate more than triple the state average. Nearly 200,000 cars travel through this interchange every day, and those numbers are expected to increase. Fourteen distinct traffic movements must be accommodated in a very confined area. There is an obvious need to address this problem; indeed this is one of the fundamental purposes of the entire project.

In attempting to address the problem, however, transportation engineers are faced with significant constraints. The area in which these fourteen traffic movements must be accommodated is surrounded by the Boston Garden, the Charles River dam, numerous industrial and commercial properties, an above ground commuter rail and a below ground transit line, historic buildings, established neighborhoods such as Charlestown and East Cambridge, existing and potential future parkland, and two rivers, one of which is active with recreational and commercial boats. These constraints pose extraordinary challenges for transportation engineers and designers.

Any solution to this snarl of roads, waterways, parks, buildings, neighborhoods, and railroad tracks will have significant impacts. Yet, unless the traffic is to be directed west of Hopkinton, the headwaters of the Charles River, any solution will necessitate crossing the river. The challenge we face as a community is to arrive at a solution that provides the greatest possible protection to the environment while meeting the transportation and economic needs of the region. The best solution will have to involve some compromise on all fronts. We cannot solve our transportation problems perfectly if the price to the environment is too great. And similarly, the legitimate need to correct very real transportation problems may mean the loss of some potential parkland or the alteration of some natural resource.

It is the responsibility of the DPW to determine how and where this road should be built. That agency has put a tremendous amount of effort into trying to make that determination. And they have chosen, after review of some 31 alternatives, what they consider to be the solution that best balances the transportation and environmental values that must be considered. It is DPW's obligation to provide, in an environmental impact report, sufficient information to analyze the various options. It is my responsibility to ensure that the report provides adequate information for the public and the permitting agencies to understand how the project will affect the environment; whether the project proponent has adequately studied reasonable alternatives to the project; and what the project proponent intends to do to minimize the environmental impacts of the preferred alternative.

It is my judgment that, through the FSEIR, DPW has satisfied its obligation under the MEPA statute and regulations to provide an adequate analysis of these issues including their preferred alternative for the crossing. The alternatives analysis in Part IIB of the report has substantially laid out the relative transportation advantages of the alternative crossings, the relative amounts of parkland that will be affected by those crossings, the noise and visual impacts of the alternatives, the construction related difficulties and impacts, and the variety of other impacts this project, and alternatives to it, will have.

However, although I am finding the FSEIR adequate as a matter of law, I strongly recommend that the DPW continue to review modifications and options to Scheme Z, which might reduce the environmental and aesthetic impacts of the crossing, and to embark on a process to better inform the public about the crossing. I do not intend for this to stop or impede this project in any way, nor does anything in this certificate do so, either expressly or by implication. The project may now go forward. But, in my judgment, it would be wise to acknowledge the high degree of public concern about and opposition to certain aspects of the crossing. Some of this opposition is based on self-serving motives, some of it is well motivated, and much of it, especially for the general public, is based on sheer confusion. The approach I am recommending has the potential for creating greater public understanding about Scheme Z, as well as the potential for further improving the project, and it is consistent with DPW's right and, I would maintain, obligation, to continue to try and improve the project as a whole, and the crossing in particular, so as to further minimize environmental harm.

One mechanism for doing so is the Bridge Design Review Committee, which I called for in August. This Committee is charged with the responsibility of identifying improvements in the Scheme Z crossing so as to make it as aesthetically pleasing and environmentally harmless as possible—in short, to develop a beautiful, architecturally significant, and environmentally friendly bridge structure.

The Design Review Committee should contain citizen members from East Cambridge, Charlestown, and the North End, representatives from the Boston Society of Architects, the Charles River Watershed Association, the Conservation Law Foundation, the Boston Preservation Alliance, the Metropolitan District Commission, the Massachusetts Historical Society, the Boston Redevelopment Authority, the Boston Transportation Department, the Boston Parks and Recreation Department, Move Massachusetts 2000, 1000 Friends of Massachusetts, the Building Trades Council, the Artery Business Committee, the Committee for Regional Transportation, three representatives appointed by the Secretary of Environmental Affairs, and other members, especially from the engineering and architectural communities, as deemed appropriate by DPW. The Committee should be appointed and convened no later than February 1, 1991.

In the work of the Bridge Design Review Committee, as well as the DPW's continued review of other options for the river crossing, it is imperative that this project and the substantial public benefits which it represents not be jeopardized through endless review and delay.

Since the Draft SEIR, the DPW has made improvements to Scheme Z. Two hundred feet of river bank have been freed from shadow by the stacking of the outer loop ramps. Elimination of the Traverse Street on-ramp means a decrease in the amount of parkland affected and in the number of lanes that cross the river. The latter improvement has made possible the consideration of several potentially attractive bridge designs. This, in turn, has enabled the DPW to reduce the number of piers needed in the river, greatly improving the impact on navigation and the visual impact.

However, elimination of the Traverse Street ramp appears to have ramifications for a number of intersections in Boston and even Somerville as cars that would have entered the highway by that ramp find alternate routes. The DPW should continue to evaluate this aspect of the project and work toward the goal of reducing the number of bridge lanes without simply shifting the problem to another part of the city. If the Traverse Street ramp is reincorporated into Scheme Z, a Notice of Project Change must be filed.

I commend Stephen Kaiser for a remarkable effort in designing an all tunnel option. The possibility of this highway going under rather than over the river is very appealing to a number of people, including myself. However, in my view, the analysis presented in the FSEIR supports the DPW's position that even though some of the early obstacles have been overcome, this option is impracticable. The great difficulty of constructing a tunnel adjacent to the Orange Line tunnel and an underground highway next to the commuter rail tracks, the enormous traffic problems at Leverett Circle and on the mainline, and the all-tunnel's incompatibility with the already approved and under construction CANA project are major problems. In addition, the excessive cost of the all-tunnel option, which is at least $1 billion more than the DPW's preferred alternative, appears to make it a highly unlikely proposition, especially in light of the fact that federal transportation officials have informed the state that there is virtually no prospect that the Federal government would assume even a modest portion of this increased cost.

Parkland Mitigation

The Charles River crossing will clearly have enormous impacts on the river, the riverbank, and the nearby residential neighborhoods, including impacts on navigation, public access, aesthetics, and existing and planned parklands. Extensive measures are needed to mitigate these impacts. I am extremely pleased to see that the DPW has committed to all of the nearly two dozen park, recreational and cultural improvements I called for in my Certificate on the Draft SEIR in August and intends to begin construction of much of the parkland even in advance of the construction of the road itself.

The numerous specific park improvements are listed at pages IIB-86 through IIB-89 of the FSEIR. They include: restoration and expansion of the Paul Revere Landing Park, construction of a commuter ferry and riverside pedestrian space at Lovejoy Wharf, design and construction of parks in the Nashua Street and GSA parcels, acquisition of land to connect the GSA parcel to parkland created by the CANA project, construction of continuous paths for pedestrians and bicyclists on both sides of the river to link the Esplanade with Boston Harbor as well as pedestrian and bicycle paths across the river immediately upstream of the Science Museum, and creation of outdoor recreational amenities such as a public skating area in the Storrow Lagoons and an Olympic size swimming pool between the Science Museum and the Hatch Shell. In short, with this impressive and unprecedented mitigation package, this project will go a long way toward enhancing the recreational and open space opportunities in the metropolitan parks system.

Although not explicitly listed in the mitigation section of the FSEIR, there is one additional mitigation measure that I direct the DPW to undertake: the restoration and lighting of the Longfellow Bridge, which runs from Kendall Square in Cambridge to Charles Circle in Boston. The impacts of construction at Leverett Circle will lead to greater reliance on the Longfellow Bridge. This reliance will likely necessitate repairs on the Bridge. Those repairs should include cleaning and the installation of decorative lighting. This architecturally and historically significant structure can once again become a bright symbol of our community's commitment to an attractive and pleasing urban landscape. Along with the agreement that we have recently reached with Harvard University to light the Larz Anderson Bridge, this improvement will provide an elegant and gracious link between the two cities that border the lower Charles and further enhance the visual beauty of our capitol and university cities.

The mitigation agreement of November 9, 1990 between the DPW and the MDC represents a firm commitment on behalf of the project proponent to carry out these important parkland mitigation measures. I hereby approve it as part of this project, with the following clarification. The Citizens Advisory Committee described in section I.E. of the MOU shall be established by the Secretary of Environmental Affairs and the seven representatives from the design and environmental community shall be appointed by the Secretary. In addition, the MDC shall submit the results of the master planning process (sections I.B. and I.D. of the MOU) to the Secretary of Environmental Affairs for review through a process to be determined by the Secretary. In addition to the acquisition and development of parkland, the DPW has agreed to other mitigation measures. One of these is the establishment of the Bridge Design Review Committee I called for last August. As I indicated earlier in this Certificate, it is essential that this be done and that this Committee be given the necessary resources to fulfill their very important obligation.

This mitigation package, while generous, is no less than what is required to mitigate the significant impacts of this project.

The Joint Development Process

The sudden appearance of 27 acres of open space in one of the country's oldest and most historically significant cities is an urban planner's and an urban dweller's dream. The Joint Development Process is the process by which the Department of Public Works, in cooperation with the City of Boston, will plan for use of these parcels of real estate that will be created by depression of the existing Central Artery. The Boston Redevelopment Authority (BRA) has assumed the lead role in developing a plan for these parcels, with the active participation of individuals, neighborhood organizations and citywide groups concerned with future development along the corridor. They have done a very able job.

In my Certificate on the DSEIR, I called for considerably more information on the Joint Development Process being undertaken by the DPW and the BRA, including information on ownership of the parcels that will be created, proposed plans for design of the tunnel box and how that will preclude or foster potential development, size of the surface roadways, and the process itself. I also established several principles to guide the development process: maximize the amount of open space, with a goal of 75% of the land to be maintained as publicly accessible open space; limit the surface roadways to no more than three lanes in either direction, if feasible, and eliminate street parking; provide for continuous pedestrian/bicycle paths along the entire length of the corridor; and explore the development of a Winter Garden in the area of Russia Wharf.

My directions in the August Certificate were explicit and ambitious because of the extraordinary opportunity this project presents to alter the look and feel of the City of Boston. This kind of opportunity will never come again. The FSEIR significantly provides the information I requested and provides me with a level of comfort that the Joint Development process will work and work well.

The Joint Development Appendix indicates that the DPW continues to work with the BRA on conceptual plans for the air rights development. The BRA plan, which is referenced in the SFEIR and attached to the BRA's comments, is an excellent plan, and with a very few exceptions adopts the principles I laid out in August. BRA Director Stephen Coyle, DPW Commissioner Jane Garvey and the staff responsible for development of the BRA plan deserve great praise for this initial work.

The BRA plan proposes mixed use and residential development for the Bulfinch Triangle area and restoration of the original urban plan conceived by Charles Bulfinch by rebuilding the original street plan. Residential development planned for the North End is appropriate and needed. Equally necessary residential development is proposed for the parcels located in Chinatown. In my view, these housing proposals are essential to the neighborhoods for which they are proposed and I strongly encourage the Secretary of EOCD to make funding for these parcels a priority. I emphasize that in all aspects of the air rights plan, but particularly those where development is envisioned, close coordination with the MHC, Boston Landmarks Commission, Boston Preservation Alliance, and other historic preservation groups should be maintained. It is essential that the scale and design of these development parcels be consistent with the historic and architectural character of the area.

For the remainder of the corridor, the BRA plan achieves the goal I set of 75% open space. While I am disappointed that the FSEIR did not provide as much information on the proposed design of the tunnel box as I had hoped for, I consider the inclusion of the BRA plan in the FSEIR to be a firm commitment to 75% open space, regardless of how the tunnel box is designed. Further information regarding tunnel box design will be provided in the future environmental review of the land use corridor, as I will outline below. I emphasize, however, that the 75% open space component of this project is an essential mitigation measure and, by virtue of my Certificates of today and last August, must be considered as an established part of the Central Artery project. Any change that would decrease the amount of open space in the land use corridor would require a Notice of Project Change for the Central Artery project and review and approval by the Secretary of Environmental Affairs. In addition, in order to meet this requirement and thus maximize the amount of open space and maintain a continuous open corridor, the BRA plan should be revised to eliminate the hotel now proposed for Parcel 17. I am pleased to see that the BRA plan does not include a proposal for a parking garage in the boulevard corridor and the ENF filed for the corridor should reflect that prohibition. Every effort should be made to eliminate from the BRA plan the parking facility planned for Parcel 20.

The Joint Development Appendix states that the surface artery is planned to be no more than three lanes in either direction, except in a few areas where additional width is needed to accommodate on and off moves from the highway or for turning storage lanes. I agree with this proposal. However, the BRA plan proposes that the surface artery contain three travel lanes and one parking lane in each direction, one more than I called for in August. I recognize that parking lanes may in some cases improve the overall quality of a roadway for pedestrians, by providing a buffer between sidewalk and street and a psychological deterrent to excessive speed. The BRA and the Boston Transportation Department (BTD) have spent considerable time working with neighborhood groups and local businesses in an attempt to design the surface artery in a way that is sensitive to the needs of the people who live, work, and walk there. I commend them for the inclusive and participatory process they have pursued. However, the continuing goal should be to minimize macadam and maximize grass. If the determination is made to dedicate one of the three lanes for parking, I have no objection. However, any proposal to increase the total number of lanes beyond three in each direction, not including turning storage lanes and what is necessary to accommodate on and off moves from the highway, is outside the clear parameters upon which this FSEIR approval is conditioned and thus would have to be filed as a Notice of Project Change for the Central Artery project. As the BRA and DPW continue to refine the land use plan, I also direct that more attention be paid to making the roadway accessible to bicyclists.

I am very pleased at the progress that has been made with respect to development of a Winter Garden by the Massachusetts Horticultural Society. It is fitting that the gift of new open space that the Central Artery project presents should be used in part to provide a first class cultural facility, which will also serve as an indoor oasis of green and living things for the public.

Given the amount of disruption in city streets and neighborhoods that will be caused by the tunneling of the new road and dismantling of the old, it is clear that substantial mitigation is in order. The corridor parcels that are designated as open space must not be left as dusty open lots, but must be fully developed as parks and recreational space by the project proponent, including landscaping, plantings, paths, lights, benches, and sidewalks in accordance with BRA design standards, that will make the area an attractive urban open space. Highway ramps on parcels 6, 12, and 18 must be covered to mitigate their impacts on the surface environment. DPW must use every effort to secure federal highway funding to finance these activities as well as substantial contributions to the capital costs of the Winter Garden. Absent Federal funding, it will nevertheless be the responsibility of the DPW to finance these activities.

As the FSEIR recognizes, state actions and ownership with respect to the parcels contained in the corridor give the DPW a continuing responsibility to ensure proper environmental review by MEPA of any proposed development. The DPW has thus committed to the filing of an Environmental Notification Form for the 27 acre corridor. Given that the City and its agencies are properly the entities most concerned with how their city is developed, it is appropriate that the DPW and BRA be co-proponents in this future MEPA process.

The DPW and BRA should therefore prepare and file on or before June 1, 1991, an ENF for the land use of the entire length of the corridor. This ENF should include, to the extent they are available, details about the proposed design of the tunnel box. The design of the tunnel boxes should be consistent with the open space requirements of this approval and, as modified by this Certificate, not allow for greater development than the BRA plan allows. The ENF shall reflect elimination of the proposed hotel and reduction of the number of lanes to three in each direction on the surface roadway. It shall also include all feasible measures, such as conservation restrictions, to ensure that parcels designated as parkland and open space will stay that way.

In addition, construction in South Boston will create additional areas for development. Plans for these areas, with appropriate attention given to open space and recreational improvements, should also be addressed in a separate ENF.

Air Quality and Transportation

A principal benefit of this project is the improvement of air quality. In my judgment and that of the Department of Environmental Protection, this project, subject to the conditions I will set forth later in this Certificate, will accomplish this purpose. The FSEIR demonstrates that by 2010, emissions of carbon monoxide will decrease by 12%, hydrocarbons by 4%, nitrogen oxides by 2% and particulate matter by 7%. Ambient air toxic concentrations will be reduced by 62%. Additional improvements in auto emission controls and gasoline formulation mandated by the new federal Clean Air Act will contribute to even greater decreases. The sensitivity analysis I called for in my Certificate indicates that even if traffic growth exceeds the DPW's basic forecasts for the proposed project, conditions will still be markedly better than without the roadway improvements. The sensitivity analysis also demonstrates the significant difference that effective parking freezes will have on traffic congestion and air quality.

The Department of Environmental Protection is finalizing a regulation that will govern the six tunnel vent stacks that will be constructed by the DPW to discharge auto emissions from the depressed Artery and Third Harbor Tunnel. The regulation will require the project proponent, prior to construction, to certify to DEP that operation of the roadway and appurtenant vent stack system will not cause or exacerbate a condition of air pollution.

Once the vent stacks are operational, the DPW is required to certify that they are operating in conformance with air quality standards as set forth in the regulation, and must recertify compliance every five years for the life of the project. If discharges from the stacks exceed ambient air quality standards, NO_2 guidelines, or result in increases in total non-methane hydrocarbons (ozone precursors), DPW must mitigate those emissions by implementing additional traffic control measures. The vent stack regulation has the full force and effect of law, and DPW is subject to the full range of DEP administrative sanctions including penalties of up to $25,000 per day for non-compliance.

I have already noted that the project will provide a 62% reduction in air toxics concentrations. Although the DPW's air quality analysis shows that in a few instances localized impacts for four of the vent buildings would be higher under build than no-build conditions, even those localized impacts would be less than what exists today. City-wide and regional pollution will be far less. The recently passed Massachusetts Auto Emissions Act and the Federal Clean Air Act will bring even further improvements in air quality to our region. The proponent has adequately addressed the air pollution impacts of the project generally and the vent stacks specifically, and I am pleased with the outcome.

Several of the commenters have suggested that DPW has failed to consider the option of controlling vent stack emissions with air pollution monitoring and control technology devices. I disagree. The DPW and the DEP agree that given the high volume of air and thus the low concentration of pollutants passing through the ventilation buildings, no technology that would be effective for the purpose is available at the present time. However, I hereby order the DPW, as part of its operating certification and at each five year recertification, to report on the state of the art in air quality monitoring and control technology with an eye towards incorporating the best available control technology if and when it becomes available.

In terms of the location and aesthetics of the vent stacks, the DPW conducted an evaluation and screening analysis which started with one hundred and eighty nine potential sites and ultimately selected six. I accept as a matter of common sense DPW's contention that the siting parameters are defined by the configuration of the Artery and Tunnel and that the specific locations are further limited by a variety of subsurface constraints. I have reviewed the Ventilation Building Site Report and find it adequate.

My primary concern goes not so much to the location of the stacks, but to their visual and aesthetic impacts on the City and its environs. I agree with the spirit of those who note the "discrete and tasteful" ventilation stacks along the southwest corridor, although I recognize that those stacks are constructed for MBTA car emissions, not for automobile traffic.

It is vitally important that, with a full understanding of the technical parameters, the stack buildings be designed and constructed with the utmost sensitivity to their natural and built surroundings and to the communities to which they are proximate. I hereby order the DPW to work closely with the BRA, the DEP, the Mass Historical Commission, the Boston Preservation Alliance and the Boston Society of Architects to ensure that the stacks are drawn and constructed in the most visually unobtrusive and aesthetically pleasing way possible, and I order further exploration of the feasibility, if any, of employing multiple small scale vent stacks as are used along the Southwest Corridor.

As I have already emphasized in this Certificate, improved traffic flow and, therefore, air quality are the twin goals of this project. In my Certificate on the Draft SEIR, I stated that the project must be seen in the larger context of reducing our reliance on the automobile as the primary means of transportation and therefore must be integrated with a panoply of other measures and strategies designed to get people out of their cars and into trains, buses and the subway.

In my August Certificate, I called for the DPW to include in the FSEIR a number of traffic mitigation measures necessary to mitigate the air pollution impacts of the project. Principally these measures were 1) parking management strategies, and parking freezes in particular; 2) an expanded HOV system accompanied by programs to facilitate and encourage the use of HOVs; and 3) clearer commitments to the mass transit improvements that were assumed in the traffic analysis (or a demonstration that the project would work, from a traffic and air quality standpoint even if those improvements were not implemented as expected). The FSEIR, incorporating as it does the Massachusetts Transportation Agenda put forth by the 1000 Friends of Massachusetts and adopted as a policy by the Secretary of EOTC and myself and supported by the Regional Administrator of the EPA, indicates the commitment of the project proponent to these essential mitigation measures.

The FSEIR lists a number of capital improvement projects that are anticipated in the public transportation system. These include extensions of the Blue
and Green Lines and commuter rail system, station improvements for rail and
rapid transit, new bus terminals at South Station and in Lynn, and a significant
increase in the number of parking spaces at rapid transit and commuter rail
stations. The Transportation Agenda represents the commitment of EOTC to
implement these improvements. Moreover, both the record of the Massachusetts Bay Transit Authority (MBTA) for improving and expanding the region's
public transportation system and the City of Boston's commitment to mass transit are excellent.

The HOV system presented in the FSEIR expands upon that contained in the
DSEIR. The Transportation Agenda addresses some kind of HOV support system that I called for, in terms of financing incentives, employee education and
the like. I understand that DPW has committed to expand the HOV system in this
proposed highway project even further. This expansion must be documented and
submitted as a Notice of Project Change. I note the commitment of the DPW to
implementation and enforcement of aggressive parking freezes contained in the
FSEIR. Additional details should be provided in the hearings that DEP will conduct on the State Implementation Plan (SIP) revisions voted in late December
by the Metropolitan Planning Organization (MPO) which include, among other
measures, strengthening of the parking freezes in Cambridge and Boston. The
SIP revisions in no way undermine the interim parking freeze agreement completed in 1990 with the City of Cambridge and establish the parameters for the
City's current efforts to reform its parking management system for inclusion in
the SIP.

In addition, I have received with comments filed by the Conservation Law
Foundation a copy of an agreement between CLF, the Secretary of EOTC, and
the Commissioner of the DPW that expands on many of the mitigation measures
I called for last August. While not a formal part of the FSEIR, the agreement
is nevertheless helpful in that it provides additional detail about the proposed
measures and, more importantly, indicates DPW's commitment to pursue these
transportation measures and agreement that they are necessary mitigation measures for this project.

It is critical that all of these mitigation measures be implemented. My approval
of this FSEIR is based on my finding that the mitigation measures I called for,
as further detailed in the CLF Memorandum of Understanding, are substantially
contained in the SFEIR. I want to be clear that they are absolutely necessary to
achieve greater air quality improvements in metropolitan Boston and are a condition of approval for this project. The commitments memorialized in the MOU,
or their substantial equivalent in terms of emission reductions, are required as
"feasible measures . . . to avoid or minimize" adverse impacts on traffic volumes
and air quality. G.L., c. 30, s. 61. The commitments are needed under the range of
conditions that may prevail in 2010, the analysis year used by the DPW. They are
all the more necessary for later years—and I consider it abundantly appropriate to

require a project that will take nearly a decade to build to have beneficial impacts for more than a decade after completion. In my view, MEPA also requires that adverse impacts be minimized or avoided and benefits maximized over the long term—that is, for considerably more than a decade.

I am pleased that the package of mitigation measures contained in the CLF agreement has already been approved by the Metropolitan Planning Organization and submitted to DEP for adoption into the SIP. The SIP establishes an enforceable strategy for reducing air pollution so as to bring the state into compliance with the federal Clean Air Act. Failure to comply with elements of the SIP can lead to enforcement action by the EPA and/or the withholding of federal highway funds. The MBTA Board of Directors has also voted to adopt this package, thus providing additional assurance that the identified capital improvements will be implemented.

In addition, there will need to be a lengthy public process before some of these measures become a reality. I recognize that in order to accomplish many of these commitments, DPW will require the approval and active support of numerous other government agencies and public bodies. I in no way intend to bind other agencies or require them to take steps to meet these conditions. As I have discussed, the mitigation measures contained in the CLF agreement are not new. For example, the capital improvements identified in Section 5 were all mentioned in the Draft SEIR and endorsed as a requirement for approval by me in August. Strengthened parking freezes are something that state and local environmental and transportation agencies have been working on for years. I recognize the concern expressed by several public agencies and groups about the lack of their involvement in developing the details of the agreement. Frankly, I find it curious that groups whose publicly stated goals are the enhancement of public transportation have criticized what in my view is a first rate mitigation package that will make Boston first among the nation's cities in its commitment to and reliance upon mass transit. However, I agree that an opportunity for public review of the agreement is appropriate.

The Current SIP revision process being conducted by DEP provides that opportunity. The public hearing that is required as part of this process should be noticed by DEP in the *Environmental Monitor.* On or before July 1, 1991, DPW should submit to this office a report describing the results of the public review process and further details on how the specific elements of the MOU will be implemented. If the agency's commitment to the measures in the MOU changes in any material way as a result of the public review process or, indeed at any point subsequent to this Certificate, a Notice of Project Change for the Central Artery Project will have to be filed. Any changes in the proposed mitigation plan, moreover, will have to demonstrate that they will be equivalent to those presented in the FSEIR and contained in the MOU in terms of achieving the goals of improved transportation and air quality. The report should also identify any elements of the MOU, such as the HOV lanes north and south of the City, that were not contained in the FSEIR, so that those can be formally adopted into the environmental record and made a condition of approval for this project as they should be. Semiannual updates on the further planning and progress toward implementation of these strategies should also be submitted to this office, on January 1 and July 1 of each year.

I stress that neither the FSEIR nor the list of commitments contained in the CLF agreement should be considered the limit for transportation related mitigation measures. The DPW should continue to think creatively and aggressively about further measures to reduce automobile use. Possibilities that should be pursued include the development of a comprehensive master plan for all forms of transit, a strategy for maximizing fleet use of alternative fuel vehicles, an efficiency study of the entire MBTA system, and additional dedicated parking for HOVs. I encourage the DPW to keep thinking when it comes to traffic mitigation and to enlist the help of other public agencies as well as the general public as it does so.

There has been considerable comment regarding the desirability of a rail link between North and South stations. At first blush, this sounds like a very logical step and one that ought not be precluded even if it were not going to be pursued as part of this project. However, the material in the FSEIR and a number of comments on this issue seem to suggest that this link does not merit a high priority, especially when weighed against the extensive public transit improvements that are committed to as part of the CA/TIIT project.

There is little serious dispute that engineering such a rail link would be extremely difficult, with or without the CA/THT project. Existing facilities, such as the Red, Orange and Blue Lines, and physical aspects of the Central Artery corridor appear to make construction of such a connection impracticable if not impossible. Because of the gentle grades that would be needed for a potential rail tunnel, trains would have to surface well to the north of North Station and well to the south of South Station. Moreover, the likely demand for such a connection appears to be low. Commuters entering Boston from the south have ready access to the Back Bay and downtown financial district. Planned improvements to the Orange and Green Line platforms at North Station will make rail to transit connections much more convenient for commuters from the north. Because of the limited capacity in such a rail connector tunnel, many trains would continue to operate to the existing commuter rail terminals and many commuters who live south of the city and work north of the city would have to change trains at least once in any event.

Thus the rail link would not provide any real service improvement or ease of travel for the vast majority of rail commuters. The South Station to Logan rail study which the DPW and the MBTA have agreed to undertake offers the promise of a much more beneficial rail improvement for both the metropolitan area and the Northeast corridor.

Given the fact that from an engineering standpoint this connection is at worst impossible and at best impracticable, and given the relatively low demand that would be expected for this service, and given the expense of such a connection,

I do not consider this project a mitigation measure necessary for the CA/THT project. I do feel however that the DPW, MBTA, BTD and others should continue to explore ways to provide an easier connection between North and South Stations and that a study of demand for such a service be undertaken to determine more definitively the number of commuters that would benefit from and use the north-south rail link. DPW should provide subsidized water shuttle service between Lovejoy Wharf and South Station as part of this effort. I also encourage study of light rail or bus connections between the two stations.

Materials Disposal

The materials disposal program that was described in the Draft SEIR has been significantly revised in the response to agency and public comment. Most significantly, the proponent now relies heavily on a program of "beneficial reuse." Materials will be used as backfill, landfill daily cover and final capping material, and to close the existing dump and provide for a park at Spectacle Island. Through the increased use of material as backfill, the disposal requirements have also been reduced.

The FSEIR provides sufficient information about this program at the conceptual level. Obviously, there are many details to be worked out at permitting level, but I feel that state and local permitting agencies have the issues well in hand, and that the remaining issues and details will be adequately addressed.

I am concerned that the reduction of the amount of fill to be barged to Spectacle Island means a corresponding increase in the amount of fill that will have to be transported over land. The FSEIR indicates that the peak removal rate will generate 1200 truckloads a day (one million truckloads for the entire project). Every effort should be made to minimize the impacts that this would cause. I therefore call upon the DPW to make the greatest possible use of rail transportation and direct the DEP to require rail transport wherever possible.

Spectacle Island

The FSEIR presents a plan for a significantly smaller park on Spectacle Island. Nevertheless, the plan still accomplishes the goals of capping the landfill and creating what we all expect to be a stunning addition to the Boston Harbor Islands park system. The DPW has committed to construct a finished, usable park on Spectacle Island, and I consider this commitment to be necessary mitigation for the impacts of this project and a condition of this approval.

As noted in the comments of the Boston Parks and Recreation Department and the Summary Volume of the FSEIR, final layout and design of the park will be carried out by the DPW with the guidance and concurrence of the DEM and the City, working in conjunction with the Spectacle Island Park Advisory Committee. Park preparation will commence within two years of the substantial completion of excavated disposal on the Island, and would be completed one year later. In order to ensure that this park truly provides the desired and anticipated recreational opportunity to the public, one third of the park should be designed for active recreational uses, as the very helpful comments from the Boston Parks and Recreation Department suggest.

Bremen Street Buffer Park

In my Certificate on the Draft SFEIR, I stated that the creation of a park between the Bremen Street residential neighborhood and the McClellan Highway was a necessary measure to mitigate the potentially adverse impacts of this project, and I reaffirm that here. I am pleased to see the DPW's commitment to this needed recreational space in an area that has historically had to bear a heavy burden of regional transportation infrastructure. This open space should be formally designated as Article 97 parkland by way of a conservation restriction or other mechanism. The DPW should continue to work closely with the Boston Transportation Department and the Parks and Recreation Department to resolve any outstanding issues related to ramp locations, park design, community involvement and the like, as well as with EOTC in their study of the South Station to Logan rail link, in order to properly design the conservation restrictions to allow for the appropriate transportation facilities.

Marginal Road Ramp

The FSEIR properly documents the air quality and traffic impacts of the exit ramp from the Southeast Expressway and Seaport Access Road in Chinatown. This ramp would convert what is now a low-volume local street to a major connection for traffic coming from Logan Airport, the Southeast Expressway, and the proposed South Boston bypass road. I understand that the DPW is consulting with the Boston Transportation Department and representatives of Chinatown to come to a resolution that makes every attempt to reduce this significant impact. If these discussions lead to a change in the ramp design, a Notice of Project Change must be filed.

Additionally, if the Massachusetts Turnpike Authority moves forward expeditiously with plans for the proposed Berkeley Street off-ramp, the concerns of the Chinatown community can be further mitigated. I understand they plan to do so and I encourage that. As a first step, an Environmental Notification Form for this project should be filed with the MEPA Unit on or before March 1, 1991.

Water Quality

Because of the recent release of EPA's new Stormwater Management Regulations, the FSEIR did not present a detailed plan for managing the runoff from the roadway structures. This plan should be developed with the EPA, DEP, MWRA and BWSC to ensure that existing facilities have the capacity to handle the discharges from this system and that pretreatment is provided where appropriate. This stormwater management plan should be developed as part of DEP's permitting activities.

The Haul Road

The FSEIR has provided additional information on the relationship of the South Boston Bypass and Haul Roads. As noted in my August Certificate, the Haul Road is a long sought after traffic mitigation measure designed to take trucks off local South Boston streets. In this way, it will act as a construction period mitigation measure for the CA/THT project, but that is by no means its exclusive purpose. I am pleased that the proponent has removed HOVs from the Haul Road, thus further enhancing the benefits to South Boston by keeping trucks off of neighborhood streets. Remaining issues regarding roadway design should be resolved with the BTD.

Wetlands Impacts

The project clearly will have impacts on resources protected by the Wetlands Protection Act in the areas of Fort Point Channel, the Charles River crossing, and Spectacle Island, among others. The proponent will be expected to comply with applicable regulations and to provide comprehensive mitigation for wetlands that will be altered.

I recognize that there is an ongoing dialogue between the proponent and the City of Cambridge regarding the type and extent of wetlands resources on the north bank of the Charles. There are established regulatory procedures for resolving any issues that might exist. These procedures will be followed to address and resolve the wetlands related issues.

Waterways

This project can now move forward to the licensing process pursuant to G. L. c. 91. The project site includes filled or flowed tidelands in East Boston, South Boston, Fort Point Channel, downtown Boston, and the Charles River crossing area. During the licensing process, DEP will establish specific conditions and parameters for the project activities. I urge upon DPW and DEP two goals in this process: the minimization of fill to the greatest extent possible, and the enhancement of public access to the waterfront.

Procedural issues

Lastly, I will address several procedural points that have arisen in the course of this review.

This project has generated an extraordinary amount of public comment. As is the case with many MEPA reviews, a number of comments have been submitted after the formal deadline of December 26, 1990. I have the authority, pursuant to 301 CMR 11.09(3), to accept late filed comments, provided they are received by the date of this Certificate. I hereby am exercising that authority, and accept into the MEPA record any and all comments filed with me or the MEPA Unit on or before January 2, 1990.

Two procedural criticisms that have been made relate to the ability of the public to participate in this review process in a meaningful way. The first of these criticisms is that the MEPA and NEPA reviews have been bifurcated. The second is that the DPW did not file a Notice of Project Change when Scheme Z was identified as the preferred alternative. Neither of these criticisms has sufficient merit to make credible the argument that members of the public or public agencies have been denied a fair opportunity to exercise their very necessary and important responsibilities to inform this process.

It is a simple fact that there has been an extraordinary amount of opportunity for public participation. An interagency draft of the DSEIR was circulated beginning in the fall of 1989, more than a year ago. The DSEIR itself was filed in May, 1990, and was subject to a nearly three month public comment period rather than the usual thirty days. There have been numerous public hearings and hundreds of meetings—nearly 200 in the last 6 months alone—sponsored by the DPW and FHWA.

The criticism that the state and federal environmental reviews should not have been bifurcated at this stage of the process is unfounded. While parallel review is often encouraged by this office, neither the MEPA statute nor its regulations require a joint review process. Moreover, I do not see that the delay of federal review detracts from my ability or the ability of the public to review this document on the merits and reach a judgment as to its adequacy pursuant to MEPA.

A Notice of Project Change need not have been filed when the DPW initially identified Scheme Z as its preferred option for the river crossing portion of this roadway. It is the nature, and indeed one of the goals, of the MEPA review process that proposed projects will develop and change as review proceeds. Phased environmental review through draft and final reports allows and encourages proponents to refine and adjust their projects in response to identified environmental concerns. In some cases those adjustments amount to radical change or redesign of the project.

MEPA does not require that a Notice of Project change be filed in every such instance. The additional scoping that can be but is not necessarily triggered by the filing of a Notice of Project Change can also be accomplished by the Secretary's Certificate on a draft or final impact report that incorporates and addresses project changes. In this case, the former Assistant Secretary for Environmental Impact Review made the judgment that requiring DPW to file a Supplemental Draft EIR, preceded by an extended interagency review period, would be more useful than a Notice of Project Change. He came to this conclusion because a Notice of Project Change would necessarily have been more general and cursory in its discussion of the project revisions than a supplemental

EIR and that the public would be better able to comment in an informed way if they had a full report before them.

Therefore, while a Notice of Project Change arguably might have provided one more opportunity for public review and comment on this project, in light of the extensive public process, it is my judgment that the filing of a Notice of Project Change would not have led to a materially different or improved public review process for this project.

Conclusion

The DPW is responsible for making comprehensive Section 61 Findings concerning this project. Those Findings should be submitted to the Secretary of Environmental Affairs for public review and comment.

Date: January 2, 1991
John DeVillars, Secretary
Executive Office of Environmental Affairs
Commonwealth of Massachusetts

Glossary of Acronyms

BRA	Boston Redevelopment Authority
BTD	Boston Transportation Department
BWSC	Boston Water and Sewer Commission
CA/THT	Central Artery/Third Harbor Tunnel
CLF	Conservation Law Foundation
DEP	Department of Environmental Protection
DPW	Department of Public Works
ENF	Environmental Notification Form
EOTC	Executive Office of Transportation and Construction
EPA	Environmental Protection Agency
FEIS/R	Final Environmental Impact Statement/Report
FHWA	Federal Highway Administration
FSEIR	Final Supplemental Environmental Impact Report
HOV	High Occupancy Vehicle
MBTA	Massachusetts Bay Transportation Authority
MEPA	Massachusetts Environmental Policy Act
MOU	Memorandum of Understanding
MPO	Metropolitan Planning Organization
MWRA	Massachusetts Water Resource Authority
NEPA	National Enviromental Policy Act
SDEIR/S	Supplemental Draft Environmental Impact Review
SFEIR	Supplemental Final Environmental Impact Review
SIP	State Implemental Plan

From Commonwealth of Massachusetts, Executive Office of Environmental Affairs. Boston, MA, January 2, 1991.

BIBLIOGRAPHY

Abel, Deryck. *Channel Underground: A New Survey of the Channel Tunnel Question.* London: Pall Mall Press, 1961.

Achenbach, Joel. *The Grand Idea: George Washington's Potomac and the Race to the West.* New York: Simon and Schuster, 2004.

"Act for Establishing a Temporary and Permanent Seat of the Government of the United States." *Congressional Record*, First Congress. Sess. II, Ch. 28, 1790: 130.

"Act to Amend 'An Act for Establishing the Temporary and Permanent Seat of the Government of the United States.'" *Congressional Record*, First Congress, Sess III, Ch. 16, 17, 1791: 214, 215.

"Act to Incorporate the Inhabitants of the City of Washington, in the District of Columbia." *Congressional Record*, Seventh Congress, Sess. I, Ch. 53, 1802: 195–197.

Adams, Michael. *Suez and After: Year of Crisis.* Boston: Beacon Press, 1958.

Adewale, Toyin, ed. *25 New Nigerian Poets.* Berkeley, CA: Ishmael Reed Publishing, 2000.

Aguda, T. Akinola. *Report of the Committee on the Location of the Federal Capital of Nigeria.* n.p. [no publisher] December 20, 1975.

Aliosi, James A., Jr., and Robert J. Allison. *The Big Dig (New England Remembers).* Boston: Commonwealth Editions, 2004.

Allen, W.B., ed. *George Washington: A Collection.* Indianapolis: Liberty Classics, 1988.

Almedingen, E.M. *My St. Petersburg: A Reminiscence of Childhood.* New York: Norton, 1970.

Ambrose, Stephen E. *Nothing Like It in the World: The Men Who Built the Transcontinental Railroad, 1863–1869.* New York: Simon and Schuster, Touchstone Books, 2000.

Andréossy, Comte Antoine-François. *Histoire du Canal de Midi ou Canal de Languedoc considéré sous les rapports de'invention, d'art, d'administration, d'irrigation et dans ses relations avec les étangs de l'intérieur des terres qui l'avoisinent.* Paris: Impr. De Crapelet, 1804.

Anisimov, Evgenii Victorovich. *The Reforms of Peter the Great: Progress through Coercion in Russia.* Translated by John T. Alexander. Armonk, NY: M.E. Sharpe, 1993.

Badian, E., and Robert Sherk, eds. *The Roman Empire: Augustus to Hadrian (Translated Documents of Greece and Rome).* Cambridge: Cambridge University Press, 1988.

Bahura, G.N., and Chandramani Singh. *Catalogue of Historical Documents in Kapad Dwara, Jaipur.* Jaipur, India: Jaigarh Public Charitable Trust, 1988.

Baker, A.J. *Suez: The Seven Day War.* New York: Praeger, 1965.

Banaja, A.A. *Red Sea, Gulf of Aden and Suez Canal: A Bibliography on Oceanographic and Marine Environmental Research.* Edited by Selim A. Morcos and Allen Varley. Compiled by A.A. Banaja, A.L. Beltagy, and M.A. Zahran, with scientific contributions by M. Kh. Ed-Sayed. Jeddah, Saudi Arabia: Alexco-Persga and UNESCO, 1990.

Barbero, Alessandro. *Charlemagne: Father of a Continent.* Translated by Allan Cameron. Berkeley: University of California Press, 2004.

Beaufré, André. *The Suez Expedition.* 1956. Reprint, New York: Praeger, 1969.

Bechtel Corporation, Brown & Root, Inc., and Morrison-Knudsen Company, Inc. *The Channel Tunnel: Design and Construction of a Channel Tunnel as Recommended by Three Engineer-Constructors.* November 1959. Available in the Channel Tunnel archives at the Historical Collections, Baker Library, School of Business Administration, Harvard University, Cambridge, Massachusetts.

Bedini, Silvio A. *The Life of Benjamin Banneker.* Rancho Cordova, CA: Landmark Enterprises, 1984.

Beebe, Lucius. *Trains in Transition.* New York: Bonanza Book, 1941. Reprint, New York: Hawthorn Books, 1976.

Beebe, Lucius, and Charles Clegg. *When Beauty Rode the Rails: An Album of Railroad Yesterdays.* Garden City, NY: Doubleday, 1962.

Beekman, A.A. *Het dijk–en waterschaprecht in Nederland vòòr 1795.* Gravenhage, Netherlands: Martinus Nijhoff, 1905.

Begley, W.E., and Z.A. Desai, comp. and trans. *Taj Mahal: The Illumined Tomb; An Anthology of Seventeenth-Century Mughal and European Documentary Sources.* Seattle: University of Washington Press, 1989.

Bénard, André. "Financial Engineering of Eurotunnel." In *Macro-engineering: MIT Brunel Lectures on Global Infrastructure* Edited by Frank P. Davidson, Ernst G. Frankel, and C. Lawrence Meador. Horwood Series in Engineering Science. Chichester, England: Horwood Publishing, 1997.

Bergman, Charles. *Red Delta: Fighting for Life at the End of the Colorado River.* Golden, CO: Fulcrum Publishing, 2002.

Bernstein, Peter L. *Wedding of the Waters: The Erie Canal and the Making of a Great Nation.* New York: Norton, 2005.

Berton, Pierre. *The Last Spike: The Great Railway, 1881–1885.* 1971. Reprint, Toronto: Anchor Canada, 2001.

Berton, Pierre. *The National Dream: The Great Railway, 1871–1881.* Toronto: Anchor Canada, 1970.

Bijker, Wiebe E., Thomas P. Hughes, and Trevor J. Pinch, eds. *The Social Construction of Technological Systems: New Directions in the Sociology and History of Technology.* Cambridge, MA: MIT Press, 1987.

Biswas, Asit K. "Aswan Dam Revisited: The Benefits of a Much Maligned Dam." *D+C* (*Development and Cooperation*). Monograph no. 6 (November/December 2002): 25–27.

Blackman, Deane R., and A. Trevor Hodge, eds. *Frontinus' Legacy: Essays on Frontinus' De Aquis Urbis Romae*. Ann Arbor: University of Michigan Press, 2001.

Boedler, Christine, trans. "Federal Capital." *Diário Oficial*. Year 95, no. 217, September 20, 1956.

Bonnaud, Laurent. *Le Tunnel sous la Manche: Deux Siècles de Passions*. New York: Hachette, 1994.

Bonner, Raymond. "Mysterious Libyan Pipeline Could Be Conduit for Troops." *New York Times*, December 2, 1997, A1, Col. 4.

Boorstin, Daniel J. *The Americans: The Democratic Experience*. New York: Random House, 1965.

Bowie, Robert Richardson. *Suez. 1956*. Reprint, London and New York: Oxford University Press, 1974.

Browning, Larry D., and Judy C. Shetler. *SEMATECH: Saving the U.S. Semiconductor Industry*. Austin: Texas A&M University Press, 2000.

Buckley, Charles Burton. *An Anecdotal History of Old Times in Singapore, from the Foundation of the Settlement under the Honourable the East India Company on February 6th, 1819 to the Transfer to the Colonial Office as Part of the Colonial Possessions of the Crown on April 1st, 1867*. Kuala Lumpur: University of Malaya Press, 1965.

Butt, John J. *Daily Life in the Age of Charlemagne*. Westport, CT: Greenwood Press, 2002.

Calame, Claude. *Myth and History in Ancient Greece: The Symbolic Creation of a Colony*. Translated by Daniel W. Berman. Princeton: Princeton University Press, 2003.

California Water Atlas. Sacramento: State of California, 1978.

Callahan, North. *TVA: Bridge over Troubled Waters*. Cranbury, NJ: Cornwall Books, 1981.

Canadian Pacific Railway Archives, Windor Station, P.O. Box 6042, Station Centre Ville, Montreal, Quebec, Canada H3C 3E4.

Cathcart, Robert. *Planet Earth Renewed: Macroprojects and Geopolitics*. Monograph. Fifth revision, 1984. Contact author at: rbcathcart@msn.com or http://geographos. blogspot.com.

Chandler, W. U. *The Myth of TVA: Conservation and Development in the Tennessee Valley, 1933–1983*. Boston: Ballinger, 1984.

Chinweizu. *Decolonising the African Mind*. Lagos, Nigeria: Pero Press, 1987.

Chinweizu. *The West and the Rest of Us*. New York: NOK Publishers, International, 1978.

Ch'len, Ssu Ma (Kuang). *The Grand Scribe's Records: The Basic Annals of Pre-Han China*. Bloomington: Indiana University Press, 1994.

Christy, Jim. *Rough Road to the North: Travels along the Alaska Highway*. Toronto: Doubleday Canada, 1980.

Clark, Ronald W. *Works of Man*. New York: Viking, 1985.

Clark, Wilson. *Energy for Survival*. New York: Anchor Press, 1974.

Clarke, Arthur C. "Extra-Terrestrial Relays." *Wireless World*, October 1945, 305–8.

Clarke, Arthur C. *The Exploration of Space*. New York: Harper, 1951.

Coates, Kenneth, and William R. Morrison. *The Alaska Highway in World War II: The U.S. Army of Occupation in Canada's Northwest*. Toronto: University of Toronto Press, 1992.

Collis, Brad. *Snowy: The Making of Modern Australia*. Palmerston, Australia: Tabletop Press, 2002.

Committee on Science and Astronautics, U.S. House of Representatives, 87th Cong., 1st sess. on HR 6874, P.L. 87–98 (May 9, 1961): 1033–34.

Congressional Record, 70th Cong., 2nd sess. (1928), chap. 42: 1057–66.

Congressional Record, 73rd Cong., 1st sess., H.R. 5081 (May 18, 1933), chap. 21.

Cooke, Edward William. *Old and New London Bridge: A Selection of Drawings Reproduced from Originals in the Possession of the Guildhall Art Library*. London: Topographical Society, 1970.

Creswell, K.A.C. *Early Muslim Architecture*. London: Oxford, 1940.

Culver, J.C., and J. Hyde. *American Dreamer: A Life of Henry A. Wallace*. New York: Norton, 2000.

Cunningham, Michelle. *Mexico and the Foreign Policy of Napoleon III*. Basingstoke, England: Palgrave, 2001.

Danigelis, Alyssa. "The Man Behind the Big Dig." *Technology Review* (July/August 2004).

Davidson, Frank P. "An Action Plan for Creating an Interconnected Network of Trails across America." Report of a conference held at the Rensselaerville Institute, Rensselaerville, NY, October 1986.

Davidson, Frank P. "An Express of the (Near) Future." *Air and Space* (December 1995/January 1996): 22–24.

Davidson, Frank P., and John S. Cox. *Macro: A Clear Vision of How Science and Technology Will Shape our Future*. New York: William Morrow and Co., 1983.

de Beauvoir, Simone. *La Force des Choses*. Paris: Gallimard, 1963.

de La Lande, M. "Edit d'Octobre 1666 pour le Canal de Languedoc." In *Des Canaux de Navigation et Spécialement du Canal de Languedoc*. Paris: Chez la Veuve Desaint, 1778.

de Lesseps, Ferdinand. *Recollections of Forty Years*. Translated by C.B. Pitman. 2 vols. London: Chapman and Hall, 1887.

Delfs, R. "Arteries of the Empire." *Far Eastern Economic Review* (March 15, 1990): 28–29.

Dempsey, Jack, and Robert E. Lyons. "Semiconductors and SEMATECH: Rebirth of a Strategic Industry?" Washington, DC: National Defense University, Industrial College of the Armed Forces, 1993.

de Roquette-Buisson, Odile, and Christian Sarramon. *The Canal du Midi*. London: Thames & Hudson, 1983.

Dixon, Pierson. *Farewell, Catullus*. London: Hollis and Carter, 1953.

Dmitriev-Mámonov, A.I., Aleksandr Ippolitovich, and Anton Feliksovich Zdziárski, eds. *Guide to the Great Siberian Railway with 2 Photographs, 360 Photo-gravurs, 4 Maps of Siberia and 3 Plans of Towns*. Revised by John Marshall. Translated by L. Kukol-Yasnopolsky. St. Petersburg: Artistic Printing Society, 1900. Reprint, Newton Abbott, England: David and Charles, 1971.

Dryzek, John S., and David Schlossberg. *Debating the Earth: The Environmental Politics Reader*. Oxford: Oxford University Press, 1998.

Durant, Will, and Ariel Durant. "Cesar and Christ." In *The Story of Civilization*. New York: Simon and Schuster, 1944.

Dyson, Freeman. *Weapons and Hope*. New York: Harper and Row, 1984.

Efimova, A., O. Turkina, and V. Mazin. *Layers: Contemporary Collage from St. Petersburg, Russia*. Catonsville: Fine Arts Gallery, University of Maryland, 1995.

Ellis, Joseph J. *His Excellency George Washington*. New York: Knopf, 2004.

Evans, Harry B. *Water Distribution in Ancient Rome: The Evidence of Frontinus*. Ann Arbor: University of Michigan Press, 1994.

Ezzell, Patricia Bernard. *TVA Photography: Thirty Years of Life in the Tennessee Valley*. Jackson: University Press of Mississippi, 2003.

Fahim, Hussein M. *Egyptian Nubians: Resettlement and Years of Coping*. Salt Lake City: University of Utah Press, 1983.

Failure, Jacob M. *The Everything Middle East Book: The Nations, Their Histories, and Their Conflicts*. Avon, MA: Adams Media, 2004.

Ferling, John. *A Leap in the Dark: The Struggle to Create the American Republic*. Oxford: Oxford University Press, 2003.

Fetherston, Drew. *Chunnel: The Amazing Story of the Undersea Crossing of the English Channel*. New York: Random House, 1997.

Fletcher, Banister. *Sir Banister Fletcher's a History of Architecture*. Revised by J.C. Palmer. New York: Charles Scribner's Sons, 1975.

Fox, J. Ronald, and Donn B. Miller. *Challenges in Managing Large Projects*. Fort Belvoir, Virginia: Defense Acquisition University Press, 2006.

Freeman, Kathleen. *Greek City-States*. New York: Norton, 1950.

Frontinus, Sextus Julius. *Frontinus: De Aquaeductu Urbis Romae*. Edited by R. H. Rodgers. Cambridge: Cambridge University Press, 2004.

Frontinus, Sextus Julius. *Strategems and the Aqueducts of Rome*. Translated by Charles Bennett. Cambridge, MA: Harvard University Press, 1961.

Frontinus, Julius Sextus. *The Aqueducts of Rome*. In *Frontinus: The Strategems and the Aqueducts of Rome*. Cambridge, MA: Harvard University Press: William Heinemann, 1950, 339–347.

Furber, H. *John Company at Work*. Cambridge, MA: Harvard University Press, 1948.

Galison, Peter, and Bruce Hevly, eds. *Big Science: The Growth of Large-Scale Research*. Stanford, CA: Stanford University Press, 1992.

Garber, Paul Leslie. "A Reconstruction of Solomon's Temple." Cambridge, MA: Archaeological Institute of America, 1952. Stand-alone monograph reprinted from the original article published in *Archaeology* 5, no. 3, 1952.

Gardner, Brian. *The East India Company: A History*. New York: Dorset Press, 1971.

Garrels, Anne. *Naked in Baghdad: The Iraq War as Seen by NPR's Correspondent*. New York: Farrar, Straus and Giroux, 2003.

Gerard, P. *Le Voyage de Thomas Jefferson sur le Canal du Midi*. Portet-sur-Garonne, France: Loubatières, 1995.

Gewertz, Ken. "A Long View of the Big Dig." *Harvard University Gazette*, September 27, 2001.

Gies, Joseph. *Adventure Underground: The Story of the World's Great Tunnels*. New York: Doubleday, 1962.

Glaser, Peter E., Frank P. Davidson, and Katinka I. Csigi. *Solar Power Satellites: A Space Energy System for Earth*. Chichester, England: John Wiley and Sons/Praxis Publishing, 1998.

Gorlov, Alexander. "Tidal Power." *Encyclopedia of Ocean Sciences*. London: Academic Press, 2001.

Gorlov, E. S. *The Founding of St. Petersburg as Recounted in the Journal (Diary Notes) of Tsar Peter the Great*. The Institute of Railroad and Transportation Engineering. Pub. Art. 1777, part 1, p. 76. Translated and adapted from the archaic Russian script of the eighteenth century.

Gutfreund, Owen D. *Twentieth Century Sprawl: Highways and the Reshaping of the American Landscape*. New York: Oxford University Press, 2004.

Haney, Louis H. *A Congressional History of Railways*. 2 vols. 1908–10. Reprint, New York: Augustus M. Kelley, 1968.

Hansen, R. D. "Water and Waste Water in Imperial Rome." *Water Resources Bulletin* 19 (1983).

Harding, Sir Harold. *Tunnelling History and My Own Involvement*. Toronto: Golder Associates, 1981.

Harrington, Lyn. *The Grand Canal of China*. Chicago: Rand McNally, 1967.

Harris, S. *The Tallest Tower: Eiffel and the Belle Époque*. Washington, DC: Regnery Gateway, 1989.

Hawkes, Nigel. *Structures: Man-Made Wonders of the World*. London: Marshall Editions, 1990.

Hawthorne, Nathaniel. "The Canal Boat." *New-England Magazine*, no. 9 (December 1835): 398–409.

Hecht, Roger W., ed. *The Erie Canal Reader, 1790–1950*. Syracuse, NY: Syracuse University Press, 2003.

Hickel, Walter J. *Crisis in the Commons: The Alaska Solution*. Anchorage: Alaska Pacific University, 2002.

Hickel, Walter J. "The Alaskan Pipeline Is Essential." *New York Times*, March 24, 1971.

Hickel, Walter J. *Who Owns America?* Englewood Cliffs NJ: Prentice Hall, 1971.

Hodge, A. T. "Lead Pipes and Lead Poisoning." AJA 85 (1981): 486–91.

Hodge, A. T. *Roman Aqueducts and Water Supply*. London: Duckworth, 2002.

Hope, Valerie M., and Eireann Marshall, eds. *Death and Disease in the Ancient City*. London: Routledge, 2000.

Horn, Paul. *Inside Paul Horn: The Spiritual Odyssey of a Universal Traveler*. New York: HarperCollins, 1990.

Hourani, Albert. *A History of the Arab Peoples*. New York: Warner Books, 1991.

Hubbard, P. J. *Origins of the TVA: The Muscle Shoals Controversy, 1920–1932*. Nashville, TN: Vanderbilt University Press, 1961.

Huber, Thomas Patrick, and Carole J. Huber. *The Alaska Highway: A Geographical Discovery*. Boulder: University Press of Colorado, 2000.

Hughes, Thomas P. *American Genesis: A Century of Invention and Technological Enthusiasm*. New York: Penguin Books, 1989.

Huisman, Pieter. *Water in the Netherlands: Managing Checks and Balances*. Utrecht: Netherlands Hydrological Society, 2004.

Huisman, Pieter. *Water Legislation in the Netherlands*. Delft: Delft University Press, 2004.

Hunt, Donald. *The Tunnel: The Story of the Channel Tunnel, 1802–1994*. Malvern, England: Images Publishing, 1994.

Irwin, Douglas A., and Peter J. Klenow. "SEMATECH: Purpose and Performance." *Proceedings of the National Academy of Science* 93 (November 1996): 12739–42.

Jackson, Peter. *London Bridge*. London: Cassell, 1971.

James, William. "The Moral Equivalent of War." In *American Youth: An Enforced Reconnaissance*. Edited by Thacher Winslow and Frank P. Davidson. Cambridge, MA: Harvard University Press, 1940.

Juvenal, Decimus Julius. *Satire*. Translated by Jerome Mazzaro. Ann Arbor: University of Michigan Press, 1965.

Kapad Dwara Collection, K.D. No 176/R. Descriptive List of Documents, Kapad Dwara Collection, Jaipur, National Register of Private Records, No. 1, Park I. Delhi: National Archives of India, 1971.

Karabell, Zachary. *Parting the Desert: The Creation of the Suez Canal*. New York: Vintage, 2004.

Kazaanxhi, Fouad. "A History of Baghdad." *Baghdad Bulletin*, July 16, 2003.

Keck, Margaret E. "Amazônia and Environmental Politics." In *Environment and Security in the Amazon Basin*. Edited by Tulchin, Joseph S. and Heather A. Golding. Washington, DC: Woodrow Wilson International Center for Scholars, 2002.

Keeble, John. *Out of the Channel: The Exxon Valdez Oil Spill in Prince William Sound*. Cheney: Eastern Washington University Press, 1999.

Kent, Hollister. "Vera Cruz: Brazil's New Federal Capital." PhD diss., Cornell University, 1956.

Khaldûn, Ibn. *The Muqaddimah: An Introduction to History*. Translated by Franz Rosenthal. Edited by N. J. Dawood. Princeton, NJ: Princeton University Press, 1967.

Kirk-Greene, Anthony, and Douglas Rummer. *Nigeria Since 1970: A Political and Economic Outline*. New York: Africana Publishing, 1981.

Kohlhepp, Gerd. *Itaipú: Basic Geopolitical and Energy Situation; Socio-economic and Ecological Consequences of the Itaipú Dam and Reservoir on the Rio Paraná*. Braunschweig, Germany: F. Vieweg, 1987.

Krause, Jeanne. "The Erie Canal: Macro-engineering When the World Was Still Simple." In *How Big and Still Beautiful? Macro-engineering Revisited*. Edited by Frank P. Davidson, C. Lawrence Meador, and Robert Salkeld. Boulder, CO: Westview Press, 1980.

Krech, Shepard. *The Ecological Indian: Myth and History*. New York: Norton, 1999.

Krylov, V. V., ed. *Noise and Vibration from High-Speed Trains*. London: Thomas Telford, 2001.

Kuhn, Thomas W. *The Structure of Scientific Revolutions*. Chicago: University of Chicago Press, 1970.

Kunz, Diane B. *The Economic Diplomacy of the Suez Crisis*. Chapel Hill: University of North Carolina Press, 1991.

Lambright, W. Henry. *Powering Apollo: James E. Webb of NASA*. Baltimore: Johns Hopkins University Press, 1995.

Landis, John. "Operation Mulberry: A Floating Transportable Harbour for World War II Normandy Invasion." Chapter 2: 25–52. In: *Macroengineering: MIT Brunel Lectures on Global Infrastructure*. Edited by F. P. Davidson, E. G. Frankel, and C. L. Meador. Chichester, England: Horwood Publishing Ltd., 1997.

Lassner, Jacob. *The Topography of Baghdad in the Early Middle Ages: Text and Studies*. Detroit, MI: Wayne State University Press, 1970.

Laws of New York, 90th sess. (1867), chap. 399.

Laws of New York, 40th sess. (1817), chap. 262: 361–65.

Lee, Kuan Yew. *The Singapore Story: Memoirs of Lee Kuan Yew*. New York: Prentice Hall, 1999.

Le Strange, Guy. *Baghdad during the Abbasid Caliphate from Contemporary Arabic and Persian Sources*. Oxford: Clarendon Press, 1940.

Lehmann, Joseph H. *All Sir Garnet: A Life of Field-Marshal Lord Wolseley*. London: Jonathan Cape, 1964.

Lemer, Andrew C. "Foreseeing the Problems of Developing Nigeria's New Federal Capital." In *Macro-engineering and the Future: A Management Perspective*. Edited by Frank P. Davidson and C. Lawrence Meador. Boulder, CO: Westview Press, 1982.

Lemer, Andrew C. "Old Cities and New Towns for Tomorrow's Infrastructure." In *Macro-engineering: MIT Brunel Lectures on Global Infrastructure*. Edited by Frank P. Davidson, Ernst G. Frankel, and C. Lawrence Meador. Horwood Series in Engineering Science. Chichester, England: Horwood Publishing, 1997.

Lemoine, Bertrand. *Le Tunnel sous La Manche*. Paris: Le Moniteur, 1994.

Leonard, Jane Kate. *Controlling from Afar: The Dauguang Emperor's Management of the Grand Canal Crisis, 1824–1826*. Ann Arbor: Center for Chinese Studies, University of Michigan, 1996.

Lévy, Bertrand. "Le tunnel du Mont-Blanc." *Travaux*, no. 780 (November 2001).

Lipsey, Roger. *Have You Been to Delphi? Tales of the Ancient Oracle for Modern Minds Including an Afterword with Lobsang Lhalungpa on Tibetan Oracles*. Albany: State University of New York Press, 2001.

Little, P. E. *Abundance Is Not Enough: Water-Related Conflicts in the Amazon River Basin*. Brasília: University of Brasília, 2002.

Litwin, George H., John J. Bray, and Kathleen Lusk Brooke. *Mobilizing the Organization: Bringing Strategy to Life*. London: Prentice Hall, 1996.

Liu, Ruifang. *The History of Chinese Emperors*. Beijing: National Defense University Publishing House, n.d., 614.

Lloyd, Selwyn. *Suez 1956: A Personal Account*. New York: Mayflower Books, 1978.

Lopresti-Essex, Michael. *Exploring the New River*. London: Brewin Books, 1997.

Louveau, L. "Le tunnel du Mont-Blanc. 'Déjà dix-huit années de bons et loyaux services.'" *Travaux*, February 1984, No. 585.

Lusk Brooke, Kathleen, and George H. Litwin. "Organizing and Managing Satellite Solar Power." *Space Policy* 16 (2000): 145–56.

Macaulay, David. *Building Big*. Boston: Houghton Mifflin, 2000.

MacKaye, Benton. "An Appalachian Trail: A Project in Regional Planning." *Journal of the American Institute of Architects* 9 (October 1921): 325–30.

Mann, Elizabeth. *The Brooklyn Bridge: A Wonders of the World Book*. New York: Mikaya Press, 1996.

Massie, Robert K. "The Founding of St. Petersburg." In *Peter the Great: His Life and Work*. New York: Knopf, 1980.

McCullough, David. *The Great Bridge*. New York: Simon and Schuster, 1972.

McCullough, David. *The Path between the Seas: The Creation of the Panama Canal*. New York: Simon and Schuster, 1977.

McDonnell, Greg. *Canadian Pacific: Stand Fast, Craigellachie!* Erin, Canada: Boston Mills Press, 1954.

McInerney, Thomas. "The Command Tactical Information System: Military Software for Macro-engineering Projects." In *Macro-engineering: MIT Brunel Lectures on Global Infrastructure*, edited by Frank P. Davidson, Ernst G. Frankel, and C. Lawrence Meador. Horwood Series in Engineering Science. Chichester, England: Horwood Publishing, 1977.

McNichol, Dan. *The Big Dig: The Largest Urban Construction Project in the History of the Modern World*. New York: Silver Lining Books, 2000.

Michel, Prince of Greece. *The Empress of Farewells: The Story of Charlotte, Empress of Mexico*. Translated by Vincent Aurora. New York: Atlantic Monthly Press, 2002.

Michener, James A. *Alaska*. New York: Fawcett Crest, 1988.

Moore, John R., ed. *The Economic Impact of TVA*. Knoxville: University of Tennessee Press, 1967.

Morita, Teruo. *Intercity High Speed Ground Transportation: History and Prospects*. Master's thesis, June 1984, Massachusetts Institute of Technology, Center for Transportation and Logistics.

Morris, Edmund. *Theodore Rex*. New York: Random House, 2001.

Morris, Peter W. G. *The Management of Projects*. London: Thomas Telford, 1994.

Mueller, Milton. *Universal Service: Competition, Interconnection, and Monopoly in the Making of the American Telephone System*. Cambridge, MA: MIT Press, 1997.

Mufson, Steven. "The Yangtze Dam: Feat or Folly?" *Washington Post*, November 9, 1997, A-01.

Murphy, Kathleen J. *Macroproject Development in the Third World*. Boulder, CO: Westview Press, 1983.

Mutting, Anthony. *No End of a Lesson: The Story of the Suez*. New York: C. N. Potter, 1967.

Navarra, John Gabriel. *Supertrains*. Garden City, NY: Doubleday, 1976.

Neal, Laura, ed. *It Doesn't Snow Like It Used To: Memories of Monaro and the Snowy Mountains*. Ultimo, Australia: Stateprint, 1988.

Needham, Joseph. *Science and Civilisation in China*. Cambridge: Cambridge University Press, 1954.

Needham, Joseph. *The Development of Iron and Steel Technology in China*. London: Newcomen Society for the Study of History of Engineering and Technology, Science Museum, 1958.

New China News Ltd. and New China Pictures Co. *The Grand Canal of China*. Hong Kong: South China Morning Post, New China News, 1984.

Newby, Eric. *The Big Red Train Ride*. New York: St. Martin's Press, 1978.

Newhouse, Elizabeth, ed. *The Builders: Marvels of Engineering*. Washington, DC: National Geographic Society, 1992.

Niemeyer, Oscar. *Oscar Niemeyer: Notebooks of the Architect*. Brussels: Edition CIVA, 2002.

Nitze, Paul. H. *From Hiroshima to Glasnost: At the Center of Decision; A Memoir*. New York: Grove Weidenfeld, 1989.

"No Official Word Yet on Taj Minaret." *Times of India*, October 21, 2004.

Overseas Food Corporation. *Annual Reports*. London: H. M. Stationery Office, 1948.

Parkins, Helen M., ed. *Roman Urbanism: Beyond the Consumer City*. London and New York: Routledge, 1997.

Pearse, Innes H., and Lucy H. Crocker. *The Peckham Experiment: A Study in the Living Structure of Society*. London: Allen and Unwin, 1943. Reprinted, Edinburgh: Scottish Academic Press, 1985.

Persico, Joseph E. *Roosevelt's Secret War: FDR and World War II Espionage.* New York: Random House, 2001.

Petroski, Henry. *Pushing the Limits: New Adventures in Engineering.* New York: Knopf, 2004.

Pierce, John Robinson. *The Beginnings of Satellite Communications.* History of Technology Monograph. Berkeley: San Francisco Press, 1968.

Poor, Henry Varnum. *Manual of the Railroads of the United States for 1870–71.* New York: H. V. and J. W. Poor, 1870.

Public Law 87–624, Communications Satellite Act of 1962 (August 31, 1962).

Public Papers of the Presidents of the United States: John F. Kennedy, Containing the Public Messages, Speeches, and Statements of the President, January 20 to December 31, 1961. Washington, DC: United States Government Printing Office, 1962, 403–4.

Ramos, A. R. *The Predicament of Brazil's Pluralism.* Brasília: University of Brasília, 2002.

Reaka-Kudla, Marjorie, Don Wilson, and Edward O. Wilson. *Biodiversity II: Understanding and Protecting our Biological Resources.* Washington, DC: Joseph Henry Press, 1997.

Redman, Charles. *Human Impact on Ancient Environments.* Tucson: University of Arizona Press, 1999.

Reischauer, Edwin O. *Ennin's Travels in T'ang China.* New York: Ronald Press Company, 1955.

Ritter, Joyce N., comp. "Development of the Interstate Program." In *America's Highways: 1776–1976.* Washington, DC: U.S. Department of Transportation, Federal Highway Administration, 1976.

Robertson, Terence. *Crisis: The Inside Story of the Suez Conspiracy.* New York: Atheneum, 1965.

Rose, Mark H. *Interstate Express Highway Politics 1941–1989.* Revised ed. Knoxville: University of Tennessee Press, 1990.

Rosenstock-Huessy, Eugen. *Out of Revolution.* Norwich, VT: Argo Books, 1993.

Rozell, Ned. *Walking My Dog, Jane: From Valdez to Prudhoe Bay along the Trans-Alaska Pipeline.* Pittsburgh: Duquesne University Press, 2000.

Rudden, Bernard. *The New River: A Legal History.* Oxford: Clarendon Press, 1985.

Sagiv, Tuvia. "What If This Isn't the Western Wall." *Jerusalem Report,* November 23, 1998.

Salvucci, Frederick, and Michael Hintlian. *Digging: The Workers of Boston's Big Dig.* Boston: Commonwealth Editions, 2004.

Sandström, Gösta E. *Man the Builder.* New York: McGraw-Hill, 1970.

Scarre, Chris, ed. *Seventy Wonders of the Ancient World: The Great Monuments and How They Were Built.* London: Thames and Hudson, 1999.

Schonfield, Hugh J. *The Suez Canal in Peace and War: 1869–1969.* Coral Gables, FL: University of Miami Press, 1969.

Seasholes, Nancy. "Landmaking and the Process of Urbanization: The Boston Landmaking Projects, 1630s–1888." PhD diss., Boston University, 1994.

Seely, Bruce E. *Building the American Highway System: Engineers as Policy Makers.* Philadelphia: Temple University Press, 1987.

Sellers, Wallace O. "Financing 'Orbital Power & Light, Inc.'" In *Solar Power Satellites: A Space Energy System for Earth.* Edited by Peter E. Glaser, Frank P. Davidson, and Katinka Csigi. Chichester, England: John Wiley and Sons/Praxis Publishing, 1998.

Shaeffer, John R. and Leonard A. Stevens. *Future Water*. New York: Morrow & Co., 1983.

Shepherd, C. W. *A Thousand Years of London Bridge*. London: John Baker, 1971.

Shors, John. *Beneath a Marble Sky: A Novel of the Taj Mahal*. Kingston, NY: McPherson, 2004.

Shultz, George P. "The Abrasive Interface." In *Business and Public Policy*. Edited by John T. Dunlop. Cambridge, MA: Harvard University Press, 1980.

Silverman, Brian S., Arvids A. Ziedonis, and Rosemarie Ham Ziedonis. "Diffusion of Research from SEMATECH: Evidence from Patent Citations." Working paper, January 2002. Paper presented at the Second Annual Wharton Technology Mini-Conference, the Emerging Technologies Management Research Program, The Wharton School, University of Pennsylvania, April 12–13, 2002.

Slater, Humphrey, and Correlli Barnett, in collaboration with R. H. Géneau. *The Channel Tunnel*. London: Allan Wingate, 1957.

Smiles, Samuel. *Lives of the Engineers with an Account of Their Principal Works Comprising Also a History of Inland Communication in Britain*. 3 vols. London: John Murray, 1861.

Song, Lian, comp. *The History of Yuan*. Shanghai: Chinese Book Bureau, 1976.

Spier, Peter. *Of Dikes and Windmills*. New York: Doubleday, 1969.

Ssu Ma Kuang. *The Grand Scribe's Records: The Basic Annals of Pre-Han China*. Bloomington: Indiana University Press, 1994.

Ssu Ma Kuang, comp. *Tzu-chi t'ung chien (Comprehensive Mirror for Aid in Government)*. N.p.: Yueh Lu Books Association, n.d.

Steele, James. "The Effect of the Aswan High Dam upon Village Life in Upper Egypt." IASTE 2nd International Conference, "First World-Third World: Duality and Coincidence in Traditional Dwellings and Settlements," University of California at Berkeley, CA, USA, October 1990.

Stein, Charles. "Be It Ditch or Dot.com, a World of Growth." *Boston Globe*, January 23, 2005.

Stephenson, Richard M. "*A Plan Whol[l]y New*": *Pierre Charles L'Enfant's Plan of the City of Washington*. Washington: U.S. Government Printing Office, 1993.

Steward, J. S., ed. *The Collections of the Romanovs: European Art from the State Hermitage Museum*. London and New York: Merrell, 2003.

Stillman, Laurie. "Health and the Big Dig." *Boston Globe*, January 19, 2005.

Stine, Jeffrey K. *Mixing the Waters: Environment, Politics, and the Building of the Tennessee-Tombigbee Waterway*. Akron, OH: University of Akron Press, 1993.

Straszak, A., ed. *The Shinkansen Program: Transportation, Railway, Environmental, Regional and National Development Issues*. Laxenburg, Austria: Institute for Applied Systems Analysis, 1981.

Sutherland, Lucy. *The East India Company in 18th Century Politics*. Oxford: Oxford University Press, 1952.

Swartzwelter, Brad. *Faster than Jets: A Solution to America's Long-Term Transportation Problems*. Kingston, WA: Alder Press, 2003.

Taniguchi, Mamoru. *High-Speed Rail in Japan: A Review and Evaluation of the Shinkansen Train*. Berkeley: University of California, Institute of Urban and Regional Development, 1992.

Taylor, Eugene, and Robert H. Wosniak, eds. *Pure Experience: The Response to William James*. St. Augustine, FL: St. Augustine Press, 1996.

Teale, J., and M. Teale. *Life and Death of the Salt Marsh*. New York: Ballantine Books, 1969.

Temple, Robert. *The Genius of China*. New York: Simon and Schuster, 1986.

Tesson, Thierry. *Ferdinand de Lesseps*. Paris: J.-C. Lattès, 1992.

Thompson, Slason. *A Short History of the American Railways*. Chicago: Bureau of Railway News and Statistics, 1925.

Tobin, James. *Great Projects: The Epic Story of the Building of America, from the Taming of the Mississippi to the Invention of the Internet*. New York: Free Press, 2001.

Toynbee, Arnold, ed. *Cities of Destiny*. London: Thames and Hudson, 1976.

Treaty Series: Treaties and International Agreements Registered or Filed and Recorded with the Secretariat of the United Nations. Vol. 923. New York: United Nations, 1981.

Tsipis, Yanni K. *Boston's Central Artery: Images of America*. Mount Pleasant, SC: Arcadia Publishing, 2001.

Tuma, J. *The Pictorial Encyclopedia of Transport*. Translated by Alena Emhornova. London: Hamlyn, 1979.

Turner, Frederick J. *The Frontier in American History*. New York: Henry Holt, 1921.

Twain, Mark. *Roughing It*. 1892. Reprint, New York: Penguin Classics, 1985.

Twitchell, Heath. *Northwest Epic: The Building of the Alaska Highway*. New York: St. Martin's Press, 1992.

U.S. Department of the Interior. National Park Service. *National Trails Assessment*. 1986.

U.S. Department of State. *The Suez Canal Problem, July 26–September 22, 1956: A Documentary Publication*. Washington, DC: U.S. Department of State, 1956.

U.S. Statutes at Large 70 (1956), chap. 462: 374–402.

U.S. Statutes at Large 42 (1921), chap. 119: 212–19.

U.S. Statutes at Large 39 (1916), chap. 241: 355–59.

Unger, Margaret. *Voices from the Snowy: The Personal Experiences of the Men and Women Who Worked on One of the World's Great Engineering Feats; The Snowy Mountains Scheme*. Kensington, Australia: NSWU Press, 1989.

Van den Brink, Erwin. "The Netherlands: A Country That Engineered Itself into Existence Is Tapping into Its Centuries-Old Expertise in Handling Water." *Technology Review* 108, no. 4 (2005).

Vanderwarker, Peter. *The Big Dig: Reshaping an American City*. Boston: Little, Brown, 2001.

Van de Ven, G. P., ed. *Man-Made Lowlands: History of Water Management and Land Reclamation in the Netherlands*. Utrecht, Netherlands: Stichting Matrijs, 2004.

Van Heinigen, H. *Tussen Maas en Waal*. Zutphen: De Walburg Pers, 1972.

Van Veen, J. *Dredge, Drain, Reclaim: The Art of a Nation*. The Hague: Martinus Nijhoff, 1962.

Verne, Jules. *De la terre à la lune* [From the Earth to the Moon]. 1865. Translated by New York: Bantam Classic, 1993.

Von Hagen, Victor Wolfgang. *The Roads That Led to Rome*. Cleveland, OH: World Publishing Company, 1967.

Vranich, Joseph. *Supertrains: Solutions to America's Transportation Gridlock*. New York: St. Martin's Press, 1991.

Waldman, Neil. *The Two Brothers: A Legend of Jerusalem*. New York: Athenaeum Books for Young Readers, 1997.

Ward, Evan Ray. *Border Oasis: Water and the Political Ecology of the Colorado River Delta, 1940–1975*. Tucson: University of Arizona Press, 2003.

Waterman, Leroy. "The Damaged 'Blueprints' of the Temple of Solomon." Chicago: University of Chicago Press, 1943.

Whalen, David J. *The Origins of Satellite Communications, 1945–1965*. Washington, DC: Smithsonian Institute Press, 2002.

Whiteside, Thomas. *The Tunnel under the Channel*. London: Rupert Hart-Davis, 1962.

Whitney, Asa. *A Project for a Railroad to the Pacific*. New York: George Ward, 1849.

Whittington, Dale, and Giorgio Guariso. *Water Management Models in Practice: A Case Study at the Aswan High Dam*. Amsterdam and New York: Elsevier Scientific Publishing, 1983.

Wiet, Gaston. *Baghdad: Metropolis of the Abbasid Caliphate*. Norman: University of Oklahoma Press, 1971.

Wilkie, Richard W., and Jack Tager. *Historical Atlas of Massachusetts*. Amherst: University of Massachusetts Press, 1991.

Wilson, David Gordon. "Guided Transportation Systems: Low-Impact, High-Volume, Fail-Safe Travel." In *Macro-engineering: MIT Brunel Lectures on Global Infrastructure*, edited by Frank P. Davidson, Ernst G. Frankel, and C. Lawrence Meador. Horwood Series in Engineering Science. Chichester, England: Horwood Publishing, 1997.

Wilson, David Gordon. "Palleted Automated Transportation (PAT)—A View of Developments at the Massachusetts Institute of Technology." Tokyo: *IATSS Research* 13, no. 1 (1989): 53–60.

Wilson, Richard Guy. "American Modernism in the West: Hoover Dam." In *Images of an American Land*. Edited by Thomas Carter. Albuquerque: University of New Mexico Press, 1997.

Winslow, Thacher and Davidson, Frank P., editors. *American Youth: An Enforced Reconnaissance*. Cambridge, MA: Harvard University Press, 1940.

Wood, Alan. *The Groundnut Affair*. London: Bodley Head, 1950.

Wood, Michael. *The Road to Delphi: The Life and Afterlife of Oracles*. New York: Farrar, Straus, and Giroux, 2003.

Yao, Han-yüan. *The Grand Canal: An Odyssey*. Edited by Liao Pin. Beijing: Foreign Languages Press, 1987.

Yen, Ching-Hwang. *A Social History of the Chinese in Singapore and Malaya 1800–1911*. Singapore: Oxford University Press, 1986.

Ziedonis, A.A., R.H. Ziedonis, and B.S. Silverman. "Research Consortia and the Dissemination of Technological Knowledge: Insights from the SEMATECH." Working paper, Stephen M. Ross School of Business, University of Michigan, and the Joseph L. Rotman School of Management, University of Toronto, 2005.

Zielinski, A. *Tajemnice polskich templariuszy* [The Polish Knights Templar]. Warsaw: Dom Wydawn, Bellona, 2003.

Internet

The editors wish to advise those who utilize the Internet references provided in the chapters and the bibliography, that we have made every effort to confirm the validity of each website as of this edition's publication date.

Abbasids. See http://www.saudiaramcoworld.com/issue/196209/building.of.baghdad. htm.

Abuja, Murtala Mohammed speech. See http://www.dawodu.com/murtala1.htm.

Aida, commission and creation. See http://www.r-ds.com/opera/verdiana.aida.htm.

Alaska native people's music, including CDs by Sylvestor Ayek, John Pingayak, Evelyn Alexander, Shxat'kwaan Daners, and Ethan Petticrew and the Atka Dancers, including sound clips. See http://www.oyate.com/.

Alaska Pipeline, history and facts. See http://www.alyeska-pipe.com/pipelinefacts. html.

Alaska Pipeline history, by the Statewide Library Electronic Doorway: Information Resources for, about, and by Alaskans. See http://sled.alaska.edu/akfaq/aktaps. html.

Alpine tunnel, 1475–1480. See http://www.ltf-sas.com/pages/articles.php?art_id=265.

Arctic National Wildlife Refuge. See http://www.anwr.org/.

Aswan Dam. See http://www.thebestlinks.com/Aswan_High_Dam.html.

Aswan Dam, statistics on increased acreage and a chart illustrating land reclamation as a result of the Aswan Dam. See http://www.fao.org/ag/agl/swlwpnr/reports/y_nf/ egypt/e_lcover.htm.

Aswan Dam, with a drawing showing both the Low Dam and the High Dam, as well as the cooperative efforts between Egypt and the Soviet Union. See http://carbon. cudenver.edu/stc-link/aswan/organi.htm.

Bailey Bridges. See http://www.fact-index.com/b/ba/bailey_bridge.html.

Benjamin Banneker, contributions. See the review by Michael Levine at http://dcpages. com/History/Planning_DC.html.

Bicycle, history, including the first Japanese production bike, and the Miyata Gun Factory. See http://www.cycle-info.bpaj.or.jp/.

Bicycling, road planning and development. See http://www.itdp.org/ (includes a link to the Global Bicycle Fund for mobility in developing areas), http://www.transalt. org/, http://www.carfree.com/, and http://www.patternlanguage.com/.

Big Dig website. See http://www.masspike.com/bigdig/index.html

Brasília, and the public archive of the Federal District. See http://www.geocities.com/ thetropics/3416/minis_ic.htm.

Brasília, history of, and Costa's pilot plan. See http://www.infobrasilia.com.br/.

Brasília's water problems. See http://www.tierramerica.net/2001/0603/iacentos2. shtml.

Brooklyn Bridge, overview with illustrations, including caissons. See http://www.endex. com/gf/buildings/bbridge/bbridgefacts.htm.

Canadian Pacific Railway. See the company's website at http://www8.cpr.ca/.

Canal du Midi, background on the concept of the canal through history and the leadership of Pierre-Paul Riquet. See http://www.canalmidi.com/anglais/historgb.html.

Canal du Midi, engineering and technical aspects of the building process, with illustrations of the tunnels and locks. See http://www.canalmidi.com/anglais/ouvraggb.html.

Canal du Midi, information on UNESCO and its designation of the Canal as a World Heritage Site. See http://whc.unesco.org/sites/770.htm.

Canals, a time line of major canals from the beginnings to 1749. See http://home.eznet.net/~dminor/Canalto1749.html.

Cats and grain storage in Europe. See http://www.isabellevets.co.uk/new_cat/newcat.htm.

Channel Tunnel, records related to planning in the 1950s and 1960s. See Technical Studies, Inc. Records from 1957–1994, Historical Collections, Baker Library, School of Business Administration, Harvard University, at http://www.library.hbs.edu/hc/additions/tsi.shtml.

Channel Tunnel, summary of. See http://www.pbs.org/wgbh/buildingbig/wonder/structure/channel.html.

Charlemagne and Alcuin, audio program including quotes and a bibliography on Charlemagne. See http://www.uh.edu/engines/epi797.htm.

Charlemagne scholarship. See http://www.rwth-aachen.de/zentral/charlemagne_scholarship.htm.

Clifton, L. J., and B. J. Gallaway, "History of Trans Alaska Pipeline System." See http://tapseis.anl.gov/documents/docs/Section_13_May2.pdf.

Clinton Engineer Works (forerunner to Oak Ridge). See http://www.vcdh.virginia.edu/HIUS316/mbase/docs/oak.html.

Crane, Hart, work on the Brooklyn Bridge. See http://www.poets.org/poems/Poempmt.cfm?45442B7C000C07070F77.

Cyrene and cultural sites in the Libyan Arab Jamahiriya. See http://whc.unesco-org/sites/190.htm, and http://www.geocities.com/Athens/8744/unesco.htm.

Cyrene. See http://www.galenfrysinger.com/cyrene_libya.htm.

Dalia, Wendy, and Robert Stevenson. "Boston and Its Environs," "Conservation New England: Past Present & Future," Conservation Biology of New England, Biology Department of the University of Massachusetts, Boston. See http://site.www.umb.edu/conne/wendy/Boston.html.

Dams and the environment. See http://www.unep.org/dams.

Disraeli, Benjamin, speech on the acquisition of the Suez Canal shares, Feb. 21, 1876, see: http://www.historyhome.co.uk

Edmondson, Brad. "In the Driver's Seat." American Demographics, March 1998. http://www.findarticles.com/p/articles/mi_m4021/is_n3_v20/ai_20375004.

Eiffel Tower. See http://www.tour-eiffel.fr/.

Eiffel Tower, webcam views. See http://www.abcparislive.com.

Eiffel Tower, 3D model and images. See http://www.greatbuildings.com/buildings/Eiffel_Tower.html.

Equestrian trails in Canada. See http://www.equinecanada.ca/.

Erie Canal, overview. See http://www.eriecanal.org/ and http://www.nycanal.com/nycanalhistory.html.

Eurotunnel, information about the company's history, management, and board and tunnel traffic. See http://www.eurotunnel.com/.

Genghis Khan. See http://news.nationalgeographic.com/news/2003/02/0214_030214_genghis.html.

Glass, Philip. *Itaipú*. See http://www.philipglass.com/html/compositions/itaipu.html.

Global Infrastructure Fund. See http://www.ecdc.net.cn/partners/gif.htm.

Grand Canal, illustrated guide. See http://www.chinapage.com/canal.html.

Grand Canal, images. See http://www.altavista.com/image/results?pg=qdstype=simage&imgset=2&q=The%20Grand%20Canal%20of%20China&Avkw=aaps.

Grand Canal, influence on America's development of canals. See http://members.tripod.com/~american_almanac/canal.htm.

Grand Canal, the "artificial Nile" did for China what the real Nile did for Egypt. See http://library.thinkquest.org/20443/grandcanal.html.

Groves, Leslie. See http://www.nuclearfiles.org/.

Handy, W.C., music, and the W.C. Handy Festival. See http://www.wchandyfest.com/.

Hansen, Oskar, and Gordon Kaufmann. See http://www.usbr.gov/lc/hooverdam/history/articles/rhinehart1.html

Hermitage. See http://www.hermitage.ru/.

Herodotus. See http://www.fordham.edu/halsall/ancient/630cyrene.html.

Hise, Troy, a history with pictures and text of Hise's lament, Morris Communications, 1998. See http://www.themilepost.com/history.html.

Hoover Dam, and its construction. See http://www.arizona-leisure.com/hoover-dam-building.html and http://www.arizona-leisure.com/hoover-dam-diversion-tunnels.html.

Hoover Dam, as a curved gravity structure. See http://www.pbs.org/wgbh/buildingbig/wonder/structure/hoover.html

Hoover Dam, official website. See http://www.hooverdam.com.

Howe, Jeffery. "A Digital Archive of American Architecture: A Study Guide for FA 267—From Saltbox to Skyscraper: Architecture in Boston." See http://www.bc.edu/bc_org/avp/cas/fnart/fa267/bos_fill2.html.

Hyperbaric chamber and its history dating to 1662. See http://www.tbims.org/combi/ubb/Forum4/HTML/000030.html.

Iditarod National Historic Trail. See http://www.iditarod.com.

Institute of the North and its founder, Walter J. Hickel. See http://www.institutenorth.org/hickel.html.

International Research Institute for Climate and Society. See http://iri.columbia.edu/ and http://wwics.si.edu/topics/pubs/Amazon.pdf.

International Rivers Network, report on the World Bank's role in funding dams. See http://www.irn.org/pubs/wp/damming.html.

International Space Station. See http://www.shuttlepresskit.com/ISS_OVR/.

International time zones at the International Meridian Conference, October 1884. See http://wwp.greenwich2000.com/millennium/info/conference-finalact.htm.

Islamic art, architecture, and culture, course syllabi from professors worldwide. See http://archnet.org/courses/.

Itaipú, key facts. See http://www.pbs.org/wgbh/buildingbig/wonder/structure/itaipu.html.

Itaipú project, general information. See http://www.itaipu.gov.br/, http://ce.eng.usf.edu/pharos/wonders/Modern/itaipu.html, and http://www.sovereign-publications.com/itaipu.htm.

Japan Rail website. See http://www.japanrail.com.

Japanese railway timeline, from 1872 to the present. See http://www.rtri.or/jp/.

Jefferson, Thomas, letters and papers at the Library of Congress. See rs6.loc.gov/ammem/collections/jefferson_papers.

Jones, Tyler, for information on Panama Canal. See http://www.june29.com/Tyler/nonfiction/pan2.html.

Kerouac, Jack. "The Brooklyn Bridge Blues." See http://www.geocities.com/yesterdayswine/BrooklynBridgeBlues.html?200415.

Kuzmichenko, Svetlana, website that explores strategies for doing business in Russia and Eurasia, including extending the Trans-Siberian Railway to Japan and Korea. See http://www.bisnis.doc.gov/bisnis/bisdoc/010921rail.htm/s

Kyotani, Yoshihiro, electrical engineer who helped develop the Shinkansen. See http://www.ieec.org/organizations/history_center/ oral_histories/transcripts/kyotani.html

Lake Baikal spur, with pictures and maps. See http://baikal.irkutsk.org/railway.htm.

Land reclamation in the Netherlands. See http://www.knag.nl/pagesuk/geography/engels/news99engelstekst.html.

L'Enfant and McMillan Plans. See http://www.cr.nps.gov/nr/travel/wash/lenfant.htm.

Library of Congress. Country Study on Libya. See http://workmall.com/wfb2001/libya/libya_history_cyrenaica_and_the_greeks.html.

London Bridge, view in 1540. See http://www.bbc.co.uk/history/3d/bridge.shtml.

"London Bridge is Falling Down," nursery rhyme and legend behind. See http://www.rhymes.org.uk/london-bridge-is-broken-down.htm.

MacAdam, John L. and Thomas Telford. See http://www.greatachievements.org/greatachievements/ga_11_2.html.

Maglev trains. See http://www.rtri.or.jp/rtri/future_E.html

Maglev trains are an advanced form of guided transport. How **maglev** trains work. See travel.howstuffworks.com/**maglev**-train.htm

Maglev trains. For maglev research on linear synchronous motor (LSM) technology developed by Richard Thornton and Associates for government and industry. See http://magnemotion.com.

Main-Danube Canal. See http://www.deutsches-museum.de/ausstell/dauer/wasserle_wass2.htm.

Manhattan Engineer District organization chart. Society for the Historical Preservation of the Manhattan Project's Web site. See http://www.childrenofthemanhattan-project.org/COTMP/MED_chart_Printable.htm.

Marconi-Tesla U.S. Supreme Court decision to resolve the proper patentholder for wireless transmission (radio). See http://www.justia.us/us/320/1/case.html.

Marib Dam. See http://www.saudiaramcoworld.com/issue/197802/a.dam.at.marib.htm and http://www.yementimes.com/98/iss52/lastpage.htm.

McGurk, Tom. *Treasures of the World* (soundtrack). Music from the PBS series *Treasures of the World,* with tracks for the Taj Mahal and Borobudur. See http://www.pbs.org/treasuresoftheworld/ and http://www.flopsy.com/.

Metros and subways around the world, selected songs ranging from Duke Ellington's "Take the A-Train" to the songs composed for Berlin's subway. See http://www. urbanrail.net/metro-song.htm.

Modern wonders of the world, American Society of Civil Engineers (ASCE) list. See http://www.asce.org/history/7_wonders.cfm.

Mont Blanc Tunnel, chronology, with bibliography before and after the 1999 fire. See the International Database and Gallery of Structures at http://www. structurae.net/.

Mont Blanc Tunnel, fire. See http://news.bbc.co.uk/1/hi/world/europe/1856504.stm.

Morgan, Morgan, and Lilienthal, the triumvirate of the TVA. See http://newdeal.feri. org/tva/tva05.htm.

Mulberry artificial harbors. See http://www.valourandhorror.com/DB/BACK/Mulberry. htm

NASA, history and the Apollo mission. See http://history/nasa.gov.

National Canal Museum. See http://www.canals.org/.

National Council for Science and the Environment. See http://www.cnie.org/.

National Trails System overview, with authorization dates and public laws of trails. See http://www.americantrails.org/resources/feds/NatTrSysOverview.html.

National Trails System, background and history. See http://usparks.about.com/library/ weekly/aa060599.htm.

Nationwide Personal Transportation Survey (NPTS). See http://www.fhwa.dot.gov/ ohim/1983/1983page.htm.

New River, aqueducts and tunnels. See http://www.citiesofscience.co.uk/go/London/ ContentPlace_2525.html.

New River, overview. See http://www.waterinschools.com/newriver/story.html.

New River, walking route, with lists of publications, including *The New River Path: A Walk Linking Hertford and Islington*. See http://www.ramblers.org.uk/info/ paths/newriver.html.

New River, water use and treatment. See http://www.riverlee.org.uk/.

Nigeria, and the city of Abuja. See http://www.nigeria.gov.ng/ and http://www. abujacity.com.

Nigeria, popular culture. See http://www.mothernigeria.com/music.html.

Nigeria, socio-political issues. See http://www.dawodu.com.

Norris, Tennessee. See http://tva.gov/heritage/norris/.

Nuclear age in the 1940s, an overview. See http://www.nuclearfiles.org/.

Nuclear time line in the 1940s, including links to video and audio clips from December 7, 1941, from CBS News. See http://www.radiochemistry.org/history/nuclear_ timeline/40s.html.

Nuclear treaties and agreements from 1963 to 1966. See http://library.thinkquest. org/17940/texts/timeline/treaties.html.

Oak Ridge Laboratories. See http://www.ornl.gov/ornlhome/about.shtml.

Panama Canal, access to legal documents: http://www.pancanal.com/eng/legal/law/ index.html.

Panama Canal, coverage of the American role. See http://www.pbs.org/wgbh/amex/tr/ panama.html.

Panama Canal Authority, future and activities. See http://www.riverdeep.net/ current/2000/12/120800_panama.jhtml.

Peter the Great, a psychological portrait and biography. See http://mars.wnec.edu/~grempel/courses/russia/lectures/12peter1.html.

Polders and dikes. See http://geography.about.com/library/weekly/aa033000a.htm.

Population growth in western Europe. See http://www.tulane.edu/~august/H303/handouts/Population.htm.

Precipitation/evaporation ratios worldwide. See the glossary of the American Meteorological Society at http://amsglossary.allenpress.com/glossary/browse?s=p&p=54.

Ragtown, and the story of Murl Emery. See http://www.arizona-leisure.com/hoover-dam-men.html.

Rail connections from Europe to the Middle East. See http://www.seat61.com/Syria.htm.

Rails-to-Trails Conservancy. See http://www.railtrails.org/.

Rail tunnel in Switzerland, and background on Alpine tunnels. See http://www.geotimes.org/feb03/feature_tunnel.html.

Reader's Companion to American History. Houghton Mifflin Online Study Center. See http://college.hmco.com/history/readerscomp/rcah/html/rc_021900_crditmobilie.htm.

Rhine River. See http://www.public.asu.edu/~goutam/gcu325/rhine.htm.

Rijksmuseum, works by famous Dutch artists. See http://www.rijksmuseum.nl/.

Rijkswaterstaat, home page: http://www.rijkswaterstaat.nl/en/.

Ritter, Joyce N., how highway statistics were developed as the basis for the U.S. interstate system. See http://www.fhwa.dot.gov/ohim/1994/text/history.htm.

River and waterways engineering. See http://www.deutsches-museum.de/ausstell/dauer/wasser/wasser.htm.

Roman aqueducts, an overview. See http://www.dl.ket.org/latin3/mores/aqua/.

Roman army and its engineering corps. See http://www.roman-empire.net.

Romans in the Netherlands. See http://www.livius.org/ga-gh/germania/woerden.html.

Roosevelt, Franklin D., including on TVA. See http://newdeal.-feri.org/.

Roosevelt, Franklin D., letters. Roosevelt Library, Marist College, Poughkeepsie, NY. See http://www.fdrlibrary.marist.edu/fdrbx.html.

Rosenstock-Huessy, Eugen. See http://www.valley.net/~transnat/erh.html.

Russian czars. See http://killeenroos.com/4/Czar1800.htm.

Salinity and irrigation. See http://web.bryant.edu/~langlois/ecology/pollution.html.

São Paulo's population. See http://www.demographia.com/db-mumsao.htm.

Satellite, broadband, cable, broadcasting, and multimedia industries. See http://www.satnews.com.

SEMATECH. See http://www.sematech.org/.

SEMATECH, interview with former CEO Mark Melliar-Smith and CEO Robert Helms. See http://www.micromagazine.com/archive/01/06/leadnews.html.

SEMATECH North. See http://www.albany.edu/pr/ualbanymagfall02/sematech.htm.

Shinkansen, train types, statistics, dates of construction, services operated, lines used, cars per unit, maximum speed, and colors. See http://www.h2.dion.ne.jp/~dajf/byunbyun/types.htm.

Silk Road. See http://www.silkroadproject.org/.

Silphium, an image of (engraved on an ancient coin). See http://www.livius.org/ct-cz/cyrenaica/cyrenaica.html.

Singapore. See http://www.singstat.gov.sg/.

Singapore, artists. See http://www.living2000.com.sg/chenchongswee/biography.htm.

Singapore, overview. Library of Congress Country Studies. See http://www.workmall.com/wfb2001/singapore and http://memory.loc.gov/frd/cs/sgtoc.html.

Singapore, pictures of ancient temples and mosques. See http://www.orientalarchitecture.com/singapore.htm.

Six Companies, Inc. See http://www.constructioncompany/historic-construction-projects/hoover-dam/.

Snowy Mountain, transition to present times. See http://www.users.bigpond.net.au/snowy/overview.html.

Snowy Mountain, website. See http://www.snowyhydro.com.au/

Solomon's Temple. See http://www.templemount.org/.

Solomon's Temple, Phoenician influence on the design and building of, including a bibliography. See http://phoenicia.org/temple.html.

South Korea's proposed new capital. See http://news.bbc.co.uk/1/hi/world/asia-pacific/3554296.stm.

Space elevator concept. See http://www.spaceelevator.com.

Space travel in popular culture. Web sites for the television show *Star Trek* and the comic strip *Buck Rogers in the 25th Century*. See http://www.startrek.com/ and http://www.buck-rogers.com/.

St. Petersburg, antique photos. See http://www.alexanderpalace.org/petersburg1900/2.html.

St. Petersburg, current news and information. See http://www.sptimes.ru/.

St. Petersburg, population and changes. See http://www.fact-index.com/s/sa/saint_petersburg.html.

St. Petersburg University. See http://www.cityvision2000.com/city_tour/university.htm.

Statistics relating to international injury and fatality statistics. See http://www.safecarguide.com/exp/statistics/statistics.htm and http://www.scienceservingsociety.com/p/141.htm.

Suez Canal crisis, a select bibliography. Dwight D. Eisenhower Presidential Library in Abilene, Texas. See http://www.eisenhower.utexas.edu/suez.htm.

Taj Mahal, background on the love story. See http://www.pbs.org/treasuresoftheworld/taj_mahal/tmain.html.

Taj Mahal, details on the unique plumbing and water systems. See http://www.tajma-halindia.net/taj-mahal-garden-water-devices.html.

Taj Mahal, photo gallery of images. See http://www.greatbuildings.com/buildings/Taj_Mahal.html.

Taylor, Robert W. "Urban Development Policies in Nigeria: Planning, Housing, and Land Policy." Upper Montclair, NJ: Department of Environmental, Urban, and Geographic Studies, Montclair State University, September 2000. See http://alpha.Montclair.edu/~lebelp/CERAFRM002Taylor1998.pdf.

Thames Water Authority. See http://www.thames-water.com/.

Time-use survey, University of Maryland. See http://www.webuse.umd.edu/.

Transatlantic cable and Morse Code. See http://www.oldcablehouse.com/cablestations/history.html.

Trans-Atlantic tunnel, using interactive 3-D options. See http://dsc.discovery.com/convergence/engineering/transatlantictunnel/interactive.html.

Trans Canada Trail. See http://www.tctrail.ca/.

Trans-Siberian Railway, building process by the 27th Infantry Regimental Historical Society. See http://www.kolchak.org/History/Siberia/Trans-Siberian%20Railroad.htm.

Trans-Siberian Railway, original *Scientific American* article (August 26, 1899), with photos. See http://www.travelhistory.org/siberia/.

Tunnel fires in New York. See http://www/panynj.gov/pr/116–99.html.

Tunneling companies. See http://www.bouygues-construction.com/en/.

Tunnels, notable rail and vehicular. Pearson Education website. See http://www.infoplease.com/ipa/A0001340.html.

TVA. See http://www.tva.gov/abouttva/history.htm.

TVA, a library of cartoons. See http://newdeal.feri.org/toons/toon10.htm.

TVA, logo. See http://newdeal.feri.org/tva/index.htm.

TVA, pictures, maps, news releases. See http://www.encyclopedia.com/html/section/TennVA_Bibliography.asp.

U.S. Arctic Research Commission and Mead Treadwell. See http://www.arctic.gov/mtreadwell.htm.

U.S. Highways, detailed information, with hyperlinks. See http://www.fhwa.dot.gov/fhwaweb.htm.

U.S. Highways, details on concrete, slag, and other road-construction techniques. See http://www.cement.org/pavements/pv_cp_highways.asp.

Udall, Morris, 1970–84 archives of the TVA, with a complete listing of the files and notations from all reports and committee documents. See http://dizzy.library.arizona.edu/branches/spc/udall/udallfindingaid/.

Ventura, Fihlo, Altino. "Itaipú: A Binational Hydroelectric Power Plant, Its Benefits and Regional Context." See http://www.dams.org/kbase/submissions/showsub.php?rec=ins237.

Villanueva Siasoco, Ricco. "Red, White, and Blue Highways: The Story of the U.S. Interstate." See http://www.infoplease.com/spot/interstate1.html.

Wallenberg, Christopher. "Digging Deep." *Panorama*. See http://www.panoramamagazine.com/panoramamagazine/articles/digging_deep.asp.

Water in times of war, and as a regional water supply, Raker Bill of 1913. See http://www.sfmuseum.org/hetch/hetchy10.html.

Weingraff, Richard F. Federal Highway Administration, history of the legislative process, with historic photos. See http://www.tfhrc.gov/pubrds/summer96/p96su10.htm.

Whalen, David J., "Communications Satellites: Making the Global Village Possible." See http://www.hq.nasa.gov/office/pao/History/satcomhistory.html.

Windmills. See http://www.kinderdijk.nl/.

Wireless, history of. See http://www.pbs.org/tesla/res/res_radtime.html.

World War II, costs of traditional military war versus the Manhattan Project. See http://www.brook.edu/fp/projects/nucwcost/manhattn.htm.

Y-12 information. See http://www.y12.doe.gov/.

Music

Armstrong, Louis. *Louis Armstrong Plays W. C. Handy*. Sony 64925, 1997.

Beethoven, Ludwig van. Piano Sonata No. 14 in C Sharp Minor, Opus 27, No. 2, "Moonlight." Rudolf Serkin, pianist. Sony 37219, 1990. Debussy, Claude. "Clair de lune." From *Suite Bergamasque*.

Brody, Martin. *Earth Studies Sound Recording*. Recorded live at the Duncan Theatre, January 12–13, 1996. Janice Felty, mezzo-soprano; William Hite, tenor; James Maddalena, baritone; Demetrius Klein Dance Company; Mary Forcase, narrator. Sound cassette. Cambridge, MA: Harvard University, Loeb Music Record Coll., Call No. AC 33342. Track II Cyrene, Ode of Pindar, Greece, 5th Center BC Published by S.I.: s.n., 1996.

Cole, Nat King. "Route 66." Written by Bobby Troup. Audio clip available at http://members.cox.net/jdmount/route_66.aif.

Edwards, Gus. "In My Merry Oldsmobile" Lyricist: Vincent Bryan. New York: M. Witmark & Sons, 1905. http://libraries.mit.edu/music/sheetmusic/childpages/inmymerryoldsmobile.html

Glass, Philip. *Itaipu*. Libretto translated from Guarani Indian text by Daniela Thomas, copyright 1988 Dunvagen Music Publishers, Inc. New York: Sony 46352, 1993.

Horn, Paul. *Inside the Taj Mahal*. Kuck Kuck Records #11062, 1991.

John, Elton, and Tim Rice. *Aida*. Island Records 524628, 1999.

Lightfoot, Gordon. "Canadian Railroad Trilogy." *The Way I Feel*. United Artists, BH22967, 1976.

Ma, Yo Yo, and the Silk Road Ensemble. *Silk Road Journeys: When Strangers Meet*. Sony Classical 089782, 2002.

Ma, Yo Yo, and the Silk Road Ensemble. *The Silk Road: A Musical Caravan*. Smithsonian Folkways Recording SFW40438, 2002.

McPartland, Marian. *Chick Corea: Marian McPartland's Piano Jazz*. Jazz Alliance, TJA120402, 2002.

Norman, Jessye. "The Holy City," by Stephen Adams. *Sanctus: Sacred Songs*. With the Ambrosia Singers and the Royal Philharmonic Orchestra, Sir Alexander Gibson, conductor. Philips 400019, 1992.

Respighi, Ottorino. *Fountains of Rome*. Conducted by Malcolm Sargent. Everest Records CD #9018, 1995.

Songs of London. Capital Radio 194, HALC3 CD UK, 1979.

Tiso, Wagner. *Memorial*. Biscoito Fino, BF-519, 2002.

Verdi, Guiseppe. *Aida*. Claudio Abbado, conductor; with Placido Domingo, Martina Arroyo, and others. 2 discs. CD Opera D'Oro 1167, 1998

Williams, Ralph Vaughan. "As I Walked over London Bridge." Six Studies in English Folk Song, No. 6. CD. *Bassoon Bon-bons*. White Line, Cat. No. 2052, Track #16, 1991.

Williams, Ralph Vaughan. "A London Symphony" No. 2. The London Symphony. Conducted by Bryden Thomson. Chandos 8629, 1988.

Film and Television

2001: A Space Odyssey. Directed by Stanley Kubrick. Burbank, CA: Warner Studios, 1968.

Brooklyn Bridge. Videocassette. Directed by Ken Burns. Boston: PBS Home Video, 1997. In *The National Dream*. Eight part series. Toronto: Canadian Broadcasting Company TV series 1975.

Building Big. Videocassette box set. Hosted by David Macaulay. Boston: WGBH, PBS Television, 2000.

Building the Alaska Highway. Produced and directed by Tracy Heather Strain. Boston: WGBH Educational Foundation, PBS Television, 2005.

Casablanca. Directed by Michael Curtiz. Burbank, CA: Warner Studios, 1942.

Destination Tokyo. Directed by Delmer Daves. Burbank, CA: Warner Studios, 1943.

Dr. Strangelove, or How I Learned to Stop Worrying and Love the Bomb. Directed by Stanley Kubrick. Culver City, CA: Columbia TriStar, 1964.

Easy Rider. Directed by Dennis Hopper. Culver City, CA: Columbia TriStar Studios, 1969.

Enola Gay. Directed by Tim Curran. New York: A&E Entertainment, 1995.

Erie Canal: Albany to Buffalo. DVD. Cicero, NY: Media Artists, 2002.

Eurotunnel—Two Years On. Videocassette. Pasadena, CA. Association of American Railroads, 1993.

Extreme Engineering: Transatlantic Tunnel. DVD. Discovery Channel. Powderhouse Productions, Somerville, MA, 2003.

Hiroshima: Decision to Drop the Bomb. New York: A&E Home Video, 2001.

Modern Marvels: Hoover Dam. A&E Home Video. Orland Park, II: M.P.I. Media Group, 1999.

Modern Marvels: The Eiffel Tower. Videocassette. Directed by Larayne Decoeur, Noah Morowitz. New York: Modern Marvels, 1994.

Modern Marvels: The Erie Canal. Videocassette. New York: A&E Television, History Channel, 2000.

Russian Ark. Sokurov, Alexander. Fox Lorber, #720917538228. DVD. 2002.

Secrets of Lost Empires: Roman Baths. Boston: WGBH TV, 2/22/02.

Thelma and Louise. Directed by Ridley Scott. Culver City, CA: MGM-UA Studios, 1991.

Thirty Seconds Over Tokyo. Directed by Mervyn LeRoy. Burbank, CA: Warner Studios, 1944.

Transcontinental Railroad. PBS. Boston: WGBH, 2003.

Treasures of the World. Videocassette. Seattle: Stoner Productions, 1999.

Index

AAA. *See* Automobile Association of America

Abbasid Dynasty, 64

Abu Ja, 733, 734

Abuja: books/articles about, 739–40; building of, 737–39; as Centre of Unity, 737; choosing site of, 737; cultural context of, 735; documents of authorization, 740–60; Economic Community of West African States located in, 735; energy crisis influence on, 733; ethnic groups of, 733, 735; historical importance of, 739; history of, 733–35; Internet sites, 740; Mohammed's Federal Capital speech, 740–41; naming of, 733, 734; new housing in, 738; planning of, 735–37; university founded in, 737–38. *See also* Nigeria

Abuja University, 737–38

Academy of Sciences (Washington, D.C.), 688

Act of Iguaçu, 712

Adams, P. F., 529, 530

Administrative Procedure Act (U.S.), 511

Advanced Micro Devices (AMD), 807, 809

Adventure Underground (Gies), 761

Africa, 316, 608, 737

Agriculture Department (U.S.), 369, 371, 380

Aida (Verdi), 189

Akashi Suspension Bridge (Japan), 241

Akbar, Divine Earth proposed by, 109

Alaska, 394, 395–96; Anvik Connector trail, 644; Cook Inlet, 518; Elmendorf Air Force base, 686; Fairbanks, 519; gold mining, 517; Iditarod National Historic Trail, 645; Klondike River, 517; Military Highway to, 519; North Slope, 518; oil fields, 681; oil/gas found in, 518; Pearl Harbor attack, 518; pipeline, 518; Prudhoe Bay, 681, 682, 683; Russian ships in, 517; Russia selling of, 517; Soldier's Summit, 517, 521

Alaska Highway: airfields linked by, 519, 522, 524; arguments against, 516; books/articles about, 521; building of, 515, 520–21; Canada and, 517; construction difficulties, 516; cultural context of, 517–19; dedication of, 521; documents of authorization, 522–26; film/television resources, 522; historical importance of, 521; history of, 515–17; Internet sites, 521–22; Military Highway to Alaska exchange of notes, 522–26; sleigh/dogsled routes followed by, 643; wartime maintenance of, 523

definitions, 513–14; enforcement,
512–13; fissionable material production,
495–96; General Advisory Committee,
493; general authority, 508–9; information control, 503–6; international
arrangements, 502; Joint Committee on
Atomic Energy, 511–12; Judicial Review/
Administrative Procedure, 511; materials
control, 496–500; military applications of
atomic energy, 500–501; Military Liason
Committee, 493–94; organization of,
492–94; patents/inventions, 506–8; property of Commission, 503; reports, 513;
research/development activities, 494–95;
separability of provisions, 514; utilization
of atomic energy, 501–2

Atomic Energy Commission (U.S.), 597

Atomic Energy Joint Committee (U.S.),
511–12

Augustus, edict of, 26

Australia: British/Irish ancestry of, 529;
Great Dividing Range, 528; Murray
Darling Basin, 532; New South Wales,
527, 529; Snowy Mountains of, 528;
Sydney Opera House, 531

Autobahn (Germany), 369, 373

Automobile: average minutes spent in,
820–21; Model T, 370

Automobile Association of America (AAA),
369

Babangida, Ibrahim, 738

Babylonians, 2, 66

Baghdad: Abbasid Dynasty, 64; Berlin-to-
Baghdad railway, 67–68; books/articles
about, 68–69; Bosnia-to-Baghdad technology corridor, 68; as city founded by
God, 63; as City of al-Mansur, 64; as City
of Peace, 63, 64; construction of, 63; cultural context of, 66–67; history of, 64–65;
Internet articles about, 69; markets of,
72–73; planning of, 67; roots of, 63; as
Round City, 64; Turks/Mongols and, 65

Bailey Bridge (Normandy), 75, 80

Balaq Hills, 428

Banking, Housing, and Urban Affairs
Committee (U.S.), 857

Banneker, Benjamin, 141, 144; Jefferson,
Thomas, letter from, 153–54

Barbosa, Antonio, 562

Barboza, Mário Gibson, 717

Battleships: *Nashville*, 320; *Oregon*, 320

Battos, 12, 13

Battus III, 16

Bay of Panama, 351, 353

Beards, taxation of, 134

Beauvoir, Simone de, 559

Bechtel Corporation, 246, 431, 766, 823

Belcher, Donald. *See* Donald J. Belcher &
Associates

Belgium: atom bomb project and, 479;
German invasion of, 490

Bell Labs, 623, 624

Benue River, 736

Bering Strait, 311, 517

Bering Strait Tunnel, 769

Bering, Vitus, 515, 517

Berlin-to-Baghdad railway, 67–68

Bey, Koenig, 198

Bhagavad Gita, 480

Bible (King James version): 1 Kings 5:8, 4; 1
Kings 5:10-11, 4; 1 Kings 5:12, 4; 1 Kings
6:1, 3; 1 Kings 6:1-10, 5; 1 Kings 6:7, 3; 1
Kings 6:11-14, 5; 1 Kings, Chapter 5, 7–8;
1 Kings, Chapter 6, 8–10; 2 Kings 20:8, 6;
Jer 52:12, 6

Bicycling: Paris-to-Moscow bikeway, 641,
648; sidewalks for, 373

Big Dig. *See* Central Artery/Tunnel Project

BISNIS. *See* Business Information Service for
the Newly Independent State

Blériot, Louis, 795

Blockades, Panama Canal, 326

Boleyn, Anne, 86

Bolshoi Theater (St. Petersburg), 136

Bonaparte, Napoleon, 189, 792

Bonifacio, José, 558

Bonneville Dam, 454

Books/articles: Abuja, 739–40; Alaska
Highway, 521; Baghdad, 68–69; Brasília,
563; Brooklyn Bridge, 247; Canadian
Pacific Railway, 260; Canal des Deux
Mers, 125–26; Central Artery/Tunnel
Project, 827–28; Channel Tunnel,
769–70; Charlemagne, 80; Colorado
River/Hoover Dam, 434–35; COMSAT,
629; Eiffel Tower, 296; Erie Canal, 162–
63; Federal Highway System, 377–78;
Grand Canal of China, 41–42; High
Dam at Aswan, 613; Itaipú Hydroclectric
Power Project, 716; London Bridge, 91;
Manhattan Project, 485; Mont Blanc
tunnel, 548; NASA, 586; National Trails
System, 650; Netherlands, 56; New River,

About the Authors

FRANK P. DAVIDSON was the American co-founder (1957) of The Channel Tunnel Study Group and the initiator of teaching and research in macro-engineering at the Massachusetts Institute of Technology. Davidson has a J.D. from Harvard Law School and has been appointed a *Chevalier* in the French Legion of Honor.

KATHLEEN LUSK BROOKE is the Founder and Managing Director of the Center for the Study of Success. Lusk Brooke has a Ph.D. from Harvard University and is the author of many works on policy and management, including the book *Mobilizing the Organization: Bringing Strategy to Life*.